Energy and Spectrum Efficient Wireless Network Design

Covering the fundamental principles and state-of-the-art cross-layer techniques, this practical guide provides the tools needed to design MIMO- and OFDM-based wireless networks that are both energy- and spectrum-efficient. Technologies are introduced in parallel for both centralized and distributed wireless networks to give you a clear understanding of the similarities and differences between their energy- and spectrum-efficient designs, which is essential for achieving the highest network energy saving without losing performance. Cutting-edge green cellular network design technologies, enabling you to master resource management for next-generation wireless networks based on MIMO and OFDM, and detailed real-world implementation examples are provided to guide your engineering design in both theory and practice. Whether you are a graduate student, a researcher, or a practitioner in industry, this is an invaluable guide.

Guowang Miao is an Assistant Professor in the Department of Communications Systems at KTH Royal Institute of Technology, Sweden. After receiving his Ph.D. in electrical and computer engineering from Georgia Institute of Technology, USA, in 2009, he spent two years working in industry as a Senior Standard Engineer at Samsung Telecom America. His current research interests are in the design and optimization of wireless communications and networking.

Guocong Song is currently the Principal Research Engineering at ShareThis, Palo Alto, California. He has been working in wireless communications and networks for a decade, since receiving his Ph.D. in electrical and computer engineering from Georgia Institute of Technology. He received the 2010 IEEE Stephen O. Rice Prize for the best paper in the field of communications theory, and he is recently active in the area of data science and machine learning.

Energy and Spectrum Efficient Wireless Network Design

GUOWANG MIAO

KTH Royal Institute of Technology, Sweden

GUOCONG SONG

ShareThis, Palo Alto, California

CAMBRIDGE
UNIVERSITY PRESS

CAMBRIDGE
UNIVERSITY PRESS

University Printing House, Cambridge CB2 8BS, United Kingdom

Cambridge University Press is part of the University of Cambridge.

It furthers the University's mission by disseminating knowledge in the pursuit of
education, learning, and research at the highest international levels of excellence.

www.cambridge.org
Information on this title: www.cambridge.org/9781107039889

© Cambridge University Press 2015

First published 2015

Printed in the United Kingdom by Clays, St Ives plc

A catalogue record for this publication is available from the British Library

Library of Congress Cataloguing in Publication data
Miao, Guowang.
Energy and spectrum efficient wireless network design / Guowang Miao, KTH Royal Institute of Technology,
Sweden, Guocong Song, ShareThis, Palo Alto, California.
   pages cm
ISBN 978-1-107-03988-9 (Hardback)
1. Wireless communication systems–Energy conservation.   2. Wireless communication systems–Energy
consumption.   3. Radio frequency allocation.   4. Radio resource management (Wireless communications).
5. Engineering economy.   I. Song, Guocong.   II. Title.
TK5102.86.M53 2014
621.384–dc23   2014020418

ISBN 978-1-107-03988-9 Hardback

To
Ting Ren, Eileen Miao, and Ryan Miao
Wei and Lyra

Contents

Preface

This book provides a comprehensive introduction to the theory and practice of energy and spectrum efficient design for various types of wireless networks. The concepts and technologies are presented in a unified way for both centralized and distributed networks. The principles of the designs are stressed so that they can be applied in the broader context of wireless systems. The detailed derivations and proofs from first principles are provided. They are intended for the reader who desires a more in-depth understanding of the results. For the reader not interested in the detailed derivations, the concepts and theories are self contained and can be easily understood while skipping the derivations.

Energy and spectrum are two fundamental resources in wireless networks. A network design can always choose to optimize the utilization of one resource over the other. If one resource is redundant and the other is not, the design will need to optimize the network behavior towards better efficiency using that other resource. If both are adequate, then the system can be operated for the best user experience. If both are scarce, the design has to choose between them. Energy efficiency and spectrum efficiency are equally important and there is no clear advantage of one metric over the other. Which metric is more desired depends on network needs. This book presents a comprehensive yet rigorous discussion of the relationships between wireless channel state, energy efficiency, spectral efficiency, implementation, and network resource management in various wireless environments and their corresponding optimal designs.

The material in this book is structured into parallel discussions of energy and spectrum efficient designs, both of which are also discussed in parallell for centralized and distributed wireless networks. We hope this structure will facilitate the understanding of their similarities and distinctions.

The book is divided into four parts. In Part I, we introduce the basic concepts of wireless communications, e.g. wireless channel properties, performance metrics, conventional centralized and distributed radio resource management, that serve as the foundation to understand the book. The reader that is familiar with this background knowledge can skip this part and start from Part II directly. Part II introduces cross-layer designs for networks with central controllers and Part III for networks without central controllers. Both Parts II and III are focused on spectrum-efficient designs. Part II presents a generic framework for optimal opportunistic radio resource management in centralized networks by exploiting the multi-user diversity of time and frequency in wireless channels and regulating the resource allocation through network economics. Part III covers how to optimally exploit multi-user diversity in distributed wireless networks and shows how distributed random access can be designed to achieve spectrum efficiency comparable to that of ideal centralized schedulers. In Part IV, we present optimal energy-efficient transmission and resource management for both centralized and distributed wireless networks. For example, while the Shannon capacity results tell us the tightest spectrum efficiency upper bound of point-to-point communications, we introduce the tightest energy efficiency upper bounds, named energy efficiency capacity, for various types of channels. We also introduce energy-efficient centralized scheduling and distributed medium access control (MAC) and power control. The relationships between energy efficiency, spectral efficiency, and several other network performance metrics are rigorously examined. At the end of this part, we give a thorough discussion on energy-efficient cellular network designs and also on how to implement energy-efficient designs in practice.

This book is highly recommended for graduate-level courses as the primary or alternate textbook and professional tutorials in wireless networks and resource management. It provides material both to guide novice students as well as plenty of detailed in-depth material for graduate students pursuing research in the field. The book is also a useful reference for practicing engineers, academics, and industrial researchers. The only expected background of the reader is a basic understanding of probability, optimization, and digital communications. Background in wireless networks, radio resource management, and signal processing is helpful but not required, since we develop the related material in the text.

Acronyms

3GPP	3rd Generation Partnership Project
AD	adjustment
AM	amplitude modulation
AP	access point
APA	adaptive power allocation
AWGN	additive white Gaussian noise
BER	bit error rate
BS	base station
C/I	carrier to interference
CAD-MAC	channel-aware distributed medium access control
CCI	co-channel interference
CDF	cumulative distribution function
CDMA	code division multiple access
CIA-MAC	co-channel interference avoidance MAC
CoMP	coordinated multi-point transmission
CRC	cyclic redundancy check
CRS	contention resolution slot
CSI	channel state information
CSMA	carrier sense multiple access
CSMA/CA	carrier sense multiple access with collision avoidance
CSMA/CD	carrier sense multiple access with collision detection
CTS	clear to send
DOMRA	decentralized optimization for multi-channel random access
DSA	dynamic subcarrier assignment
EMMPA	energy-efficient MU-MIMO power allocation
ESPA	exhaustive search power allocation
EXP	exponential
FCC	Federal Communications Commission
FDM	frequency division multiplexing
FDMA	frequency division multiple access
FEC	forward error correction
FFR	fractional frequency reuse
FFT	fast Fourier transform
FM	frequency modulation

FPA	fixed power allocation
FS	frequency selective
HOL	head-of-line
ICR	interference to carrier ratio
ICT	information and communication technology
IFFT	inverse fast Fourier transform
LDPC	low density parity check
LLC	logical link control
LOS	line of sight
LS	least squares
LTE	long-term evolution
MAC	medium access control
MCS	modulation and coding scheme
MDU	maximum delay utility
MIMO	multiple-input multiple-output
M-LWDF	modified largest weighted delay first
MMSE	minimum mean squared error
M-QAM	M-ary quadrature amplitude modulation
MSC	maximum sum capacity
MT	mobile terminal
MU-MIMO	multiple user MIMO
OFDM	orthogonal frequency division multiplexing
OFDMA	orthogonal frequency division multiple access
OSI	open systems interconnect
PA	power amplifier
PAPR	peak to average power ratio
PC	personal computer
PDF	probability distribution function
PER	packet error rate
PF	proportional fair
PHY	physical
PSK	phase shift keying
QoS	quality of service
RF	radio frequency
RNC	radio network controller
RTS	request to send
SDMA	space division multiple access
SIMO	single-input multiple-output
SINR	signal to interference plus noise ratio
SNR	signal-to-noise ratio
TDD	time division duplex
TDMA	time division multiple access
WFQ	weighted fair queueing
WLAN	wireless local area networks

1 Introduction

The explosive growth and widespread applications of high-rate multimedia wireless communications have prompted waves of research and standard development activities. The ultimate goal is to connect all users to their targets all the time: to the Internet, to the cloud, and to the various devices in their lives, from phones to PC to tablet and to their cars. They expect to access a rich set of services regardless of where they are: whether at home, around the office, or outdoors. This exponential growth in data traffic is inevitably limited by spectrum availability and energy consumption of mobile terminals and access networks. The future success of wireless networks thus relies on the ability to overcome the mismatch between the requested *quality-of-service* (QoS) and limited network resources.

1.1 Motivation

Among available resources, spectrum and energy are two fundamental ones that enable wireless communications. Spectrum is a natural resource that cannot be replenished and therefore must be used efficiently; so that is where the special significance of *spectral efficiency* lies. Tremendous efforts and progress have been made in past decades to improve spectrum efficiency of wireless networks. However, the increasing market for all things wireless generates more demand than supply can handle. Wireless networks therefore need to be designed in a much more spectrum-efficient way to keep up with consumer demands.

Indeed, *energy efficiency* is also becoming increasingly important. From the perspective of user experiences, small form factor mobile devices are getting more and more energy hungry, since battery technology has not kept up with the growing requirements stemming from ubiquitous multi-media applications [114]. From a global perspective, we are confronted with severe challenges of environmental protection and prevention of climate changes. Improving the energy efficiency of wireless networks will not only reduce their impacts on environments but also cut network costs and make communications more affordable for everyone [68].

Spectral efficiency and energy efficiency are two different metrics for measuring wireless network efficiency. Some design criteria optimized for improving one metric may not necessarily improve the other. For example, a single wireless user may achieve its highest spectral efficiency if it transmits with the highest radio power. However, energy efficiency will then be adversely affected due to too high or concentrated consumption of that energy resource. We can see that there is an urgent need to address spectral efficiency and energy efficiency issues together. New metrics, which are completely distinct from existing ones in literature, may be required to address this need. The spectral efficiency–energy efficiency relationship varies depending on the type of wireless networks we design. Different approaches should therefore be analyzed and better understood in order to design spectral- and energy-efficient operations for different types of wireless networks. This will be the focus of this book.

1.2   Wireless networks

1.2.1   Overview

Wireless networks refer to communication networks of any kind that are implemented by using radio communications. Wireless networks have various application requirements, including coverage, data rate, delay, error rate, mobility, functionality, and so on. No single wireless technology is capable of handling all these requirements effectively, and that is why different types of wireless networks with different topologies and coverage have been designed to handle different situations. The protocols running on different types of wireless networks also differ significantly from one another.

In general, wireless networks can be classified into two main categories, having or not having infrastructure support. Wireless networks with infrastructure support typically consist of fixed access points, e.g. base stations, and wireless mobile users (or nodes, devices, mobile stations, which we will use interchangeably). The access points are connected to the Internet via backhaul links and the mobile users communicate with the access points through wireless air interfaces. Some typical infrastructure wireless networks are cellular networks and wireless local area networks (WLANs). For those without infrastructure support, e.g. a mobile ad hoc network, the network can operate autonomously. In a mobile ad hoc network, each user can communicate with neighbors within communication range, and may operate as a source, destination, router, or relay. Because they are easy to deploy, ad hoc networks can be applied in critical situations such as battlefield communications and disaster recovery, where infrastructure networks are hard to build or maintain.

Traditionally, wireless networks have limited or no inter-operability between the various technologies. With the advances in wireless technologies, the evolution is toward a mixture of various technologies and topologies that coexist and inter-operate to provide seamless services to mobile users anywhere. Future wireless networks will consist of hybrid structures and different access technologies. For example, they may consist of both wide-area cellular networks providing high-speed mobile services to users over a large area, small cells covering local area hot spots with much higher data rates, and ad hoc direct communications among users in a vicinity. Correspondingly, mobile users in such a hybrid network are expected to operate with multiple access protocols and switch intelligently in different operating modes.

As multiple users need to share limited resources, resource management essentially guarantees the performance of individual users as well as that of the whole network. There are well-known orthogonal wireless resources in four dimensions; time, frequency, space, and code. As long as different transmissions occur in resources that have no overlap in any of the four dimensions, they do not interfere with each other. So the problem of resource management is how many orthogonal resource chunks should be assigned to each user. Another resource, that is even more fundamental, is energy, which is a non-orthogonal resource and is much more difficult to manage. When different users send data over different time, frequency, space, or code dimensions, their power can be managed independently. However, when the transmissions overlap in time, frequency, space, or code, the transmissions interact with each other through mutual interference. For example, inter-cell co-channel interference in cellular networks happens because there is overlap in the space domain and different cells are not spatially separated far enough. Inter-symbol interference takes place when there is time overlap and previous symbols are delayed and overlap with the following symbols. Similarly, inter-channel interference in frequency division multiple access (FDMA) and code of division multiple access (CDMA) result from overlap in frequency or code. Interference is controlled by transmission power levels. In addition, transmission power also determines energy consumption. Therefore, power plays a pivotal role in both network spectrum and energy efficiency.

There are many other factors affecting wireless resource management. One principal limitation is the wireless channel. With wireless communications, transmission signals are usually severely distorted by the channel and the distortions differ from user to user and from time to time. The wireless channels of different users also vary significantly in the number of paths, path delays, phase shifts, path attenuation, etc. Furthermore, the broadcast nature of wireless channels implies limitations on transmission power to avoid co-channel interference, because excessive interference can deteriorate network performance and waste scarce wireless resources. In addition, government regulations also apply strict control on the amount of power that can be transmitted, such that the radio frequency (RF) exposure levels people may be subjected to are safe. As a consequence, limited power can be used for wireless transmission to compensate the path loss. The wireless channel is therefore error-prone and highly unreliable and subject to many impairment factors that are of transient nature.

Besides these physical limitations, users also have different QoS demands, indicating the need to use different amounts of wireless resources. The interest of one user usually conflicts with that of others in resource allocation. Therefore, it is critical to design resource management schemes that can allocate wireless resources effectively, efficiently, and fairly. It is the scheduling and medium access control (MAC) protocols that allocate wireless resources to users on demand, multiplex and separate transmissions of different users, control interference, and ensure network-wide flexibility, efficiency, and fairness of resource sharing. These schemes need to consider both wireless properties and user demands. The diversity underlying channel conditions and user demands should be exploited together to enhance network performance.

Wireless resource management schemes can be divided into several categories depending on if they use central controllers or not. In centralized networks, a central scheduler collects all network information and decides the resource allocation for all users in the network. In distributed wireless networks, users in the network make decisions on resource allocation themselves and the conflicts in resource allocation among users are resolved by their autonomous behaviors. The performance of distributed approaches can be enhanced through distributed collaborations among users in the network. The centralized and distributed schemes can be combined together to further improve the spectral and energy efficiency of the network.

1.2.2   Traditional layered architecture

Traditionally, network protocols are described with a layered model and the protocols are stacked on top of each other. Each layer has specific responsibilities and knows nothing about the procedures of other layers. Each layer carries out its own tasks and delivers messages to the adjacent layer in the process. Data sent to others are passed from the highest-level protocol down to the lowest-level one and vice versa. The layered model allows network services to be defined with their functions, rather than specific implementations. The isolation of communications functions in different layers minimizes the impact of technique change on the entire protocol stack. Protocols in each layer can be updated without affecting other layers, as long as the updates use the same interfaces with adjacent layers as those used before.

A classic layered network model developed by the International Standards Organisation (ISO) is called the Open Systems Interconnect (OSI) reference model, which is frequently used to describe the structure and function of data communication protocols. The OSI model contains seven layers and each layer represents a function performed when data are transferred through the layer. In the OSI model, a layer defines the function instead of the protocols used to implement the function. Therefore, each layer may have multiple protocols. Every layer communicates only with its remote peer entity, which runs the same protocol in the equivalent layer.

The OSI model has seven layers, which are described briefly in the following:

- Layer 1: Physical layer. This layer defines the functions of the hardware necessary to transmit the data signals on a certain carrier. The main responsibilities of this layer are to send and receive information bits to or from the medium. It describes the way data are actually transmitted on the channel, but does not define the medium. There are many varieties of media for data communication, e.g. radio, cable, fibre optics, light waves, and so on. Different medium need different sets of physical layer protocols. The physical layer describes how information bits are encoded into media signals and the characteristics of the media interface.
- Layer 2: Link layer. This layer transfers data between entities in the network, detects, and possibly corrects errors that may occur in the physical layer. It formats packets, defines network-frames, and is responsible for error control on the frame level. Not all physical layer bits go into link layer frames, as

some of these bits are for physical layer functions. The link layer connects users in the same network and provides intranet address information for the physical layer in the transmitted frames. Usually the link layer is divided into two sublayers: the MAC and logical link control (LLC) layers. The MAC sublayer controls how users in the network gain access to the channel to transmit data, while the LLC layer controls frame synchronization, flow control, and error checking.

- Layer 3: Network layer. This layer transmits data and decides which route the data must follow through the whole network. The network layer maintains the quality of service provided to the upper transport layer. The network layer receives data packets from the upper layer from the sender, and transmits them by as many connections and subsystems as needed to reach the destination. The source node may reside in one type of network while the destination in a different one. This layer also controls packet delivery between intermediate stations. Therefore, the network layer manages connections across the network and isolates the upper-layer protocols from the details of the underlying physical networks. An example is the Internet Protocol (IP), which handles the addressing and delivery of data in the TCP/IP protocol.

- Layer 4: Transport layer. This layer ensures transparent transfer of data between end users and is responsible for end-to-end error recovery and flow control. It controls the reliability through flow control, segmentation/desegmentation, and error control. The transport layer may keep track of the segments of a packet and retransmit those that fail to ensure reliable delivery. Sometimes the transport layer also provides acknowledgement of successful data transmission. In TCP/IP, the function of the transport layer is performed by TCP. However, TCP/IP offers a second transport layer protocol, the User Datagram Protocol (UDP) that does not perform the end-to-end reliability checks.

- Layer 5: Session layer. This layer establishes, manages, and terminates connections between dialogues or connections of users. A dialogue is a formal conversation in which two nodes agree to exchange data. It deals with session and connection coordination. The communication can take place in full-duplex, half-duplex, or simplex mode. Sessions enable users to communicate in an organized manner. Each session has three phases: connection establishment, data transfer, and connection release. In the establishment phase, the source and destination users negotiate the rules of communication, including the protocols to be used and communication parameters. Then they exchange data in a dialogue. When the users no longer need to communicate, they engage in an orderly release of the session.

- Layer 6: Presentation layer. This layer cooperates applications to exchange data and is responsible for presenting data to the application layer. It transforms data into the form that the application layer can accept and provides freedom from compatibility problems. For example, translations could be made between ASCII and Unicode. In addition, this layer may also provide security assurance in the form of encryption and compression.

- Layer 7: Application layer. This layer supports applications that users directly interact with, as well as other processes that users are not necessarily aware of. For example, this layer provides application services for file transfers, e-mail, and other network services. Telnet, FTP, and Email are typical examples.

One advantage of the layered protocol is that they break the communication process into manageable chunks. Designing a small part of the protocol stack is much easier than designing the entire model. So it simplifies engineering. A change in one layer does not affect others and new technology can be easily introduced into the system.

1.2.3   Necessity of cross-layer optimization

As introduced in the previous section, communication networks can be modeled using the OSI standard and the networks are divided into layers that are in charge of different functionalities. For example, the physical (PHY) layer takes care of reliable and efficient bit transmission using modulation and coding techniques. The MAC layer handles resource allocation to multiple users. On the other hand, the network layer is in charge of routing. The traditional design paradigm emphasizes transparency between layers for the purpose of implementation simplicity. With this design, each layer does not need to know how adjacent layers work inside. Instead, each layer accesses only the interfaces provided by the adjacent layers and the interfaces are usually minimally designed. The layered design has some advantages from a design and implementation perspective. For example, each layer can be independently updated without affecting other layers. However, the layered design ignores coupling among adjacent layers that leads to significant information loss between layers and thus significant performance degradation. The performance loss is even more significant in wireless networks than wired ones. This is because the MAC layer is closely related to the underlying physical layer. For example, in the downlink of a wireless cellular network, the network sum capacity is maximized if, in each time slot, the user with the best instantaneous channel gain is scheduled. This is one way of exploiting the so-called multi-user diversity, and the scheduler is a channel-aware scheduler. The instantaneous channel gain is PHY layer knowledge. However, the MAC layer deals with user scheduling. With traditional layered protocol, the MAC layer will not be able to apply the channel-aware scheduler to exploit multi-user diversity because of the lack of channel information. If there are many users in the network and the channel varies significantly among users, the loss in spectral and energy efficiency would be huge without exploiting multi-user diversity.

Spectral and energy efficiency are affected by all layers of system design, ranging from silicon to applications. While the traditional layer-wise approach leads to independent design of layers and results in high design margin, cross-layer approaches exploit interactions between different layers and can significantly improve system performance as well as adaptability to service, traffic, and environment dynamics. Cross-layer optimization for throughput improvement has been a popular research theme [197, 151, 131]. Recent efforts have also been undertaken to tackle energy consumption at all layers of communication systems, from architectures [135, 119, 29] to algorithms [48, 93, 216].

The PHY layer plays a very important role in wireless communications due to the challenging nature of the communication medium. In wireless networks, the PHY layer deals with data transmission over wireless channels and consists of RF circuits, modulator, power control, channel coding units, etc. The physical layer has many tasks and some examples are listed below.

- Modulation and demodulation: Map information bits into analog signals and vice versa. Some common techniques are:

  Amplitude modulation (AM): Uses different amplitude levels of the carrier signal to represent information bits. AM is not robust to noise and interference and rarely used in wireless communications.

  Frequency modulation (FM): Uses different frequencies to represent information bits.

  Phase shift keying (PSK): Uses the phase of the carrier signal to represent information bits.

- Coding and decoding: Convert messages from their original forms, e.g. bits, into other forms that represent the messages for efficient transmission and vice versa. There are numerous encoding and decoding algorithms and the oldest one is the Morse code, which was used in the landline telegraph in the 19th century.

- Time or frequency synchronization: Enable the transmitter and receiver to agree on the frequency or time that the communications take place.

- Multiplexing and demultiplexing: Allow multiple users to transmit at the same time without interfering with each other. For example, with frequency division multiplexing (FDM), different users will use different sets of frequencies to send data at the same time.

- Carrier sensing in some MAC protocols: Detects if the carrier is in use before attempting to send data.

- Signal processing: Uses equalization, filtering, pulse shaping, channel estimation, signal detection, and so on, that are used to process signals.

- Interleaving: Reorders data such that consecutive bits are distributed over a larger sequence of data to reduce burst errors. Many error protection coding algorithms cannot correct for errors in groups, and interleaving increases their ability to correct for burst errors.

Traditional wireless systems are built to operate on a fixed set of operating points [38], e.g. no power adaptation. This results in excessive energy consumption or a pessimistic data rate for peak channel conditions. Hence, a set of PHY parameters should be adjusted to adapt the actual user requirements (e.g. throughput and delay) and environments (such as shadowing and frequency selectivity) to tradeoff energy efficiency and spectral efficiency. As wireless is a shared medium, communication performance and energy consumption are affected not only by the layers comprising the point-to-point communication link, but also by the interaction between the links in the entire network. Hence, a system approach is required.

On the other hand, in a multi-user network, the MAC layer ensures that wireless resources are efficiently allocated to maximize network-wide performance metrics while maintaining user QoS requirements. Here, pessimistic medium access strategies that allocate wireless resources to assure worst-case QoS may hurt network spectral and energy efficiency. In distributed access schemes, MAC should be improved to reduce the number of wasted transmissions that are corrupted by interference from other users, while in centralized access schemes, efficient scheduling algorithms should exploit the variations across users to maximize the overall network performance. The MAC layer manages wireless resources for the PHY layer and they both directly impact overall network performance and energy consumption.

In this book, we introduce cross-layer approaches to optimize the performance of wireless networks. More specifically, we emphasize joint physical-MAC layer designs to improve wireless spectral and energy efficiency because the two layers closely depend on each other. We will emphasize scheduling, channel access, radio power control, modulation, coding, network energy consumption, and their interactions, together with the wireless channel states.

Orthogonal frequency division multiplexing (OFDM) is a key modulation scheme for next-generation broadband wireless standards [63, 18], including digital video broadcasting (DVB) systems, WLAN standards such as American IEEE 802.11 a/g/n and the European equivalent HIPERLAN/2, the fourth-generation (4G) mobile communications such as IEEE WiMAX and 3GPP LTE. OFDM has been widely applied in wireless networks because of its high bandwidth efficiency and robustness to multi-path fading and delay channels. In addition, OFDM converts a frequency-selective fading channel into several nearly flat-fading channels by dividing the entire available spectrum into narrow-band subchannels. The high spectral efficiency is obtained by overlapping the orthogonal frequency responses of the subchannels. From a resource allocation perspective, these multiple channels in OFDM systems have the potential for more efficient MAC design since subcarriers can be assigned to different users [8] and this is usually called Orthogonal Frequency-Division Multiple Access (OFDMA). Furthermore, adaptive power allocation on each subcarrier can be applied for further improvement [9]. Therefore, the exploitation of these OFDMA properties in network resource management will significantly boost network spectral and energy efficiency. This book will emphasize joint PHY and MAC optimization to improve the spectral and energy efficiency for OFDM-based wireless networks.

1.3   Book outline

The major goal of this book is to introduce state-of-the-art cross-layer transmission and resource management designs that significantly improve both spectral efficiency and energy efficiency in wireless networks. The book is divided into four parts. In Part I, we introduce the basic concepts of wireless communications and networks that serve as a foundation for understanding the book. Readers familiar with this background knowledge can skip this part and start from Part II directly. Part II introduces cross-layer designs for networks with central controllers while Part III does so for networks without central controllers. Both Parts II and III emphasize enhancing spectral efficiency. In Part IV, energy-efficient design techniques are introduced and the relations between energy efficiency, spectral efficiency, and some other network performance metrics as well as their impact on network designs are discussed in detail.

To be more specific, Part I consists of four chapters. The first chapter talks about wireless channel properties such as path loss, shadowing, and fading for individual links. Channel state information (CSI) is essential in wireless networks for cross-layer design and we will discuss some channel estimation methods that can be used to obtain the CSI. The second chapter gives a detailed comparison of the spectral and energy efficiency metrics from both link and network levels. In the third and fourth chapters, we discuss traditional MAC protocols for different types of wireless networks. We will introduce both centralized resource management schemes and distributed access protocols.

Part II is focused on centralized cross-layer optimization assuming a centralized scheduler to manage network resources. It uses two major mechanisms in resource management to improve the spectral efficiency of wireless networks: exploiting the time variance and frequency selectivity of wireless channels through adaptive modulation, coding, as well as packet scheduling and regulating resource allocation through network economics. With the help of utility functions that capture the satisfaction level of users for a given resource assignment, Part II introduces a generic utility optimization framework for centralized OFDM networks, in which the network utility at the level of applications is maximized subject to the current channel conditions and the modulation and coding techniques employed in the network. We will introduce novel efficient dynamic subcarrier assignment (DSA) and adaptive power allocation (APA) algorithms, which are proven to achieve optimal or near-optimal performance with very low complexity. Based on a holistic design principle, we will also introduce max-delay-utility (MDU) scheduling, which senses both channel and queue information. MDU scheduling can simultaneously improve spectral efficiency and provide the right incentives to ensure that all applications can receive the different required QoS. To facilitate cross-layer design, we will also investigate the mechanisms of channel-aware scheduling, such as efficiency, fairness, and stability, using extreme value theory. Especially, we will analyze the impact of multi-user diversity on throughput and packet delay and provide a method to design cross-layer scheduling algorithms that allow the queueing stability region at the network layer to approach the ergodic capacity region at the physical layer.

In Part III, we continue the discussion of spectrum efficiency improvement using cross-layer techniques but in distributed wireless networks. We first introduce the concept of distributed multi-user diversity and its potential to improve network throughput and then the detailed technologies to exploit this diversity. The design philosophy of distributed approaches heavily depends on how different users in the network interact or interfere with others. When different communicating pairs are very close to each other, they will not be able to send data simultaneously and distributed MAC protocols are needed to avoid collisions. We will first introduce distributed MAC protocols that utilize channel state information to exploit the multi-user diversity in the wireless channels for collision resolutions. In these protocols, we will discuss opportunistic random access for single-cell cellular networks and then for any network topologies. The final goal is to introduce how to use distributed random access approaches to achieve performance comparable to that of centralized approaches when channel state information is used at the MAC layer. We will introduce an optimal channel-aware distributed MAC, which, although quite preliminary, reaches this goal. In the last chapter, as an example, we will discuss how in practice we can use these distributed approaches in cellular networks to improve spectral efficiency. Besides resolving collisions, power control is essential in determining network performance for simultaneous data transmissions. At the end of this part, we will introduce distributed power control for both real-time and elastic traffic.

In Part IV, the focus is on energy-efficient cross-layer design techniques, and the relationship between energy and spectrum efficiency in various types of wireless networks. We will use a bottom-up approach. This means we will first study simple point-to-point energy-efficient communications and then more complicated multi-user networks. We start by discussing energy-efficient wireless transmission techniques in both flat-fading and frequency-selective channels. After that, we consider a multi-user single-cell network and discuss energy-efficient orthogonal resource management schemes in different resource domains. We will introduce energy-efficient scheduling technologies both with and without fairness and show that conventional spectrum-efficient schedulers are indeed special cases of energy-efficient schedulers in specific regimes. Then we will study distributed energy-efficient communications, including both energy-efficient distributed random access and power control in interference-limited networks. We will discuss the fundamental tradeoffs in wireless resource allocation and investigate relationships between energy efficiency, spectral efficiency, and several other network performance metrics such as deployment cost, system bandwidth, etc. Finally, we move on to the whole network level and investigate system-level energy-efficient designs for both homogeneous and heterogeneous cellular networks. We will discuss implementation issues in practice, illustrating how the technologies discussed in this part can be implemented in real-world wireless networks.

Part I

Basic concepts

Wireless communications are mainly limited by the properties of radio channels, i.e. wireless channels. Wireless signals experience severe distortions in delays, amplitudes, phases, and many other aspects, and the distortions vary significantly with time, frequency, and location of the communications. Furthermore, there are other limiting factors such as system bandwidth and radiation power allowed. Therefore, wireless channels are transient, unreliable, and error prone. Wireless communications have a broadcast nature, and excessive interference in the network will cause deterioration in network performance and waste network energy and spectrum resources. A wireless channel observed by a transceiver pair is affected by both the link performance and also the network interference environment. Therefore, radio resource management and interference management are necessary in wireless networks to improve channel qualities of all users in the network and to provide acceptable QoS to users. There are many mechanisms for radio resource and interference management. Medium access control and power control are the most important ones as they grant users channel access rights and determine how different users may affect the communications of each other. In the following chapters, we will introduce the most basic concepts of wireless channel properties and modeling, wireless transceiver design, and classical existing MAC protocols to facilitate understanding of the rest of book. For readers familiar with these concepts, please skip the following discussions and move on to Part II.

2   Wireless channel properties

With wireless communications, signals are carried on electromagnetic waves that travel through wireless propagation channels. The signals will be attenuated. If a transmitter sends a signal with power p_t, the power of the received signal would be

$$p_r = p_t g, \qquad (2.1)$$

where g is the channel power gain. The gain depends on several effects: path loss, shadowing, small-scale fading, and can be written in the following format:

$$g = \alpha^2 10^{\frac{x}{10}} g(d) G_t G_r, \qquad (2.2)$$

where G_t and G_r are the power gains of the transmitter and receiver antennas respectively, $g(d)$ the path loss, x the shadowing, and α^2 the small-scale fading. Shadowing is also frequently referred to as large-scale fading.

2.1   Path loss

If the signal travels through free space, the channel gain is determined by the distance, d, between the transmitter and receiver and the carrier frequency, f, and given by

$$g = G_t G_r \left(\frac{c}{4\pi df} \right)^2, \qquad (2.3)$$

where c is the light speed in free space and $c = 3 * 10^8$ ms. The impact of G_t and G_r on g is obvious. Other than these, there are two main effects of wireless communications: $\frac{1}{4\pi d^2}$ and $\frac{c}{4\pi f^2}$. The first one is mainly because, when an antenna radiates a signal, the signal will propagate in all directions. If we look at a sphere of radius r around the antenna, the total power going through the sphere is a constant. The signal power per unit area on the sphere is the total signal power divided by the area of the sphere, $4\pi d^2$. In addition to this frequency-independent effect, the gain of $\frac{c}{4\pi f^2}$ describes how well an antenna can receive power from an incoming electromagnetic wave, which depends on the wavelength or frequency. For an isotropic antenna, the gain is $\frac{c}{4\pi f^2}$. Combining all these effects, we have the path loss model in (2.3). Note that this path loss model is accurate in the far field where we can assume that the spherical spreading does not hold when the receiver is close to the transmitter.

Let g_0 denote the channel gain at the first meter and

$$g_0 = G_t G_r \left(\frac{c}{4\pi f} \right)^2 . \tag{2.4}$$

Then

$$g = \frac{g_0}{d^2}, \tag{2.5}$$

or in decibels form,

$$10 \log_{10}(g) = 10 \log_{10}(g_0) - 20 \log_{10}(d), \tag{2.6}$$

indicating a 20 dB loss per ten meters in free space.

The signals in most wireless networks, however, do not experience free-space propagation. For non-free-space cases, the path loss is frequently assumed to be

$$g = \frac{g_0}{d^\alpha}, \tag{2.7}$$

where α is called the path loss exponent. Rewriting the above expression in decibels form, we have

$$10 \log_{10}(g) = 10 \log_{10}(g_0) - 10\alpha \log_{10}(d). \tag{2.8}$$

The values of α may vary between 2 and 6 and depend on the carrier frequency and communication environments, e.g. urban, rural, etc.

There are many other more accurate path loss models, such as Hata's model, Okumura's model, and the COST231 model, most of which are experiment based. These models are obtained by curve fitting experimental data. For more details, please refer to [79].

2.2   Shadowing

The path loss models in the previous section describe the relationship between distance and channel gain. In real wireless environments, the relationship would vary depending on the environments and locations. The path loss models indeed provide mean values of the expected channel gain when the distance is d and the actual value will vary around it. This location-dependent variation of channel gain is called shadowing.

Shadowing is mainly due to the existence of large obstacles in the propagation path of wireless signals. These large obstacles, like walls, mountains, cars, buildings, and foliage obstructions, block wireless signals and therefore shadows are formed. Shadowing causes channel variations over large areas. For paths longer than a few hundred meters, e.g. in cellular systems, shadowing can be modeled as a log-normal random variable. If we express the channel gain in dB scale, then the shadowing can be modeled by a zero-mean normal random variable x with a certain standard deviation σ^2. The standard deviation σ^2 of the shadow is an environment characteristic and typically ranges from 6 to 12 dB [194]. In addition, the standard deviation increases slightly

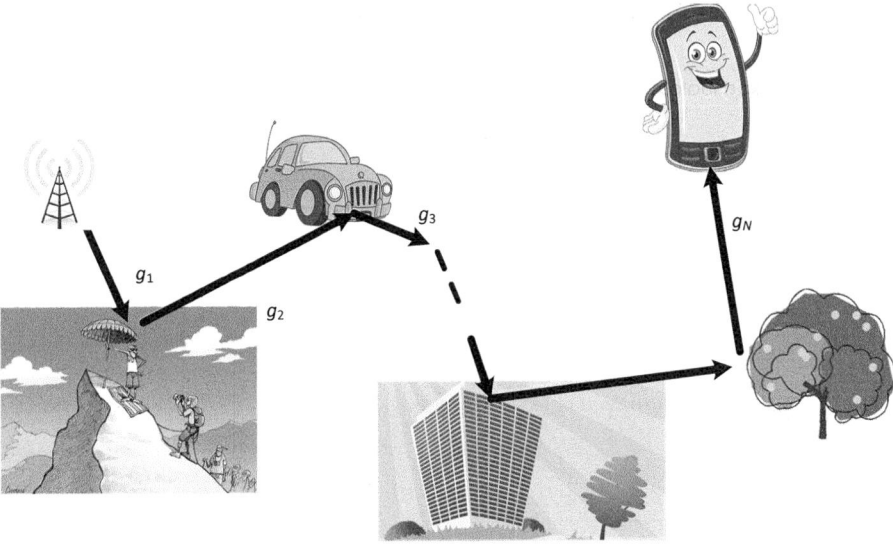

Figure 2.1 Example of signal propagation

with frequency (0.8 dB higher at 1800 MHz than at 900 MHz), but has been observed to be nearly independent of radio path length, even for distances that are very close to the transmitter [79].

To have an intuitive understanding, Figure 2.1 illustrates an example. Assume there are many obstacles in the signal propagation path and the whole path is divided into N channels, each with a random gain g_n. The overall channel gain would be

$$g = \prod_n g_n \tag{2.9}$$

which can be rewritten as

$$x = 10 \log_{10} g = \sum_n 10 \log_{10} g_n. \tag{2.10}$$

Since N is large, $10 \log_{10} g$ is approximately normally distributed according to the central limit theorem.

Shadowing can be estimated using terrain data and the standard deviation depends on the geographical resolution. With a geographical data base of infinite resolution, the standard deviation can theoretically achieve 0 dB. So the standard deviation describes how precisely the terrain is described. While the path loss depends only on distance, not the detailed terrain features, the standard deviation of shadowing is used to capture how random the features are.

2.3 Small-scale fading

The signal arriving at a receiver contains not only direct line-of-sight (LOS) signals, but also many reflected ones, as shown in Figure 2.2. In some environments, the LOS signals

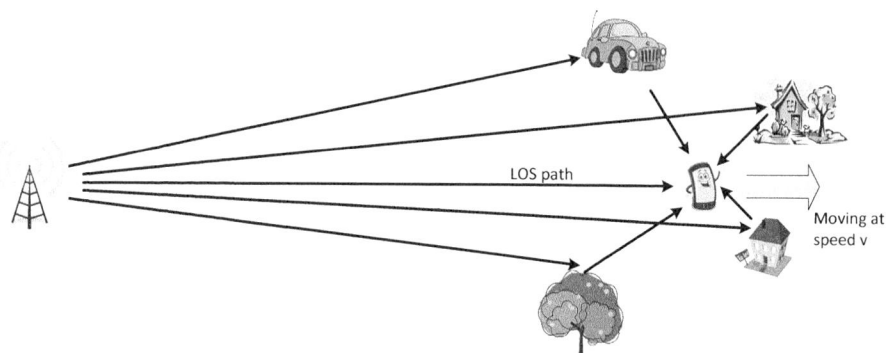

Figure 2.2 Illustration of small fading

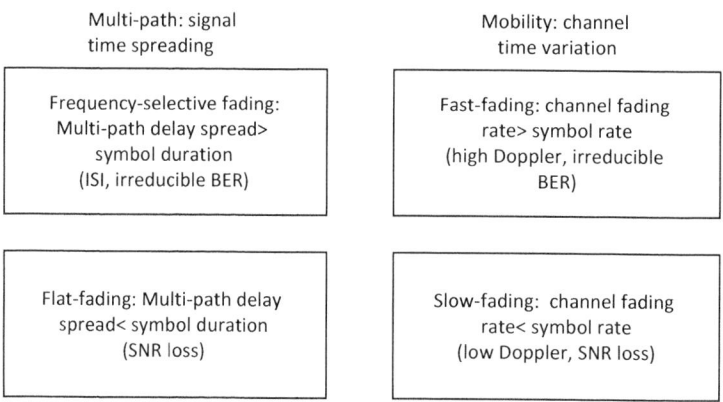

Figure 2.3 Phenomenon of small fading

may even be blocked by large obstacles. Small-scale fading is caused by interference between several versions of the transmitted signal that arrive at the receiver at different times. The multi-path waves are combined at the receiver and the resultant signal may vary widely in amplitude and phase.

There are two main effects in small-scale fading: multi-path and mobility, as shown in Figure 2.3. Multi-path waves cause fades within small distances and time spreading of signals while mobility of users or the environment causes time-variant behavior of wireless channels. Small-scale fading results in significant channel variation within a short period of time or travel distance.

In the following, we introduce several mathematical models for different types of fading channels.

2.3.1   Flat-fading channels

Consider first flat-fading channels where the differences in the delays of different paths are negligible compared to the symbol time. Therefore, it looks like only one copy of the transmitted signal is received. This happens in narrow-band wireless communications, where signal bandwidth is smaller than the channel coherence bandwidth. Define the corresponding small-fading complex channel gain to be

$$\beta(t) = \alpha(t)e^{i\theta(t)}, \tag{2.11}$$

where $\alpha(t)$ is the amplitude gain and $\theta(t)$ the phase shift. The received signal is

$$r(t) = \beta(t)x(t), \tag{2.12}$$

where $x(t)$ is the transmitted signal. $\beta(t)$ is the superposition of different copies of signals from all paths and depends on the combining algorithm. $\alpha(t)$ and $\theta(t)$ can be modeled as wide-sense stationary processes.

For slow-fading channels, $\alpha(t)$ and $\theta(t)$ almost do not vary over an observation period such as a frame or a symbol duration. An example is the block-fading channel where the channel stays static in each frame duration and the channel may vary in different frames. Over each observation period, $\alpha(t)$ and $\theta(t)$ can be modeled using random variables, α and θ. An example is the Rayleigh slow-fading channel model, which assumes a rich scattering environment without LOS signals. In this case, both the in-phase and quadrature channels are the superposition of signals from many paths and can be modeled as Gaussian random variables. Correspondingly, α is Rayleigh distributed with the probability density function

$$p(\alpha) = \frac{2\alpha}{\alpha_0} e^{\frac{-\alpha^2}{\alpha_0}}, \alpha \geq 0, \tag{2.13}$$

where α_0 is the mean of α. θ is uniformly distributed between 0 and 2π radians. Where there is a LOS signal, the channel can be modeled as a Rician fading channel and α is Rician distributed with the probability density function

$$p(\alpha) = \frac{\alpha}{\sigma^2} e^{\frac{-(\alpha^2+v^2)}{2\sigma^2}} I_0\left(\frac{\alpha v}{\sigma^2}\right), \tag{2.14}$$

where I_0 is the modified Bessel function of the first kind with order zero,

$$v^2 = \frac{K}{1+K}\omega, \tag{2.15}$$

and

$$\sigma^2 = \frac{\omega}{2(1+K)}, \tag{2.16}$$

where K is the ratio between the power in the LOS path and the power in the other paths. ω is the total power from all paths. When $v = 0$, or $K = 0$, the Rician fading channel reduces to a Rayleigh channel.

For fast-fading channels, $\alpha(t)$ and $\theta(t)$ vary in time and depend on the Doppler power spectrum. There are many factors that would affect the Doppler power spectrum, e.g. antenna heights, moving speed, polarization of electromagnetic waves, etc. For Rayleigh fading with a vertical monopole receive antenna, $\beta(t)$ can be modeled as a a zero-mean complex Gaussian process with the autocorrelation function [201]

$$R_\beta = J_0(2\pi f_d \tau), \tag{2.17}$$

where f_d is the Doppler frequency shift relative to the carrier frequency and

$$f_d = \frac{v f_c}{c}, \tag{2.18}$$

where v is the mobile speed projected on the direction to the base station, f_c the carrier frequency, and c the light speed. Correspondingly, the Doppler spectrum is the Fourier transform of the autocorrelation function and is

$$S(f) = \frac{1}{\pi f_d \sqrt{1 - \left(\frac{f}{f_d}\right)^2}} \tag{2.19}$$

for all $|f| \leq f_d$; otherwise, $S(f)$ is zero.

2.3.2 Frequency-selective fading channels

With a frequency-selective channel, the receiver will receive multiple copies of the transmitted signals with different delays that are comparable with or even longer than the symbol duration. This happens in wide-band wireless communications where signal bandwidth is larger than the channel coherence bandwidth. The received signal is frequently modeled using a tapped delay line model,

$$r(t) = \sum_{l=1}^{L(t)} \beta_l(t) x(t - \tau_l(t)), \tag{2.20}$$

where $\tau_l(t)$ is the propagation delay of the lth path at time t and

$$\beta_l(t) = \alpha_l(t) e^{i\theta_l(t)}, \tag{2.21}$$

where $\alpha_l(t)$ and $\theta_l(t)$ are the corresponding amplitude attenuation and phase shifts of the lth path or tap at time t. If the channel is experiencing slow-fading, $L(t)$, $\tau_l(t)$, $\alpha_l(t)$, and $\theta_l(t)$ are random variables in each observation duration. In the fast-fading case, they are all random processes. Usually, $L(t)$ and $\tau_l(t)$ are assumed to be fixed in the model. Each $\beta_l(t)$ follows the properties discussed in Section 2.3.1. The channel gain in the frequency domain can be expressed as

$$B(f, t) = \sum_{l-1}^{L(t)} \beta_l(t) e^{-i2\pi f \tau_l(t)}. \tag{2.22}$$

2.4 Channel estimation

The utilization of channel information not only improves the performance of receivers significantly, but also provides an additional dimension of freedom in resource management.

In channel estimation, the transmitted sequence is assumed to be known at the receiver, in which the channel information is estimated through the received distorted signals. There are two approaches to obtaining the sequence. One is pilot-based, in which a known sequence, called a pilot, sounding tone, or training sequence, is transmitted to the receiver. Most commercial wireless systems deploy the pilot scheme and send training sequences periodically for channel estimation. Although transmitting pilot sequences consumes bandwidth from data streams, it makes channel estimation simpler and estimation accuracy higher. Another approach, called decision feedback, is to use the previously detected symbols as the known sequences. However, it may suffer from error propagation. If we assume that the detection is error free, then there is no difference in channel estimation analysis between these two approaches.

The most popular performance metric of channel estimation is *mean squared error* (MSE) of pilot and estimated signals. Since the sensitivity of a receiver also depends on the modulation and coding schemes, other receiver performance metrics, such as the *bit error rate* (BER), are also useful to understand the impact of channel estimation error on the receiver.

2.4.1   Flat slow-fading channels

We start with the simplest case, in which channel fading is both flat (unchanged) on the frequency and time domains. Assume that at time slot i, x_i and y_i are the pilot and received symbols, respectively. β is the channel information to be estimated. Then the received symbol is

$$y_i = \beta x_i + n_i,$$

where n_i is the independently and identically distributed white Gaussian noise with power N_0. In order to minimize the mean squared error $E(\hat{\beta} - \beta)^2$, where $\hat{\beta}$ is the estimated value, the *least squares* (LS) estimate is just

$$\hat{\beta} = \frac{y_i}{x_i}.$$

It is straightforward to show that the estimated channel coefficient is a Gaussian random variable as follows:

$$\hat{\beta} \sim N\left(\beta, \frac{N_0}{E[x_i]^2}\right). \tag{2.23}$$

It is shown in Equation (2.23) that increasing the transmit power of the training signals can reduce the channel estimation error. Many wireless standards call this channel estimation technique as power booting.

Channel estimation error can be reduced further by averaging multiple channel measurements, such as

$$\hat{\beta} = \frac{1}{K}\sum_{k=1}^{K}\beta_k,$$

where K is the measurement number. Then

$$\hat{\beta} - N\left(\beta, \frac{N_0}{E[x_i]^2 K}\right).$$

It is shown that both increasing the power by K and measuring K times result in the same channel estimation accuracy. The difference is that multiple measurements reduce the information payload rate (bandwidth), but power boosting increases the transmission power.

2.4.2   Frequency-selective slow-fading channels

For broadband communication systems, frequency-selective fading channels have to be dealt with. Orthogonal frequency division multiplexing (OFDM) is a multi-carrier modulation technique where data symbols modulate a parallel collection of regularly spaced subcarriers. Therefore, OFDM requires to estimate a vector of subchannels rather than a single one. More details about OFDM can be found in [147]. For this chapter, however, it is sufficient to know that OFDM has multiple narrow-band subchannels, each of which can be regarded as a flat-fading one. Mathematically, the pilot symbols $x_k, k = 1, \ldots, K$, are assumed to be known, where K is the subchannel number. The received OFDM vector is given by

$$y_k = \beta_k x_k + n_k,$$

where y_k, β_k, and n_k are the received symbols at subchannel k, the channel coefficient of subchannel k to be estimated, and the white noise that is independent over all subchannels.

The LS estimator is as simple as

$$\tilde{\beta}_k = \frac{y_k}{x_k} = \beta_k + \frac{n_k}{x_k} == \beta_k + \tilde{n}_k.$$

A straightforward question is: Can we do better? The answer is: Yes. Utilizing correlation among β_k can improve the accuracy of channel estimation, which can be presented as

$$(\hat{\beta}_1, \hat{\beta}_2, \ldots, \hat{\beta}_i) = f(\tilde{\beta}_1, \tilde{\beta}_2, \ldots, \tilde{\beta}_K),$$

where f is an enhanced estimator (or filter) using all outputs from the LS estimator. Much R&D effort on studying f has been made over the decades. There are so many implementations of f. Here, we only introduce the basic ideas behind the various implementations and designs.

Assume that the correlation matrix $\mathbf{R}_\beta$ is known,

$$\mathbf{R}_\beta = \begin{pmatrix} r_\beta[0] & r_\beta[1] & \cdots & r_\beta[K-1] \\ r_\beta[-1] & r_\beta[0] & \cdots & r_\beta[K-2] \\ \vdots & \vdots & \ddots & \vdots \\ r_\beta[-K+1] & r_\beta[-K+2] & \cdots & r_\beta[0] \end{pmatrix}.$$

The elements of $\mathbf{R}_\beta$ represent the frequency-domain correlation of the channel defined as

$$r_\beta[k] = E\{\beta_{k_0+k}\beta_{k_o}^*\}.$$

The *minimum mean squared error* (MMSE) estimate of the channel is obtained by

$$\hat{\beta} = \mathbf{R}_\beta (\mathbf{R}_\beta + \rho \mathbf{I})^{-1} \tilde{\beta}.$$

The correlation matrix is fully determined by the power-delay profile of the channel. The MMSE estimator has the best performance when the statistics for the estimator matches the real channel power-delay profile. However, the performance degrades significantly when the estimator does not match the channel. Fortunately, the degree of freedom of the channel is usually a small number. This is because there only exist a few significant multiple paths. Therefore, the correlation matrix can be decomposed by low-rank matrices. The low-rank channel estimator not only achieves robust performance but also reduces the computation time.

2.4.3   Fast-fading channels

Fast-fading channels are time-selective. Similarly, time correlation information helps improve channel estimation accuracy. Due to the Doppler effect, the higher the moving speed, the lower the time correlation.

For the most general scenario in which channel fading is both time- and frequency-selective, the use of the time- and frequency-domain correlations is a natural extension for enhancing the LS estimator. For more details, please refer to [124].

2.4.4   Conclusion

Channel estimation is required for physical layer technologies, and can achieve fairly high accuracy in practice. The rest of the book will show that channel information also significantly improves network performance via cross-layer design techniques. The channel information for the upper layers usually does not require the phase of the channel. The physical layer passes required channel information to upper layers to help the upper layers improve network efficiency. Throughout the book, we assume channel information is always available and investigate the impact of utilizing this information in improving network energy and spectrum efficiency.

2.5   Other challenges

Besides the challenge from wireless channels that vary in time, frequency, and location, wireless networks also have a second major problem of interference among multiple users that communicate over a common medium. The main task of the wireless physical layer is to send bit streams reliably. Therefore, the physical layer must be able to combat varying channels, thermal noise, and interference. When the interference is too strong for the physical layer to handle, the medium access control (MAC) layer will control channel access of all users in the network such that interference can be avoided.

3 Spectral and energy efficiency of wireless networks

Wireless networks can be measured by a number of metrics from different perspectives and spectral and energy efficiency are the most important two. Spectral efficiency, defined as the system throughput for unit bandwidth, is a widely accepted metric. For example, the peak value of spectral efficiency is always one key performance indicator of 3GPP evolution. For instance, the target downlink spectral efficiency of 3GPP increases from 0.05 b/s/Hz to 5 b/s/Hz as the system evolves from GSM to long-term evolution (LTE). In contrast, energy efficiency, has previously been ignored by most research efforts and has not been considered by 3GPP as an important performance indicator until very recently. As the green evolution becomes a major trend, energy-efficient wireless networking becomes more and more important. Unfortunately, spectral efficiency and energy efficiency are not always consistent and sometimes conflict with each other. Therefore, how to balance the two metrics needs careful study.

3.1 Spectral efficiency

Wireless communications use electromagnetic signals to carry information, and the signals are characterized by their frequency bands. The signals of each user occupies a frequency band, i.e. a set of radio frequencies. If multiple users are communicating over the same frequency simultaneously, they will interfere with each other. Therefore, different wireless access technologies usually use different frequency bands to avoid interfering with each other.

Frequency spectrum is a precious natural resource of a very limited amount and is strictly regulated by national governments all over the world. For example, in the United States, the Federal Communications Commission (FCC) manages spectrum for non-Federal use, e.g. local governments, commercial businesses, and personal uses, and the National Telecommunications and Information Administration (NTIA) for Federal uses, e.g. Army and the Federal Bureau of Investigation (FBI). With spectrum management, spectrum is divided into different bands and each band is allocated for a certain application. The purpose of spectrum management is to optimize the use of the spectrum, avoid interference, mitigate radio spectrum pollution, and coordinate wireless communications of different entities. Table 3.1 gives some examples of the frequency allocations in the United States.

Table 3.1 Sample spectrum allocations in the US (data from United States radio spectrum frequency allocations chart as of 2003 [198])

Application	Frequency band
AM radio	535–1705 kHz
FM radio	88–108 MHz
TV channels 5–6	76 MHz–88 MHz
TV channels 7–13	174 MHz–216 MHz
802.11b and 802.11g	2.4 GHz
GSM	824–849 and 869–894 MHz

The demand for wireless broadband access is soaring because of the explosive growth in mobile Internet services. Though originally intended for wired communications, it is now envisioned that broadband multimedia services will also be provided in mobile wireless environments because of technology advances and user demand. However, there is only a fixed amount of spectrum resources available for wireless communications. Spectrum efficiency is the amount of information that can be transmitted per unit frequency band in a wireless network. It quantifies how efficiently the system spectrum is used by the whole network. Therefore, spectrum efficiency is crucial for all wireless networks because it allows these networks to maximize utilization of assigned frequencies to provide services for more users. One main goal of network design is to increase spectral efficiency as much as possible.

3.2 Energy efficiency

With the explosive growth of high data-rate applications, more and more energy is consumed in wireless networks to guarantee quality-of-service (QoS). Information and communication technology (ICT) plays an important role in global greenhouse gas emissions since the amount of energy consumed by ICT increases dramatically to meet the explosive growing service requirements. It is shown nowadays that the total energy used by the infrastructure of cellular networks, wired networks, and Internet takes up more than 3% of worldwide electric energy consumption [80]. In addition, this amount of energy is expected to increase rapidly in the future because of future demand. More importantly, a large portion of energy consumption originates from the operation of wireless networks. Mobile devices in wireless networks are usually battery driven. Slow development in battery technology compared to the demand of mobile applications necessitates ways of saving energy for mobile devices. Therefore, it is critical to improve the energy efficiency of wireless networks, from both the infrastructure and device perspectives.

In the past, most efforts of communication researchers have been devoted to the enhancement of network spectral efficiency. Usually high network spectral efficiency implies high energy consumption, which is not desired when we want high energy efficiency of the network. How to reduce energy consumption while meeting throughput requirements is urgent, and energy-efficient wireless design has received significant attention in both academia and industry recently. For instance, several projects and organizations, e.g. Energy-Aware Radio and neTwork tecHnologies (EARTH) [2] and Green Touch, have been set up to develop architectures, specifications, roadmaps, and techniques to increase network energy efficiency. For example, GreenTouch [3] is a consortium of leading ICT research institutes, including AT&T, Bell Labs, China Mobile, Swisscom, Tsinghua University, and so on, and is dedicated to fundamentally transforming communications and data networks and to significantly reducing the carbon footprints of ICT devices, platforms, and networks. The goal of GreenTouch is to achieve a 1000-fold improvement in the future energy efficiency of the Internet and other networks that support communications, commerce, and entertainment. The goal is bold, but achievable. Recent research suggests that key technology limits are still 10 000 times below current operating levels [3].

3.3   Link metrics versus network metrics

3.3.1   Link spectral efficiency

According to information theory, the channel capacity of a channel, denoted by C, is the maximum possible data rate that can be reliably delivered through the channel without any error and is measured in bits per second, i.e. b/s. Define the channel bandwidth by B Hz. The maximum spectral efficiency that can be achieved on this channel is

$$\eta_S = \frac{C}{B}, \tag{3.1}$$

where η_S stands for spectral efficiency with the unit bits per second per Hertz (b/s/Hz). If the channel is an additive white Gaussian noise (AWGN) channel,

$$C = B \log_2 \left(1 + \frac{Ph}{N_o B} \right)$$

and the spectral efficiency is

$$\eta_S = \log_2 \left(1 + \frac{Ph}{N_o B} \right), \tag{3.2}$$

where P is the radio transmission power, h is the channel power attenuation, and N_o is the noise spectral density. Note this is the theoretical tight spectral efficiency upper-bound that in practice cannot be achieved. In a practical wireless system, the link spectral efficiency is the net data rate, R, i.e. the useful information rate excluding physical (PHY) layer implementation overhead like coding and pilots, divided by the channel bandwidth B,

$$\eta_S = \frac{R}{B}. \tag{3.3}$$

For example, if user A uses 1 MHz bandwidth to send data to user B at a net data rate of 2 Mb/s, the spectral efficiency is 2 b/s/Hz.

Spectral efficiency may also be measured in the number of net bits, i.e. bits received without error, per channel use (b/CU). Here per channel use indicates the occupation of the system bandwidth for a certain time period. Therefore, CU is the same as $s * Hz$ and b/CU is indeed the same as $b/s/Hz$. If in each channel use only one symbol is transmitted, the spectral efficiency can also be calculated by the number of net bits transmitted per symbol.

With uncoded M-ary quadrature amplitude modulation (M-QAM) modulation, each symbol carries $\log_2(M)$ bits and the modulation efficiency is $\log_2(M)$ b/Symbol, which is usually different from spectral efficiency and spectral efficiency, is also dependent on channel, coding and other PHY implementations. Define the average bit error rate (BER) of M-QAM modulation to be p_e and assume that bit errors are independent and uniformly distributed. For example, the BER for coherently detected M-QAM with Gray mapping over an additive white Gaussian noise channel can be well approximated by [24]

$$P_e(\eta) \approx 0.2 \exp\left(-\frac{1.5G_c\eta}{M-1}\right), \tag{3.4}$$

where G_c is the coding gain indicating the SNR improvement because of channel coding. The packet error rate (PER), p_p, is

$$p_p = 1 - (1 - p_e)^L, \tag{3.5}$$

where L is the number of bits in each packet. Correspondingly, the spectral efficiency is

$$\eta_S = \log_2(M)(1 - p_p)(1 - p_o), \tag{3.6}$$

where p_o stands for the percentage of overhead, including forward error correction (FEC) coding overhead, cyclic redundancy check (CRC) bits, etc. Clearly, the spectral efficiency is upper bounded by the modulation efficiency. The spectral efficiency of M-QAM in (3.6) relies on three parameters: modulation order M, packet error rate, and implementation overhead. Higher modulation order results in higher modulation efficiency but not necessarily higher spectral efficiency because the PER may turn out to be too high. With a stronger FEC code and a lower coding rate, the packet is better protected and the PER p_p is smaller. However, it means more coding overhead and larger p_o and it is not clearer if the spectral efficiency is increased. Therefore, there is a tradeoff in choosing the modulation order and the FEC coding. In addition, it is not clear whether the spectral efficiency can be improved by adjusting one parameter solely. The modulation order and channel coding should be jointly selected considering the channel state such that the spectral efficiency is maximized.

3.3.2 Network spectral efficiency

Link spectral efficiency measures how well a link uses the spectrum resource. Extending this concept to a multi-user wireless system, i.e. a wireless network, the spectral efficiency should measure how well the whole network uses the spectrum resource. It is the sum of net data rates of all users in the network divided by the network bandwidth B, i.e.

$$\eta_S = \frac{\sum_i R_i}{B}, \tag{3.7}$$

where R_i stands for the net data rate of user i. It also has a unit of $b/s/Hz$.

Since all users share the same spectrum, the increase in the link spectral efficiency of one user may affect the link spectral efficiency of other users because of the conflict of spectrum sharing. For instance, in a multi-cell cellular network, if the base station in one cell is transmitting with higher power to achieve higher spectral efficiency, it is bringing higher co-channel interference to the neighboring cells and thus reduces their spectral efficiency. Then it is not clear if the network spectral efficiency can be improved. The allocation of spectrum resources to all users in a network is therefore essential in determining the network spectral efficiency and the allocation is mainly managed by scheduling algorithms, medium access protocols, and frequency planning.

It is easy to calculate the spectral efficiency based on (3.7) for a small-scale wireless network such as a 802.11 WiFi network, a wireless sensor network with a small coverage, a wireless body area network, etc. You only need to measure the throughput of all users, sum them up, and divide the sum by the total system bandwidth to get the network spectral efficiency. Since the system bandwidth is usually fixed, improving the sum throughput of all uses is equivalent to improving the network spectral efficiency.

For a large-scale network, e.g. a cellular network with numerous cells, it is usually difficult to determine the sum network throughput and therefore the network spectral efficiency as defined in (3.7). Instead, some equivalent definitions can be used. For example, the network spectral efficiency of a cellular network can be expressed in the form of area spectral efficiency, i.e.

$$\eta_{SD} = \frac{\sum_{i \in \{\text{users in the area of interest}\}} R_i}{B \cdot A}, \tag{3.8}$$

with the unit $b/s/Hz/m^2$, where A is the area of interest in the cellular network. For example, if the area of interest is one sector in a cell, η_{SD} is the sector spectral efficiency. Comparing η_{SD} of different areas, we can see how efficiently the spectrum is used in different locations of the network.

Link spectral efficiency is mainly a PHY-layer concept and only the PHY layer overhead should be excluded in calculating the net data rate and therefore the spectral efficiency. This is because the upper-layer protocol overhead such as the medium access control (MAC)-layer packet header are useful information that must be successfully transmitted regardless of the design and implementation in the PHY layer. Network spectral efficiency is a concept above the PHY layer and the overhead across the whole network protocol stack should be excluded in determining the net data rates of all users. This is because overheads such as PHY preambles, MAC headers, routing bits, and so on are all designed to facilitate the functioning of the network and to deliver the net bits of applications of all users in the network. Efforts across the whole network protocol stack are needed to improve the network spectral efficiency. As this book focuses on joint

PHY-MAC layer optimization, MAC-layer throughput will be used in the following to measure the net data rate of all users and thus to determine network spectral efficiency.

3.3.3 Link energy efficiency

Similar to spectral efficiency, energy efficiency quantifies how efficiently the energy resource is used by the network. Therefore, we define energy efficiency as the amount of information that can be reliably transmitted per unit of energy consumption. For link energy-efficient communications, at time t, with energy consumption $\Delta e[t]$ in the following duration Δt, a user achieves a clean data rate $R[t]$, which excludes error correcting codes and other overhead. Note that Δe is the total energy consumption of the user and so is $R[t]$. For example, if the user is sending data to multiple users at the same time, $R[t]$ should be the sum of data rates of all receivers. The energy efficiency at time t of the user is

$$\frac{R[t]\,\Delta\,t}{\Delta e[t]}, \tag{3.9}$$

which is equivalent to

$$u[t] = \frac{R[t]}{\Delta e[t]/\Delta t} = \frac{R[t]}{P[t]}. \tag{3.10}$$

$u[t]$ is called the energy efficiency of the user at time t and has a unit b/J. Similarly, the energy efficiency of the user for communications between time 0 and t is defined as

$$u[t] = \frac{\int_0^t R[t]\,\Delta\,t}{\int_0^t \Delta e[t]} = \frac{\int_0^t R[t]}{\int_0^t P[t]}, \tag{3.11}$$

which is the total amount of bits successfully delivered in this time duration divided by the total amount of energy consumed.

3.3.4 Network energy efficiency

Similar to the network spectral efficiency, one straightforward definition of network energy efficiency is the total amount of net bits of all users delivered in the whole network divided by the total amount of energy consumed in the same time duration in the whole network, i.e. at time t, the network energy efficiency is

$$u[t] = \frac{\sum_n R_n[t]}{\sum_n P_n[t]}, \tag{3.12}$$

where the two summations are over all devices in the network. This definition also has a unit b/J. The network energy efficiency for the operations between time 0 and t is

$$u[t] = \frac{\int_0^t \sum_n R_n[t]}{\int_0^t \sum_n P_n[t]}. \tag{3.13}$$

However, it is difficult to apply this metric in practice, both measurement- and design-wise. Measurement-wise, it is difficult to put power meters around each component of each user, base station, or access point to measure the power consumption. What is more, these measurement results need to be collected globally to determine final network energy efficiency and incur lots of signaling overheading. So this metric has poor scalability. From the design perspective, while this metric is a global one, each device and access point has their individual physical constraints on power. The power resources of different users can not be shared. If we want to design a system based on this metric, the power allocation that maximizes network energy efficiency may not be feasible in practice.

Because of the above concerns, the energy efficiency of each device that has an independent power source should be individually defined and the network energy efficiency should be a certain sum of the energy efficiency of individual devices. In this book, we will use the following two network energy efficiency metrics, sum and log sum of individual energy efficiency, named network energy efficiency and proportional fair (PF) network energy efficiency respectively. With the sum metric, the network energy efficiency is

$$u[t] = \sum_{n=1}^{N} u_n[t].$$ (3.14)

Maximizing the network energy efficiency is the same as maximizing the arithmetic average of the energy efficiency of all users. With the log sum metric, the PF network energy efficiency is

$$v[t] = \sum_{n=1}^{N} \log(u_n[t]).$$ (3.15)

Maximizing the PF network energy efficiency is the same as maximizing the geometric average of the energy efficiency of all users.

In the context of spectral efficiency, the arithmetic-mean metric leads to schemes for sum throughput maximization and no fairness will be assured since most users may have no throughput. The geometric-mean metric introduces proportional fairness in the achieved data rates among all users [165, 152]. Analogously, the log sum metric in (3.15) leads to the PF network energy efficiency.

Compared to the metrics in (3.12) and (3.13), these two metrics can be easily used in practice because it is much easier to measure individual energy efficiency. Also it is much easier to observe the energy efficiency of individual users in the network so that in radio resource management the network can allocate more resources to the users that are the most energy hungry.

4 Centralized resource management in wireless networks

4.1 Overview

In a centralized wireless network, a central controller has complete network information, e.g. channel states, traffic properties, and queue statuses. The controller determines the resource allocation for each user and the resources can be a set of time slots, codes, frequency bands, power, etc. The allocation can be orthogonal such that different users use resources that do not conflict with each other, e.g. different time slots. The allocation can also be non-orthogonal. In the latter case, different users may use resources that partially overlap with each other and, therefore, the communications of one user will impact those of others.

Many centralized wireless medium access control (MAC) protocols have been designed for use in infrastructure-based networks, where access points or controllers decide access to the channel or perform some other centralized coordination functions. One typical example is a single-cell cellular network where multiple mobile users are communicating simultaneously to a base station. The base station has the channel knowledge of all users as well as the quality of service (QoS) requirements of them. The base station schedules the network resources for both the downlink and uplink transmissions of all users. If the network uses frequency division multiple access (FDMA), the BS divides the system frequency spectrum into several frequency bands, called channels. These channels are separated far enough such that they do not interfere with each other. The BS then assigns each channel to one user exclusively. For example, FDMA is used in the Advanced Mobile Phone System (AMPS), which is a first-generation cellular network and is an analog mobile phone system standard developed by Bell Labs. As each user takes at least one channel, AMPS requires a large amount of system bandwidth if the number of users in the network is large. Similar examples are time slot allocation in time division multiple access (TDMA), code assignment in code division multiple access (CDMA), and so on. Centralized resource management can also be used in multi-cell cellular networks, where several base stations are controlled by a Radio Network Controller (RNC) and the RNC controls the network resource allocation of all users in all cells.

Centralized resource management can also be used in wireless local area networks (LANs). For example, the IEEE 802.11 standard features two operation modes: the distributed coordinated function and point coordinated function. The channel access time is divided into two periods: the contention free period and contention period. The point coordinated function mode resides in the contention free period, in which the access point coordinates the communications of all users in this mode. In the contention free period, the access point polls point coordinated function-capable users one by one and grants-access to the channel in the following data frame. In case the polled user does not have any frame to send, it must transmit a null frame. New users that want to get enrolled in the poll list, should send a request in the contention period.

The benefit with a central scheduler is the easiness and effectiveness in controlling the amount of resources allocated to each user and therefore it can achieve good network performance. Especially, with a central scheduler, it is likely that the operation of the network is globally optimal. However, the functioning of the network heavily depends on the central controller. If the central controller malfunctions, the entire network may break down. The central controller can also be the bottleneck of the whole network. For example, it may be computationally prohibitive at the controller when there are a large number of users to be scheduled. In addition, the collection of information for scheduling decisions may require a large amount of signaling overhead, which grows fast with the number of users in the network as the information from all users needs to be collected. Therefore, networks operated with central controllers usually have limited scalability. Further, the delay in the signaling procedure may result in inaccuracy in collecting network information and therefore incurs performance degradation.

4.2   Wireless scheduling challenges

There are certain factors that affect wireless scheduling: the set of available wireless resources to share among the terminals, the channel information available at the scheduler, the service requirements of each terminal, and the objectives to optimize in relation to the performance metric of the service provider.

Broadband mobile cellular networks need to support mixed classes of traffic with different QoS requirements. QoS differentiation and guarantees are therefore needed.

Take universal mobile telecommunications system (UMTS) as an example. UMTS defines four QoS classes, conversational, streaming, interactive, and background, as shown in Table 4.1. The conversational class has a strict requirement of the transfer time, and the transfer time should be small enough such that the time relation between information entities, e.g. transmitter and receiver, should be preserved. The delay requirement is given by how the human perceives audio or video conversations. The streaming class is the delivery of real-time audio or video services and is a one-way service. This class of services does not require very low transfer delay because it only affects response in the initial access. Besides, data buffering is usually used in the receiver side to smooth the delay jitter, and therefore delay variation can also be larger than the conversational class. The interactive class characterizes applications such as web browsing and remote server access. Here a user sends a service request and the remote server sends back the required contents. Users receiving this class of services expect to receive the content reliably within a certain time, but not necessarily immediately (not real time). So the delay requirement is much looser than the conversational and streaming classes. But the content must be preserved, meaning data should be received with very low probability

Table 4.1 QoS classes in UMTS (3GPP TS 23.107)

Traffic class	Characteristics	Example
Conversational	Preserve time relation (variation) between information entities of the stream. Conversational pattern (stringent and low delay)	voice
Streaming	Preserve time relation (variation) between information entities of the stream	streaming video
Interactive	Request response pattern. Preserve payload content	web browsing
Background	Destination is not expecting the data within a certain time. Preserve payload content	emails

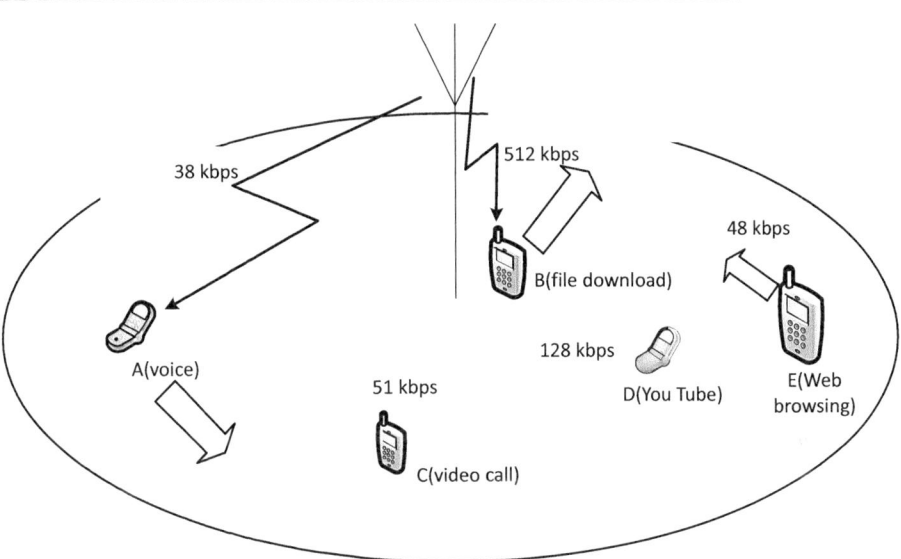

Figure 4.1 A single-cell cellular network (128KHz system bandwidth)

of error. The background class is the least time insensitive and is usually called best-effort service. This class of applications has no time requirements as long as the content can be delivered reliably.

The traffic in wireless networks therefore may be diverse in rate, delay, bit error rate (BER), service quality, and so on. An example is shown in Figure 4.1, where five users are accessing the network with different service requirements, including voice, video call, file download, You Tube, and web browsing. Besides requirement differences, users may have other limiting factors that differ from each other, e.g. peak power limits. To accommodate these different requirements, wireless networks should be flexible in allocating network resources, which can be achieved using scheduling algorithms. A simple scheduling algorithm is to select users according to a fixed cyclic order, yielding a fair transmission opportunity for every user. However, if a user has urgent packets that need to be delivered within a time limit, i.e. with a delay requirement, this algorithm may fail because the user has to wait for its turn. So, priority may need to be considered in the scheduling algorithms. Usually, the improvement in the QoS of one user will degrade the performance of other users. For example, in Figure 4.1, assume only users A and B are accessing. If the whole system bandwidth is allocated to user A, it will achieve 38 kbps throughput, while to B, it will achieve 512 kbps. User A is receiving voice service and desires only a 25 kbps stable data rate while B desires as high a data rate as possible. To obtain the highest network throughput, user B should be always scheduled, indicating no service to user A at all. Obviously, this is not a desired solution as user A will be very unhappy. When there are multiple users in the network, all with different delay and rate requirements, like the five users in Figure 4.1, the scheduling problem will be even more complicated because the resources are shared by all users and it may be difficult to satisfy all users at the same time. Therefore, while maximizing system performance, it is also critical for the scheduling to enhance the QoS performance of individual users in a balanced way.

In addition to the challenges in supporting various QoS requirements, because of shadowing, fading, noise, interference, and user mobility the quality of wireless communications is unstable, error-prone, and cannot be predicted. The capacity of almost every wireless link varies significantly over different time periods and locations. Even when the scheduler knows the QoS requirement of a user, it is still difficult to estimate the amount of resources needed to meet the requirement. An adaptive procedure is therefore needed to provide QoS assurance considering user requirements and channel variations. Good scheduling approaches should try to exploit channel conditions opportunistically to achieve better network performance. Here, the term opportunistic means the ability to schedule users based on favorable channel conditions. However, the potential of opportunistic transmission also introduces a tradeoff between efficiency in using wireless resources and level of user satisfaction. For example, as shown in Figure 4.1, a base station is communicating with several mobile users at the same time. The users are located at different places and some are very close to the base station, called cell-center users, and others far away, called cell-edge users. Cell-center users can enjoy very reliable communications with the base station, like user B, while cell-edge users can barely communicate with the base station, like user A. Furthermore, users may move, as indicated by the arrows in Figure 4.1. This mobility will make the link qualities more dynamic. Therefore, the scheduler should have certain dynamic mechanisms to deal with channel variations at different time periods and locations such that the QoS requirements can be met.

4.3   Centralized scheduling algorithms

Scheduling makes the system flexible enough to operate in adapting to the channel characteristics and QoS requirements, thereby permitting a flexible service architecture for integrating various types of services in a single air-interface. Scheduling in wireless

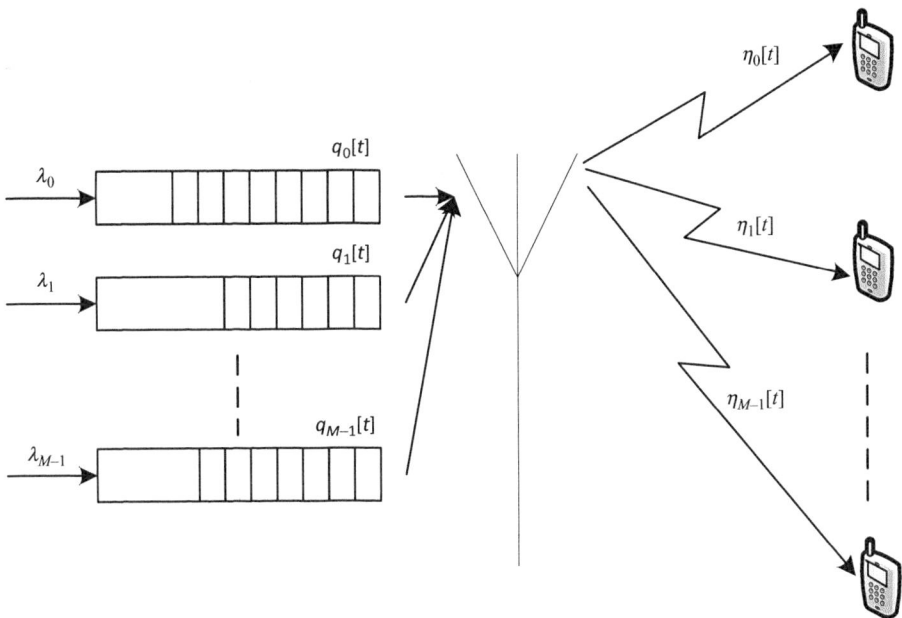

Figure 4.2 Downlink scheduling

networks, especially in cellular mobile network, can lead to significant performance improvement without more spectrum. This section presents a framework for scheduling users in an opportunistic way to achieve high network performance by exploiting time-varying channel conditions while controlling the level of QoS among users. In the following paragraphs, we will introduce how wireless resources are scheduled to maximize the network performance while meeting individual QoS requirements.

Assume a base station is scheduling M users on one single channel. In each time slot, only one user will be scheduled. An example can be the uplink or downlink communications in cellular networks, as shown in Figures 4.2 and 4.3. The coming traffic of each user is stored in an independent queue. At time t, the queue of user i changes as in the following state equation:

$$q_i[t + 1] = q_i[t] + \delta_i[t] - \eta_i[t], \tag{4.1}$$

where $q_i[t]$ is the number of bits in the queue of users i, $\delta_i[t]$ the number of bits arriving at the queue, and $\eta_i[t]$ the number of bits scheduled to transmit in this time slot. Scheduling algorithms are used to determine $\eta_i[t]$ for all users in each time slot.

4.3.1   Round-robin scheduling

Round-robin (RR) scheduling is one of the simplest scheduling algorithms. Users in a RR algorithm are scheduled in a round robin, i.e. cyclic order, manner. Mathematically,

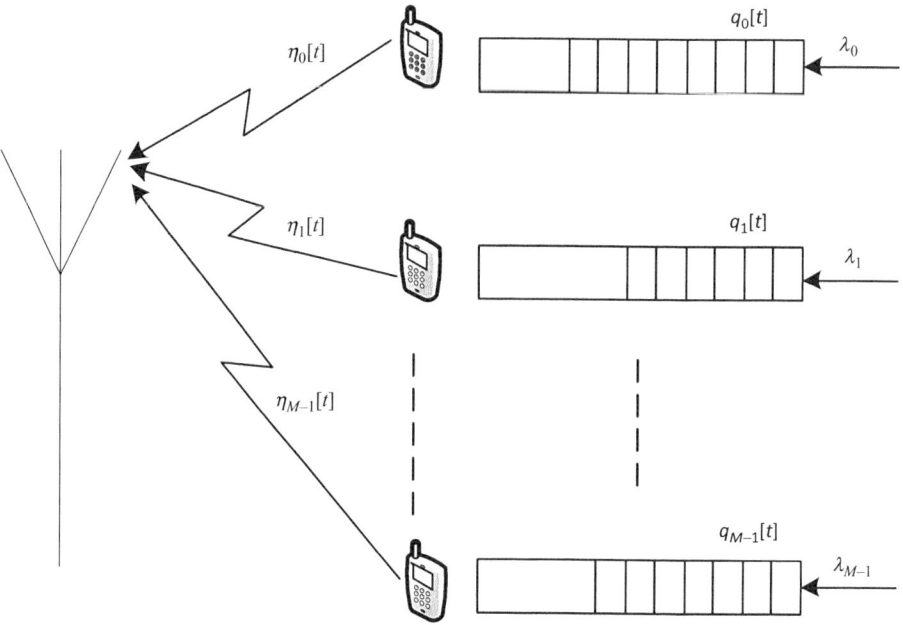

Figure 4.3 Uplink scheduling

the index, $i[t + 1]$, of the user that is scheduled at time $t + 1$ is given by

$$i[t + 1] = i[t] + 1. \tag{4.2}$$

It can be seen that the scheduler always selects the user that has not been served for the longest time. RR scheduling is fair in the sense that it gives all users the same amount of time resources. However, this algorithm does not use the user channel quality information and may suffer from low throughput. In the following, we introduce more advanced schedulers to improve network performance.

4.3.2   Max throughput scheduling

The goal of max throughput scheduling is to schedule the user in each time slot such that the total network throughput is maximized. If user i is scheduled, the expected instantaneous throughput in this slot would be

$$\hat{r}_i[t] = \frac{\hat{\eta}_i[t]}{T_s}, \tag{4.3}$$

where T_s is the slot length and $\hat{\eta}_i[t]$ is the estimated number of bits that can be successfully delivered. The total expected network throughput would be

$$\hat{r}[t] = \sum_{i=0}^{M-1} \hat{r}_i[t]I(i), \tag{4.4}$$

where $I(i)$ denotes the scheduling indicator and is 1 if user i is scheduled and 0 otherwise. The goal is to schedule the user such that the total network throughput is maximized. Therefore, the user with the highest expected throughput should be scheduled to maximize $\hat{r}[t]$.

One way of estimating $\hat{r}_i[t]$ is

$$\hat{r}_i[t] = W \log_2 \left(1 + \frac{\Gamma_i[t]}{\theta} \right), \tag{4.5}$$

where W is the frequency bandwidth, $\Gamma_i[t]$ the SINR at time, θ the signal to interference plus noise ratio (SINR) gap that defines the gap between the channel capacity and a practical coding and modulation scheme. Obviously, the user with the highest estimated throughput $\hat{r}_i[t]$ is also the user with the highest SINR, $\Gamma_i[t]$. Therefore, the max throughput scheduler is also frequently called max SINR scheduler or max C/I scheduler. Since the variations in channel conditions are used for scheduling, max SINR scheduler is a kind of channel-aware scheduler.

The max throughput scheduler is the most aggressive advanced packet scheduler, which always schedules the user with the best instantaneous channel quality. While maximizing network throughput, the main drawbacks are the unfairness and coverage limitations. With this scheduler, the users with the most favorable positions or channels will have the highest throughput, but users in unfavorable positions may never be served. If there are many users in the network, most of them will suffer from scheduling starvation because they have to wait until the user with the best channel has no more data to transfer and no other users with better channels are admitted.

4.3.3 Proportional fair scheduling

Proportional fair (PF) scheduling is a compromised scheduling policy, trying to balance the competing interests of maximizing total network throughput and providing all users with at least a minimal level of service. The objective of proportional fair scheduler is to maximize

$$\sum_{i=0}^{M-1} \ln S_i, \tag{4.6}$$

where S_i is the long-run throughput of user i. S_i may change in every time slot and define the throughput in slot $t - 1$ to be $S_i[t - 1]$. $S_i[t]$ can be predicted using an exponential low-pass filter,

$$\hat{S}_i[t] = \left(1 - \frac{1}{\tau} \right) S_i[t - 1] + \frac{1}{\tau} \hat{r}_i[t] I(i), \tag{4.7}$$

where $\tau \gg 1$. It can be shown that in order to maximize $\sum_{i=0}^{M-1} \ln S_i$, the user with the highest

$$\frac{\hat{r}_i[t]}{S_i[t - 1]} \tag{4.8}$$

should be scheduled.

The reason that the PF scheduler is proportional fair is because it meets the proportional fairness criterion. A feasible vector of throughputs $\mathbf{x} = [x_1, x_2, ..., x_n]$ is proportional fair if for any other feasible vector $\hat{\mathbf{x}}$, the sum of proportional changes is non-positive, i.e.

$$\sum_{i=1}^{n} \frac{\hat{x}_i - x_i}{x_i} \leq 0. \tag{4.9}$$

When a network has already been proportional fair and we change the scheduler such that the throughput of one user is increased by $a\%$, there will be more than $a\%$ cumulative decrease of the throughputs of all other users.

Now let's have an intuitive understanding of why the objective in (4.6) indeed is proportional fair. Assume $\delta_i \ll x_i$. Then, a feasible vector $\mathbf{x}$ satisfies

$$\sum_{i=1}^{n} \ln(x_i + \delta_i) = \sum_{i=1}^{n} \ln x_i + \sum_{i=1}^{n} \ln\left(1 + \frac{\delta_i}{x_i}\right)$$
$$\approx \sum_{i=1}^{n} \ln x_i + \sum_{i=1}^{n} \frac{\delta_i}{x_i} \tag{4.10}$$
$$\leq \sum_{i=1}^{n} \ln x_i.$$

Therefore, a proportional fair throughput vector is a vector that maximizes $\sum_{i=1}^{n} \ln x_i$.

4.3.4   Max–min scheduling

The objective of a max–min scheduler is to maximize the minimum performance of all users in the network, i.e.

$$\max \min_i S_i. \tag{4.11}$$

The scheduler result is max–min fair if and only if a further increase of throughput of one user will result in the decrease of a user that already has a smaller throughput.

The understanding can be facilitated with the following example. There are M empty cylindrical buckets (users), all with the same radius but different heights. We need to allocate a certain amount of water (resource) to the buckets. With max–min allocation, any small amount of water should be distributed equally among all the buckets that are not yet full. Repeat this process until either all the water is allocated or all the buckets are full. Figure 4.4 illustrates an allocation result and we can see that the buckets will have the same amount of water if they are not yet full. Besides, if we want to increase the water in a bucket that is not yet full, the water in another bucket that has less or an equal amount will decrease.

Now let's see who should be scheduled. Combining (4.7) and (4.11), we have the following equivalent objective:

$$\max \min_i \left(1 - \frac{1}{\tau}\right) S_i[t - 1] + \frac{1}{\tau} \hat{r}_i[t] I(i). \tag{4.12}$$

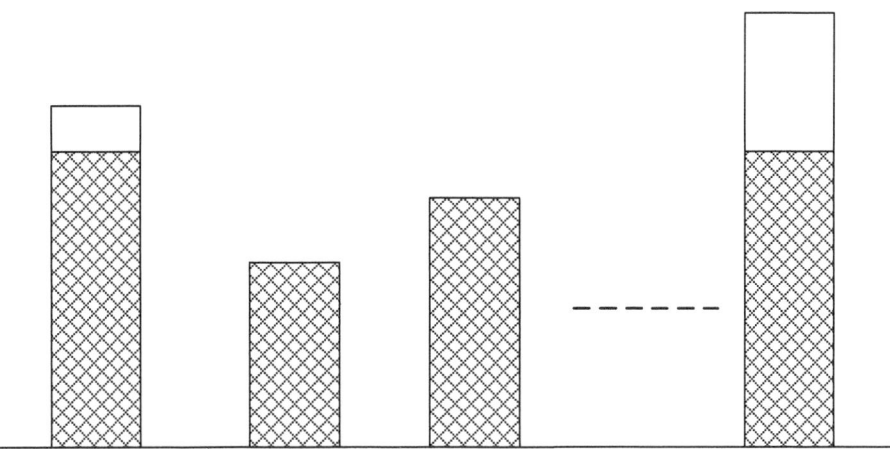

Figure 4.4 Water allocation achieving max–min fairness

Therefore, we should schedule the user with the minimum

$$\left(1 - \frac{1}{\tau}\right) S_i[t - 1], \qquad (4.13)$$

i.e. the one with the lowest throughput at time $t - 1$.

4.3.5   Max utility scheduling

The previous schedulers do not relate explicitly to the QoS users desire to get, or assume all users are demanding, best-effort data services. To fill this gap, utility-based scheduling algorithms can be used. Utility quantifies the satisfaction of each user given the allocated resources. With utility-based scheduling, the objective is to maximize the sum utility of all users, i.e. the total network satisfaction.

To model the relation between services provided by the network and how users perceive the services, utility functions can be used. Utility functions should be determined based on traffic characteristics. Figure 4.5 illustrates three popular utility functions. The first one in (a) can be used for best-effort data traffic, e.g. file transfer and e-mail. These services are elastic and users can adapt well to long delays and low throughput. More utility can be achieved if a higher data rate can be achieved, but the increase in utility slows with respect to the increased data rate. This type of utility function can be used for the interactive and background classes of services in Table 4.1. The type of utility function in subfigure (b) can be used to characterize real-time services with strict delay requirements, e.g. the conversational class. These services will not be acceptable if packet delays are beyond a certain limit or the rate is smaller than a certain amount. On the other hand, higher data rate will not help either because these applications have an almost constant rate requirement and higher allocated bandwidth will not be used. Therefore, the utility function behaves like a step function and the utility increases quickly from zero to the saturation point, beyond the desired rate. The

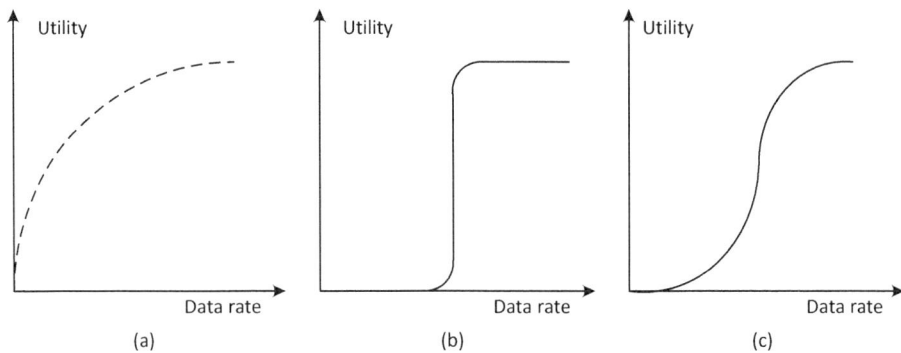

Figure 4.5 Utility of various traffic types with respect to data rate

utility function in subfigure (c) can be used to describe real-time services with looser delay requirements, e.g. the streaming class. For example, streaming video usually uses some scalable coding techniques and therefore is able to adapt to packet delay and data rate. These applications may be able to adjust the service quality, e.g. video resolution, based on channel conditions and therefore have elastic rate requirements. However, if the channel rate is smaller than the minimum requirement, the performance would be extremely poor, as users may not be able to receive any service at all. While increasing channel data rate will improve the service quality and therefore the utility, there is an intrinsic upper limit of the rate requirement. Therefore, when the data rate provided goes beyond the upper limit, the utility saturates and the additional gain diminishes very fast.

Utility-based scheduling aims to maximize the sum utility of all users in the network, i.e.

$$\max \sum_{i=0}^{M-1} U_i, \tag{4.14}$$

where U_i is the utility of user i given the allocated resources. Here we focus on the throughput performance and want to find a scheduling result that maximizes

$$\max \sum_{i=0}^{M-1} U_i(S_i). \tag{4.15}$$

Different utility functions can be designed. For example, if

$$U(S) = S, \tag{4.16}$$

the max utility scheduler is the max throughput scheduler. If

$$U(S) = \ln(S), \tag{4.17}$$

the max utility scheduler is the PF scheduler. A more generic definition of the utility function can be [103]

$$U_\alpha(S) = \begin{cases} \frac{S^{1-\alpha}}{1-\alpha} & \alpha \geq 0 \text{ and } \neq 1, \\ \ln(S) & \alpha = 1. \end{cases} \tag{4.18}$$

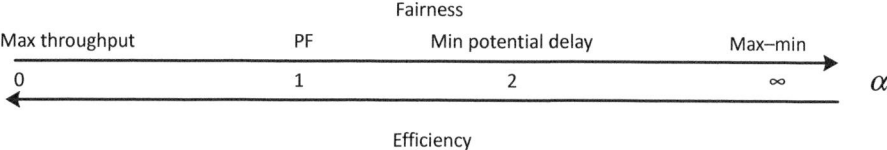

Figure 4.6 Alpha fair scheduling

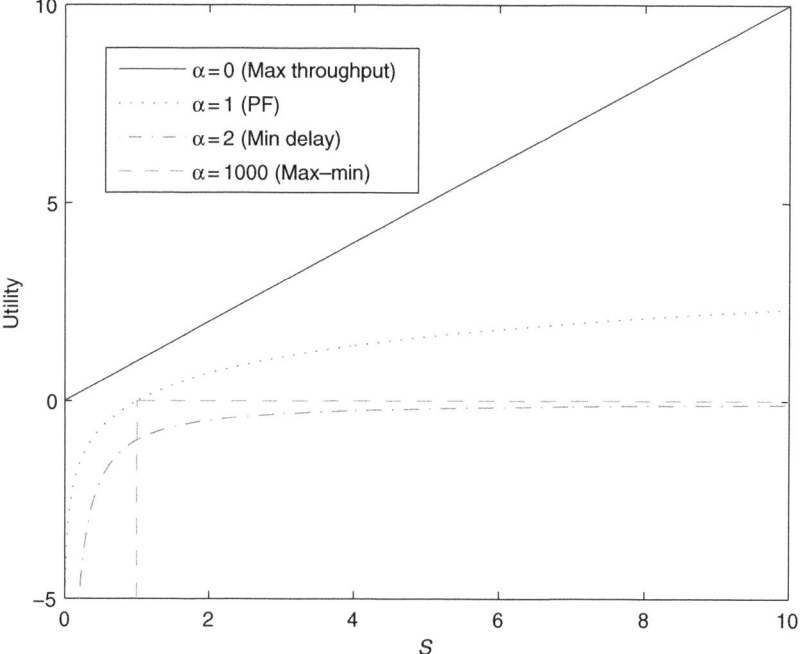

Figure 4.7 Alpha fair utility

The parameter α measures how fair the scheduler is [103] and is illustrated in Figure 4.6. The total network throughput is maximized when $\alpha = 0$, which is the max throughput scheduler, but the resource allocation is unfair. When $\alpha = 1$, it is the proportional fair scheduler. When $\alpha = 2$, the goal is to minimize

$$\sum_{i=0}^{M-1} \frac{1}{S_i}. \tag{4.19}$$

Note that the potential delay in transferring packets equals $\frac{1}{S_i}$ for user i. So the objective in (4.19) seeks a resource allocation to minimize the total potential delay of all users. In the limiting case where $\alpha = \infty$, the most fair allocation is the max–min scheduler.

The curves of $U_\alpha(S)$ when α has different values are drawn in Figure 4.7. It can be easily seen that the utility functions in (4.16), (4.17), and (4.18) are the detailed implementations of the type of utility for best-effort data traffic in subfigure (a) of

Figure 4.5. Indeed, subfigure (a) of Figure 4.5 has a concave monotonic increasing shape and so is $U_\alpha(S)$ for all α values. Other types of utility functions can also be defined, following the shapes in Figure 4.5, to reflect how users perceive the QoS and therefore assure QoS in scheduling. Different applications may have different utility functions or even different parameters. For instance, utility functions can also be defined with respect to delay, instantaneous data rate $r[t]$, etc. and the detailed scheduling algorithm may vary accordingly. Users may refer to [40] for more discussions. The focus here is on utility functions with respect to the long-run throughput, the same as other schedulers introduced in this chapter.

Now let's see how to schedule users. We want to

$$\max \sum_{i=0}^{M-1} U_i(\hat{S}_i[t]) = \max \sum_{i=0}^{M-1} U_i \left(\left(1 - \frac{1}{\tau} \right) S_i[t-1] + \frac{1}{\tau} \hat{r}_i[t] I(i) \right). \tag{4.20}$$

Note that $(1 - \frac{1}{\tau}) S_i[t-1] >> \frac{1}{\tau} \hat{r}_i[t] I(i)$ and we have the following approximation:

$$\begin{aligned} U_i &\left(\left(1 - \frac{1}{\tau} \right) S_i[t-1] + \frac{1}{\tau} \hat{r}_i[t] I(i) \right) \\ &\approx U_i \left(\left(1 - \frac{1}{\tau} \right) S_i[t-1] \right) + U_i' \left(\left(1 - \frac{1}{\tau} \right) S_i[t-1] \right) \frac{1}{\tau} \hat{r}_i[t] I(i), \end{aligned} \tag{4.21}$$

where the first portion is fixed at time t regardless of the scheduling result. The objective can be approximated by

$$\max \sum_{i=0}^{M-1} U_i' \left(\left(1 - \frac{1}{\tau} \right) S_i[t-1] \right) \frac{1}{\tau} \hat{r}_i[t] I(i). \tag{4.22}$$

Therefore, the optimal scheduling policy is to schedule the user with the largest

$$U_i' \left(\left(1 - \frac{1}{\tau} \right) S_i[t-1] \right) \hat{r}_i[t] \tag{4.23}$$

or, observing that $\frac{1}{\tau} \to 0$, the user with the largest

$$U_i' \left(S_i[t-1] \right) \hat{r}_i[t]. \tag{4.24}$$

These two have almost the same performance.

5 Distributed resource management in wireless networks

5.1 Overview

The drawbacks of centralized scheduling can be resolved by using distributed management schemes. With distributed approaches, each user determines the amount of resources to be used based on local information. Compared to centralized schemes, the main difference of distributed resource management schemes is that there are multiple decision makers in the network, whose decisions may conflict with each other, therefore resulting in performance degradation.

With distributed resource management, the network does not need infrastructure support and therefore can be easily constructed. This is very helpful in some application scenarios. For example, some sensor networks must operate in remote areas and harsh environments, without infrastructure support or maintenance. Sensor nodes must manage themselves autonomously and configure themselves to operate and collaborate with other nodes. In addition, each sensor node must be able to adapt to failure, topology, density, environment, and traffic characteristics automatically without too much overhead. Furthermore sensor networks usually have large scales, which make it not feasible to rely on centralized implementations, because centralized approaches may incur exhaustive overheads for adaptations that may affect all users in the network. Instead, sensor nodes must make local decisions without global knowledge. Therefore, it is essential to design distributed resource management schemes for sensor networks.

A more detailed example is the Request to Send (RTS)/Clear to Send (CTS) option in the IEEE 802.11 standard. In this case, a user that has packets to transmit sends a RTS frame to the destination user. The RTS frame includes a field to indicate how long the users wants to hold the channel. If the destination user receives the RTS frame successfully, it replies with a CTS. All other users that have detected the RTS or CTS frames will mark the channel as busy and refrain from transmission for a certain time to avoid collisions.

The performance of networks with distributed schedulers is in general not as good as that of networks with centralized schedulers because of the lack of centralized coordination in the contention of resource utilization. In addition, from a whole network perspective, the contention behaviors of different users may not necessarily converge to a stable state. Even if a stable state is achieved, the equilibrium state may only be a local optimum and the global optimum is in general unknown. We should also keep in mind that distributed wireless approaches have the advantages of robustness, simple implementation, good scalability of network design, and so on.

With the advances in wireless networks, network topologies are getting more and more complicated and hybrid resource management schemes that consist of both centralized and distributed protocols and algorithms are needed to make efficient use of wireless resources. The hybrid approaches exploit the advantages of both centralized and distributed schemes. With a hybrid approach, each local central controller determines some resource allocation of a subset of users and these users may further adjust their resource utilizations using distributed approaches based on the information provided by the local controller. Certain information, such as channel conditions, or queueing information, may be required at the central controllers. Message exchange among some mobile users is therefore necessary to facilitate the decisions of the central controllers. With this type of local collaboration, network performance can be improved. In the hybrid design, more collaborations with higher signaling overhead will in general result in better network performance and therefore an intrinsic tradeoff between performance and signaling overhead always exists in the network.

Contention-based access protocols allow terminals to access channel randomly when they have packets to send. With random access or contention-based methods, no terminal is superior to another station and none is assigned control over another. No terminal permits, or does not permit, another one to send. At each instance, a terminal that has data to send uses a procedure defined by the protocol to make a decision on whether or not to send.

The conventional rule of distributed medium access control (MAC) design can be summarized as the following two items:

- Removal of Idle States: An idle state takes place when some users have data to transmit but their MACs decide not to while the channel is idle and the channel capacity is wasted. The MAC protocol design is to avoid as many idle states as possible.
- Removal of Collision States: A collision state happens when several users transmit packets at the same time and the transmissions fail, which results in waste of both channel capacity and user energy. The MAC protocol design is to avoid as many collision states as possible.

The idle states happen more frequently with light network load and the collision states with with high network load. The ideal MAC protocol is the one that completely removes all idle and collision states. In the following, we will introduce different MAC protocols and see how MAC evolves to remove the two states for better network performance. We will first introduce the Aloha protocol, which is a pioneering MAC protocol in computer networks. The first version of the Aloha protocol is the pure Aloha with which each terminal sends the packet whenever it wants. The slotted Aloha improves the pure Aloha by dividing time into slots and transmission occurs only when a new slot starts. With this improvement, the chance that different transmissions collide is reduced and the network throughput is improved. Later, we will also introduce more complicated and advanced contention-based MAC protocols, e.g. carrier sense multiple access (CSMA), CSMA with collision avoidance, and IEEE 802.11 MAC.

5.2 Aloha

5.2.1 Pure Aloha

In the late 1960s, the Aloha protocol was developed at the University of Hawaii. The algorithm was devised for use in a VHF-radio system to connect remote terminals on the many islands with a central computer site. Each terminal was equipped with a radio interface as were the main computers. Thus, all the terminals belonging to a main computer were connected to the computer via a shared medium, the air interface. In the original protocol, now called pure Aloha, the stations were not synchronized. The control signaling, i.e. ACK, is sent on an independent control channel. Pure Aloha is quite simple. For each terminal in the network, it runs the following algorithm:

- Step 1: If there is a message to send, send it.
- Step 2: If the transmission succeeds, remove the message from the queue and go to Step 1. If the transmission fails, wait a random time interval, i.e. backoff randomly, and go to Step 1.

The transmission failure is mainly due to a collision. It is of paramount importance that the terminals do not wait identical time intervals after a collision, since this would almost certainly cause another collision. The quality of the backoff scheme determines the efficiency of pure Aloha and network capacity. When total traffic is low, pure Aloha works well; otherwise, performance is poor.

Assume all frames have the same length, 1. Suppose the ith frame is sent starting from t_0, as shown in Figure 5.1. In order to have a successful transmission, no other terminals should transmit any signal between t_0 and $t_0 + 1$. In addition, no other terminal should send a frame between $t - 1$ and t_0 because, otherwise, the latter part of this frame will overlap with the ith frame. So two consecutive frame durations are needed for sending one frame successfully in a network.

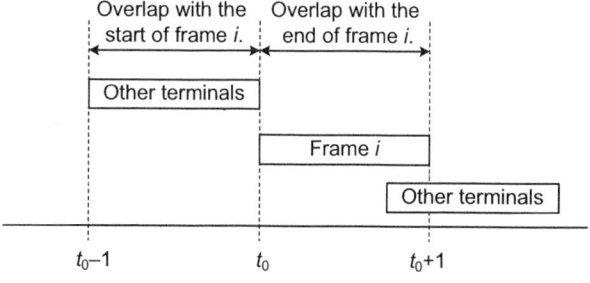

Figure 5.1 Collision in pure Aloha

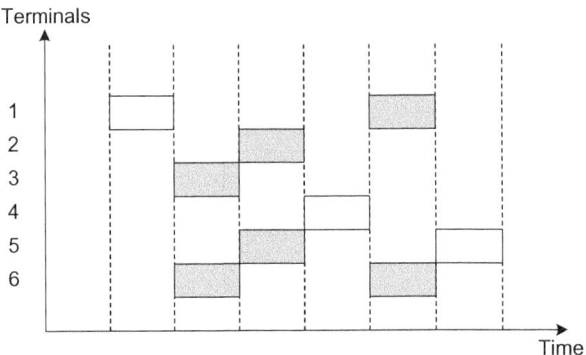

Figure 5.2 Slotted Aloha

5.2.2   Slotted Aloha

A simple improvement in the basic Aloha protocol is slotted Aloha. With slotted Aloha, all terminals are synchronized and time is assumed to be slotted. Terminals can send packets only at the beginning of slots, as shown in Figure 5.2, and therefore collisions are reduced. We can see that with slotted Aloha, collisions happen only if two terminals send frames in the same slot. So only one frame duration is needed for sending one frame successfully. The efficiency of slotted Aloha is therefore twice that of pure Aloha.

5.3   Carrier sense multiple access (CSMA)

To further improve network performance, we can abort transmission as soon as a collision is detected. This is because once two packets start colliding, it is useless to continue their following transmissions as their respective receivers will not be able to decode the packets. Otherwise, it will lead to a waste of bandwidth and energy. Hence, it is necessary to detect collision while transmitting a packet. This motivates the design of CSMA.

CSMA is popular in asynchronous (non-slotted) wireless networks with low propagation delays. Each terminal is equipped with a receiver to monitor if any transmission is in progress in the channel. The terminals simply measure the signal level to detect transmission, i.e. carrier sensing. If there is a carrier present, i.e. existing transmission in progress, the station defers its transmission attempt to a later time, randomly determined. Otherwise, it transmits its packet. Once the packet is transmitted, the terminal waits for an acknowledgement. If no acknowledgement is received later, the packet will be scheduled for future transmissions. Like Aloha, CSMA requires an independent control channel for acknowledgement transmissions.

There are three types of CSMA algorithms, non-persistent, 1-persistent, and p-persistent.

5.3.1   Non-persistent CSMA

Each terminal runs the following protocol:

- Step 1: If the channel is sensed idle, transmit a packet immediately.
- Step 2: If the channel is sensed busy, wait a random amount of time and go to Step 1.

The random backoff in Step 2 reduces the probability of collisions. If the random backoff time is too long, the channel capacity may be wasted as no other terminals may transmit either.

5.3.2   1-persistent CSMA

Each terminal runs the following protocol:

- Step 1: If the channel is sensed idle, transmit a packet immediately.
- Step 2: If the channel is sensed busy, continue sensing until the channel is idle; then transmit a packet immediately.

The protocol is called 1-persistent because a terminal will transmit with probability 1 whenever it finds the channel idle. With 1-persistent CSMA, if two or more terminals want to send packets while the channel is busy, the transmissions will always collide as they all begin transmission simultaneously as soon as the channel becomes idle.

5.3.3   p-persistent CSMA

Each terminal runs the following protocol:

- Step 1: If the channel is sensed idle, transmit a packet with probability p; with probability $1 - p$, delay its transmission by one time slot and repeat this step.
- Step 2: If the channel is sensed busy, continue sensing until the channel is idle; then go to Step 1.

1-persistent CSMA is a specific case of p-persistent CSMA. The p-persistent protocol improves the 1-persistent protocol by reducing the likelihood of packet collisions through random backoff of packet transmission.

The selection of p determines the performance of p-persistent CSMA. Suppose there are N terminals having packets to send while the channel is busy. Np is the expected number of terminals that will send packets once the channel is idle. If $Np > 1$, a collision is expected. So the choice of p must ensure that $Np < 1$ to avoid collision and $p < \frac{1}{N}$, where N is the maximum possible number of active terminals at the same time.

5.3.4   Effect of detection delay

An important parameter in a CSMA system is the detection delay. As shown in Figure 5.3, this is the time span between when a terminal decides to start a transmission

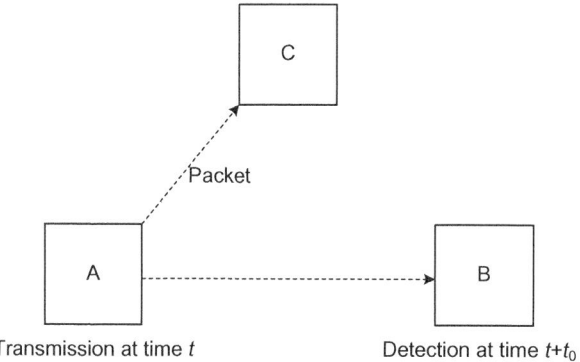

Figure 5.3 Delay in channel sensing

at time t until the instant, $t + t_0$, when all terminals become aware of this transmission (and thus avoid a collision). This delay includes the carrier measurement delay, the switch time from reception to transmission, and the propagation delay. If the detection delay is large compared to the message size, the carrier sensing information will be of little use since it is only able to describe the state of the channel some time ago. The capability to prevent collisions deteriorates and two terminals may both sense that the channel is empty and start their respective transmissions. However, if the detection delay approaches zero, almost all collisions can be avoided.

5.4   CSMA with collision detection

With CSMA, if a terminal decides to send a packet at a certain time t, it will send the whole packet. It there is a collision with other terminals, the resources, time, frequency, and energy sending the whole packet are wasted. Therefore, CSMA/CD is designed to reduce such waste.

    With CSMA/CD, each transmitting terminal will also monitor its own transmission. If it detects a collision, it stops transmission immediately and instead sends a jam signal to indicate that there has been a collision. The channel is therefore much more efficiently used since the bandwidth transmitting the entire frame is not wasted. The detection techniques depend on channel media. For example, with electrical wires, collision detections are done by comparing the transmitted packet with the received ones.

    In addition to the collision detection capability, the backoff window also grows to reduce collision probabilities. So after a collision, a terminal involved in the collision will retransmit its packet after waiting for a random time period. If another collision occurs, the time window from which the random waiting time is selected is increased step-by-step. One popular way of increasing the backoff window is called the exponential back off. In one example, in the first collision, suppose the random backoff time is chosen randomly between $[0, t]$. In the Nth collision, the random backoff time will be chosen randomly between $[0, t_N]$, where $t_N = \min(2^{N-1}t, t_{Max})$; and t_{Max} is the maximum backoff window size.

CSMA with CD (CSMA/CD) is easy to implement and is widely used in wired networks.

5.5   Carrier sense multiple access with collision avoidance (CSMA/CA)

CSMA/CD is not appropriate for wireless networks. CSMA/CD detects collisions at senders not receivers. In wireless networks, a signal should reach the receiver without any collision for successful transmission. So the sender needs to detect collisions, which is different from CSMA/CD. Further, the signal strength is almost the same for wired networks but varies dramatically in wireless networks depending on the wireless channel between the transmitter and receiver. The collision at the receiver in many cases will not be detected by the sender. More importantly, in wireless networks, the transmission power is usually much higher than the reception power and collision detection by the sender is very difficult and not possible in practice. Some more issues are described below.

5.5.1   Hidden and exposed terminal problems

The hidden terminal problem is illustrated in Figure 5.4, where the arrows stand for the traffic flows and the dashed circles the communication ranges of A and B respectively. A and B are sending packets to C, but A and B are out of the transmission range of each other. They will both sense the channel idle and send packets to C. Their packets collide at C. So A and B are hidden terminals to each other.

The exposed terminal problem is illustrated in Figure 5.5, where the dashed circles stand for the communication ranges of A and C, respectively. Obviously A and C send packets to B and D respectively at the time without any collision at the receiver sides. However, if A sends a packet to B, C will sense the channel busy and will not transmit. Half channel capacity is therefore wasted.

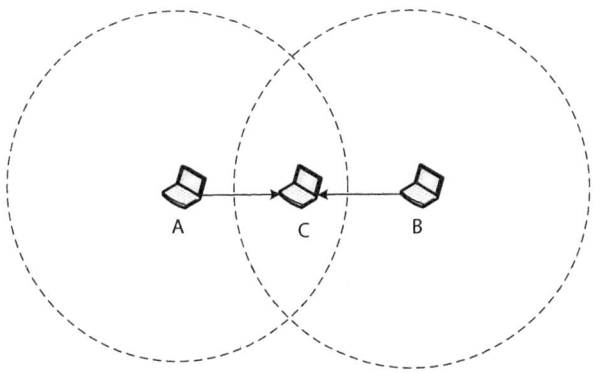

Figure 5.4 The hidden terminal problem

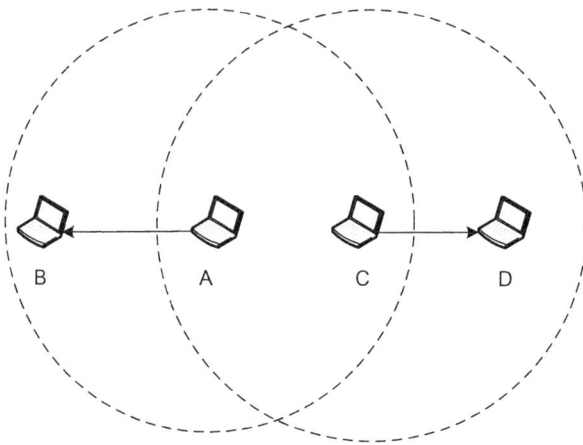

Figure 5.5 The exposed terminal problem

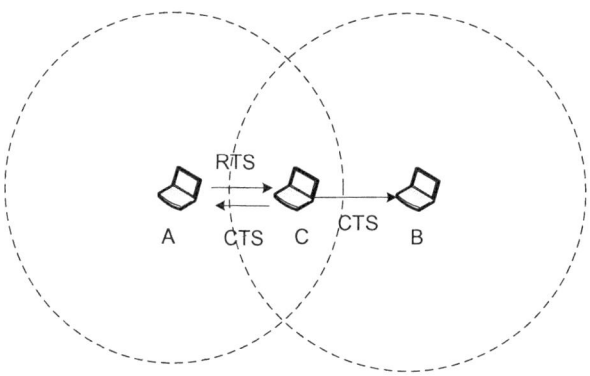

Figure 5.6 RTS/CTS Exchange

5.5.2   CSMA/CA protocol

CSMA/CA differs from CSMA/CD in that collisions are avoided. A terminal with a message to transmit will sense the channel first. If the channel is idle for a short period called the DIFS period, the terminal can transmit. Otherwise, the terminal will defer and continue monitoring the channel until the channel is idle for the DIFS period. Then a random backoff counter within the contention window will be generated before actually sending the message. The backoff counter counts down as long as the channel remains idle. The counter will pause if the terminal detects the channel is busy and resume when it detects the channel is idle for another DIFS period. The terminal will transmit immediately when its backoff counter is zero. CSMA/CA can use a binary exponential backoff algorithm to control the contention window size of each terminal and resolve collisions. Each terminal has a contention window size w which has a minimum value W_{min} and a maximum value W_{max}. Before a new transmission, the backoff counter is chosen uniformly between $[0, w - 1]$. At the first transmission attempt, $w = C_{min}$. After each failure, w is doubled until reaching W_{max}.

Optionally but almost always implemented, CSMA/CA can use a RTS/CTS exchange mechanism to avoid collisions. An example is illustrated in Figure 5.6. When a terminal, A, intends to send a message to C, it will sense whether the channel is idle and back off if the channel is busy. If the channel is idle, it will send a small RTS message to the intended receiver, C. If the receiver senses the channel is clear and receives the RTS, it immediately replies with a small CTS message to A. If A receives the CTS, it sends the data to the receiver. Both the RTS and CTS messages carry the time length to transmit the data message. After hearing the CTS, B will no longer transmit anything when A sends the package and therefore the collision is avoided. If A does not receive a CTS message, it starts the RTS procedure over again.

RTS/CTS also solves the exposed terminal problem. Consider the traffic flows in Figure 5.5. Terminal A first sends to B an RTS, which will also be heard by C. C will therefore defer its RTS transmission to D. However, later C will not receive a CTS from B and knows that B is not within its communication range. Then C can start its transmission to D.

Part II

Centralized cross-layer optimization

6   Overview

Orthogonal frequency division multiplexing (OFDM) divides an entire channel into many orthogonal narrow-band subchannels (subcarriers) to deal with frequency-selective fading and to support a high data rate. Furthermore, in an OFDM-based wireless network, different subcarriers can be allocated to different users to provide a flexible multi-user access scheme [52, 122] and exploit multi-user diversity.

There is plenty of room to exploit the high degree of flexibility of radio resource management in the context of OFDM. Since channel frequency responses are different at different frequencies and for different users, data rate adaptation over each subcarrier, *dynamic subcarrier assignment* (DSA), and *adaptive power allocation* (APA) can significantly improve the performance of OFDM networks. Using data rate adaptation [144, 83], the transmitter can send higher transmission rates over the subcarriers with better conditions so as to improve throughput and simultaneously to ensure an acceptable *bit error rate* (BER) at each subcarrier. Despite the use of data rate adaptation, deep fading on some subcarriers still leads to low channel capacity.

Channel characteristics for different users are almost mutually independent in multi-user environments; the subcarriers experiencing deep fading for one user may not be in a deep fade for other users; therefore, each subcarrier could be in a good condition for some users in a multi-user OFDM wireless network. By dynamically assigning subcarriers, the network can benefit from multi-user diversity. Resource allocation issues and the achievable regions for multiple access and broadcast channels have been investigated in [195] and [123], respectively, which have proved that the largest data rate region is achieved when the same frequency range is shared with overlap by multiple users in broadcast channels. However, when optimal power allocation is used, from [84], there is only a small range of frequency with overlapping power sharing. Thus, optimal power allocation with dynamic subcarrier (non-overlap) assignment can achieve a data transmission rate close to the channel capacity boundary. In [208], the authors have investigated optimal resource allocation in multi-user OFDM systems to minimize the total transmission power while satisfying a minimum rate for each user. The numerical optimization algorithms have been proposed in [219] for characterizing the uplink rate region achievable in OFDM with inter-symbol interference. Several algorithms have been presented in [168, 118] for subcarrier and power allocation.

As seen in previous sections, exploiting multi-user diversity can significantly improve spectral efficiency. In addition to spectral efficiency, fairness and quality of service (QoS) are crucial for resource allocation for wireless networks. Usually, it is impossible to achieve optimality for spectral efficiency, fairness, and QoS simultaneously. For instance, scheduling schemes aiming to maximize the total throughput are unfair to those users far away from a base station or with bad channel conditions. The absolute fairness may lead to low bandwidth efficiency. Therefore, an effective tradeoff among efficiency, fairness, and QoS is desired in wireless resource allocation.

The issues on efficient and fair resource allocation have been well studied in economics, where utility functions are used to quantify the benefits of usage of certain resources. Similarly, utility theory can be used in communication networks to evaluate the degree to which a network satisfies service requirements of users' applications, rather than in terms of system-centric quantities such as throughput, outage probability, packet drop rate, power, etc. [184]. The basic idea of utility-pricing structures is to map the resource use (bandwidth, power, etc.) or performance criteria (data rate, delay, etc.) into the corresponding utility or price values and optimize the established utility-pricing system.

In wireline networks, utility and pricing mechanisms have been used for flow control [116, 117], congestion control [137], and routing [33]. In wireless networks, the pricing of uplink power control in code division multiple access (CDMA) has been investigated in [181, 85, 179, 212]. Utility-based power allocation for CDMA downlinks for voice and data applications has been proposed in [127, 220, 188]. To guarantee QoS and exploit multi-user diversity, utility-pricing structures are applied in opportunistic communications [129].

In brief, network economics is becoming more and more important in modern network designs, especially for cross-layer optimization in wireless networks.

6.1   System model and problem description

The architecture of a downlink data scheduler with multiple shared channels for multiple users is shown in Figure 6.1. OFDM provides a physical basis for the multiple shared channels, where the total bandwidth B is divided into K subcarriers (subchannels), and each subcarrier has a bandwidth of $\triangle f = B/K$. Let $\mathcal{K} = \{1, 2, \ldots, K\}$ denote the subcarrier index set. The OFDM signaling is time-slotted, and the length of each time slot is T_s. The base station simultaneously serves M users, each of which has a queue to receive its incoming packets. Let $\mathcal{M} = \{1, 2, \ldots, M\}$ denote the user index set. To achieve high efficiency, both frequency and time multiplexing are allowed in the whole resource. The scheduler makes a subcarrier assignment once every slot based on each user's current channel quality and queue length.

6.1.1   Channel characteristics in OFDM

The complex baseband representation of the impulse response of a wireless multi-path channel for user i can be described by

$$h_i(t, \tau) = \sum_k \gamma_{k,i}(t) \delta(\tau - \tau_{k,i}),$$

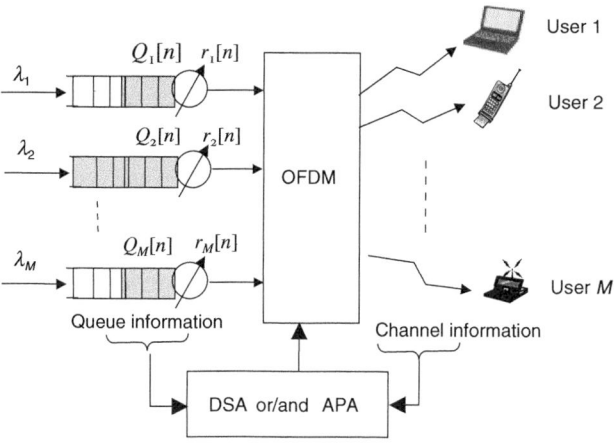

Figure 6.1 Downlink data scheduling over multiple shared channels based on OFDM

where $\tau_{k,i}$ is the delay of the kth path and $\gamma_{k,i}(t)$ is the corresponding complex amplitude at time t. The $\gamma_{k,i}(t)$ are assumed to be wide-sense stationary and narrow-band stochastic Gaussian processes, which are independent for different paths and users. The frequency response of the above channel impulse response is expressed as

$$H_i(f,t) = \int_{-\infty}^{+\infty} h_i(t,\tau)e^{-j2\pi f\tau}\,d\tau = \sum_k \gamma_{k,i}(t)e^{-j2\pi f\tau_{k,i}}.$$

For OFDM systems with proper cyclic extension and sample timing, the channel frequency response at subcarrier k at time n can be expressed as

$$H_i[k,n] \triangleq H_i(k\Delta f, nT_s).$$

Then, the channel quality of user i is given by

$$\rho_i[k,n] = \frac{|H_i[k,n]|^2}{N_i[k]},$$

where $N_i[k]$ is the noise power of user i at subcarrier k. With a power allocation $p[k,n]$, the signal to noise ratio (SNR) at subcarrier k at time n is

$$\gamma_i[k,n] = p[k,n]\rho_i[k,n].$$

There are many ways to obtain the channel state information (CSI) at the base station. In a *frequency division duplex* (FDD) system, using pilot symbols that are inserted in the downlink with a certain time-frequency pattern, the mobile terminals can effectively estimate the channel parameters $H_i[k,n]$ and $\rho_i[k,n]$ [125] and feed them back to the base station. In a *time division duplex* (TDD) system, since the symmetry of the channel characteristics for the downlink and uplink, the base station can obtain the CSI by directly measuring the uplink channels.

6.1.2   Rate adaptation in OFDM

By estimating the CSI via pilot signals and feeding it back to the base station, the achievable data transmission rate per Hz for user i at subcarrier k during time slot n, $c_i[k,n]$, can be known at the base station. Usually, the $c_i[k,n]$ are determined by the current channel SNR, the required BER, and the modulation and coding techniques that are used in the system.

If continuous rate adaptation is used, the achievable transmission rate per Hz at subcarrier k for user i can be written as a function of the current SNR, $\gamma_i[k,n]$, [158]

$$c_i[k,n] = \log_2(1 + \beta\gamma_i[k,n]), \tag{6.1}$$

where β is a constant related to a targeted BER by

$$\beta = \frac{-1.5}{\ln(5 \cdot \text{BER})}.$$

Generally, $c_i[k,n]$ can be expressed as

$$c_i[k,n] = f\left(\log_2(1 + \beta\gamma_i[k,n])\right), \tag{6.2}$$

where $f(\cdot)$ depends on the used rate adaptation scheme. For instance, if variable *M-ray quadrature amplitude modulation* (MQAM) with modulation levels $\{0, 2, 4, 6,\dots\}$ is employed,

$$f(x) = 2\lfloor\frac{1}{2}x\rfloor,$$

where $\lfloor x \rfloor$ represents the largest integer that is less than x.

6.1.3   Dynamic subcarrier assignment and adaptive power allocation

Each subcarrier in the adaptive OFDM can be dynamically assigned to any user. Let $D_i^{(n)}$ denote the set of subcarrier indices assigned to user i at time n. In the OFDM system, each subcarrier cannot be shared by multiple users, which is mathematically expressed as

$$D_i^{(n)} \bigcap D_j^{(n)} = \varnothing, \ \ \forall \, i \neq j,$$
$$\bigcup_{i \in \mathcal{M}} D_i^{(n)} \subseteq \mathcal{K}.$$

With a subcarrier assignment, the data transmission rate of user i at time slot n, $r_i[n]$, is given by

$$r_i[n] = \sum_{k \in D_i^{(n)}} c_i[k,n]\triangle f.$$

Let $\mathbf{p}[n]$ be the transmit power vector defined as $[p[1,n], p[2,n], \dots, p[K,n]]^T$, which $p[k,n]$ is the transmit power at subcarrier k at time n. If adaptive power allocation is used, the transmit powers can be adjusted but constrained by

$$\sum_{k=1}^{K} p[k, n] \leq \bar{P},$$

where $\bar{P}$ is the total power constraint.

6.1.4 Queue structure

Each connection is assumed to have a queue with infinite capacity at the base station. Let $Q_i[n]$ be the amount of bits in the queue of user i at time nT_s. During time slot n, the base station serves the queue of user i at rate $r_i[n]$. Then, the queue length evolution equation is given by

$$Q_i[n+1] = Q_i[n] - \min(Q_i[n], r_i[n]T_s) + a_i[n] \tag{6.3}$$

where $a_i[n]$ is the amount of arrival bits during time slot n.

6.1.5 Problem description

The major problem is how to effectively assign subcarriers and allocate power on the downlink of OFDM-based networks by exploiting knowledge of the wireless channel conditions and the characteristics of traffic to improve the spectral efficiency and guarantee diverse QoS.

6.2 Approach

In the joint physical and medium access control (MAC) layer optimization framework, we use two major mechanisms in resource management: exploiting the time variance and the frequency selectivity of wireless channels in network protocols through adaptive modulation, coding, as well as packet scheduling and regulating resource allocation through network economics. Besides leading to high capacity, OFDM provides fine granularity for resource allocation since different subcarriers can be assigned to different users. With the help of utility functions that capture the satisfaction level of users for a given resource assignment, we establish a utility optimization framework for resource allocation in OFDM networks, in which the network utility at the level of applications is maximized subject to the channel conditions and the modulation and coding techniques employed in the network.

Figure 6.2 illustrates the structure of Part II. In the cross-layer optimization with utility functions with respect to instantaneous data rates, we develop novel efficient DSA and APA algorithms, which provide the algorithm implementation for channel-aware scheduling and joint channel- and queue-aware scheduling. Using utility functions with respect to average data rates, we can design channel-aware scheduling desirable for best-effort traffic. We reveal a generic relationship between a specific convex utility function and a type of fairness. Based on a holistic design principle, we develop a joint channel- and queue-aware scheduling scheme that maximizes the total utility with

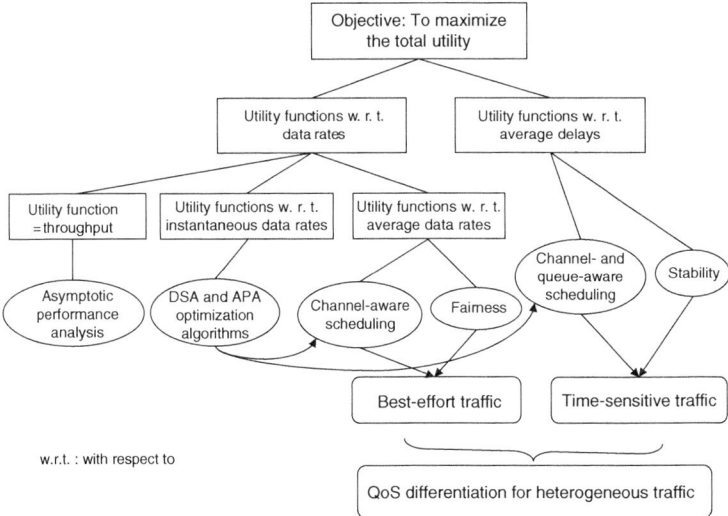

Figure 6.2 Structure of Part II

respect to average delays. The stability issue of the queueing system is comprehensively investigated because of the importance to delay-sensitive traffic. The utility-based architecture is finally proven to have the ability of QoS differentiation for heterogeneous traffic. Moreover, in the case when the utility function is just the throughput, we provide a concise asymptotic analysis for throughput and packet delay to reveal the impact of multi-user diversity.

From a traditional point of view, cross-layer design would usually seem complicated and intractable. Therefore, we pursue consistency in methodology and simplicity in results. An asymptotic approach is extensively used in analysis since it leads to elegant results. The results can not only be helpful in obtaining insights but can also be fully applied to the system design. For instance, the study of the optimization properties of the "extreme" OFDM in which the number of subcarriers is infinite directly guides development of the algorithm for practical systems. The study on the stability properties of joint channel- and queue-aware scheduling plays a crucial role in designing scheduling for heterogeneous traffic with diverse QoS requirements. The asymptotic throughput analysis of channel-aware scheduling is very accurate for typical environments and can deal with a general fading distribution. In addition, the convexity of the ergodic capacity region at the physical layer is fully exploited throughout the book, which makes most results concerning such complicated problems as fairness and stability elegant.

7 Utility-based optimization framework for OFDMA

In this chapter and Chapter 8, we do not consider the burstiness of arrival streams and investigate resource allocation and scheduling for best-effort traffic. Therefore, rate-based utility functions are used to perform cross-layer optimization and balance efficiency and fairness in this chapter. In Section 7.1, we discuss the general properties of rate-based utility functions. In Section 7.2, we investigate the optimization problems at an instantaneous time, which are formulated based on utility functions with respect to instantaneous data rates. In Section 7.2, we focus on the orthogonal frequency division multiplexing (OFDM) network that contains an infinite number of subcarriers and employs continuous rate adaptation. In the first four sections of Chapter 8, we develop efficient resource allocation algorithms for cross-layer optimization in various system configurations. In Section 8.5, we discuss channel-aware scheduling based on utility functions with respect to average data rates. The algorithms developed can be directly used for the channel-aware scheduling. In Section 8.6, we discuss efficiency and fairness issues. The relationship between a utility function and a certain type of fairness is revealed. In Section 8.7, we demonstrate performance improvement in cross-layer optimization through numerical results.

7.1 Rate-based utility functions

Utility functions are used for cross-layer optimization and balancing efficiency and fairness. A utility function maps the network resources that a user utilizes into a real number. In almost all wireless applications, a reliable data transmission rate is the most important factor to determine the satisfaction of users. Thus, the utility function $U(r)$ should be a non-decreasing function of the data rate r. In particular, when $U(r) = r$, the utility is just the throughput, which is the objective of most traditional network optimizations. Therefore, our work can be regarded as a general extension of traditional network optimizations.

Utility functions serve as an optimization objective for adaptive physical and medium access control (MAC) layer techniques. Consequently, they can be used to optimize radio resource allocation for different applications and to build a bridge among the physical, MAC, and higher layers.

When a utility function is used to capture a user's feelings, such as the level of satisfaction for assigned certain resources, it cannot be obtained only through theoretical

61

derivation. In this case, it can be estimated from subjective surveys. For best-effort traffic [108], a utility function can be described by

$$U(r) = 0.16 + 0.8 \ln(r - 0.3), \tag{7.1}$$

where r is in unit of kbps. To prevent assigning too much resource to the user with good channel conditions, the slope of the utility curves decreases with an increase in the data rate. We will discuss more on the issues of fairness and efficiency in Section 8.6.

7.2   Theoretical framework

To obtain the performance bound of cross-layer optimization, we assume in this section that there is an infinite number of orthogonal subcarriers in all frequency resources, or the bandwidth of each orthogonal subcarrier is infinitesimal, which can be regarded as an extreme situation of OFDM. In a practical OFDM system, the minimum granularity of resource allocation is one subcarrier. The OFDM system in which $\Delta f \to 0$ provides an infinitesimal granularity of resource allocation, thereby presenting the performance upper bound.

7.2.1   Problem formulation

Since we consider the "extreme" OFDM system, some parts of the system model in Section 6.1 should be modified slightly. Thus, we will briefly describe the modifications in the system model.

Because we investigate cross-layer optimization in terms of instantaneous data rates, we ignore time parameter t in all formulas in this section.

The M-user frequency-selective broadcast fading channel is shown in Figure 7.1. The channel frequency response corresponding to user i is denoted by $H_i(f)$. The quality of each user's channel can be indicated by the signal-to-noise ratio (SNR) function, $\rho_i(f)$, when the transmission power density $p(f) = 1$, which is defined as

$$\rho_i(f) = \frac{|H_i(f)|^2}{N_i(f)},$$

where $N_i(f)$ is the noise power density function of user i.

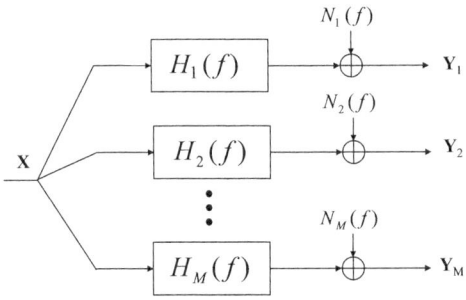

Figure 7.1 Channel model

Let $c_i(f)$ denote the achievable throughput of user i at frequency f for a given bit error rate (BER) and a transmission power density $p(f)$. When continuous rate adaptation is used, $c_i(f)$ can be expressed as [158]

$$c_i(f) = \log_2\left(1 + \frac{\beta p(f)\,|\,H_i(f)\,|^2}{N_i(f)}\right) \text{(b/s/Hz)}$$
$$= \log_2(1 + \beta p(f)\rho_i(f)). \tag{7.2}$$

In this scenario, the D_i become the frequency sets assigned to different users, which are constrained by

$$D_i \bigcap D_j = \varnothing, \quad i \neq j, \tag{7.3}$$

$$\bigcup_{i=1}^{M} D_i \subseteq [0, B]. \tag{7.4}$$

Besides dynamically assigning the frequency sets, the transmission power density at different frequencies can also be adjusted to improve network performance with a total transmission power constraint by

$$\frac{1}{B}\int_0^B p(f)\,df \leq 1. \tag{7.5}$$

The achievable transmission efficiency of user i in the continuous-frequency case is given by

$$c_i(f) = \log_2[1 + \beta p(f)\rho_i(f)],$$

and the transmission throughput of user i can be expressed as

$$r_i = \int_{D_i} c_i(f)\,df. \tag{7.6}$$

Let the utility function of user i be $U_i(\cdot)$. If user i has a data rate r_i, the user's utility is $U_i(r_i)$. The utility-based cross-layer optimization is to assign wireless resources (including frequency band and power density) to maximize the average utility of the network, which can be expressed as

$$\frac{1}{M}\sum_{i=1}^{M} U_i(r_i). \tag{7.7}$$

In the next several sections, we will discuss *dynamic subcarrier assignment* (DSA), *adaptive power allocation* (APA), and joint DSA and APA, respectively.

7.2.2   Dynamic subcarrier assignment

In this section, we investigate DSA to improve the performance of an OFDM-based network when the transmission power is uniformly distributed over the entire available frequency band, i.e., $p(f) = 1$, then the achievable throughput at frequency f, $c_i(f)$, can be expressed as

$$c_i(f) = \log_2(1 + \beta\rho_i(f)).$$

Thus, the DSA problem is to maximize

$$\frac{1}{M}\sum_{i=1}^{M} U_i(r_i) = \frac{1}{M}\sum_{i=1}^{M} U_i\left(\int_{D_i} c_i(f)\,df\right), \tag{7.8}$$

subject to

$$\bigcup_{i=1}^{M} D_i \subseteq [0, B], \tag{7.9}$$

$$D_i \bigcap D_j = \emptyset, \quad i \neq j \text{ and } i, j = 1, 2, \ldots, M. \tag{7.10}$$

We first present the results for a network with two users and then extend to general networks.

Network with two users

Assume a network with only two users sharing the bandwidth $[0, B]$. Define

$$\bar{D}_1(\alpha) = \left\{ f \in [0, B] : \frac{c_2(f)}{c_1(f)} = \frac{\log_2(1 + \beta\rho_2(f))}{\log_2(1 + \beta\rho_1(f))} \leq \alpha \right\}, \tag{7.11}$$

and

$$D_1(\alpha) = \left\{ f \in [0, B] : \frac{c_2(f)}{c_1(f)} = \frac{\log_2(1 + \beta\rho_2(f))}{\log_2(1 + \beta\rho_1(f))} < \alpha \right\}. \tag{7.12}$$

Similarly, we can define $\bar{D}_2(\alpha)$ and $D_2(\alpha)$ as the regions where $\frac{c_2(f)}{c_1(f)} >= \alpha$ and $\frac{c_2(f)}{c_1(f)} > \alpha$, respectively. It can be easily seen that

$$\bar{D}_2(\alpha) \cup D_1(\alpha) = \bar{D}_1(\alpha) \cup D_2(\alpha) = [0, B],$$

and

$$\bar{D}_2(\alpha) \cap D_1(\alpha) = \bar{D}_1(\alpha) \cap D_2(\alpha) = \emptyset.$$

The following theorem is proved in Appendix A.1 and it determines the optimal subcarrier assignment for cross-layer optimization:

THEOREM 7.1   *For a network with two users, if the subcarrier assignment, $\{D_1^*, D_2^*\}$, is optimal, then D_1^* and D_2^* satisfy*

$$D_1(\alpha^*) \subseteq D_1^* \subseteq \bar{D}_1(\alpha^*), \quad D_2^* = [0, B] - D_1^*, \quad \alpha^* = \frac{U_1'(r_1^*)}{U_2'(r_2^*)},$$

and

$$r_i^* = \int_{D_i^*} c_i(f)\,df = \int_{D_i^*} \log_2(1 + \beta\rho_i(f))\,df, \text{ for } i = 1, 2,$$

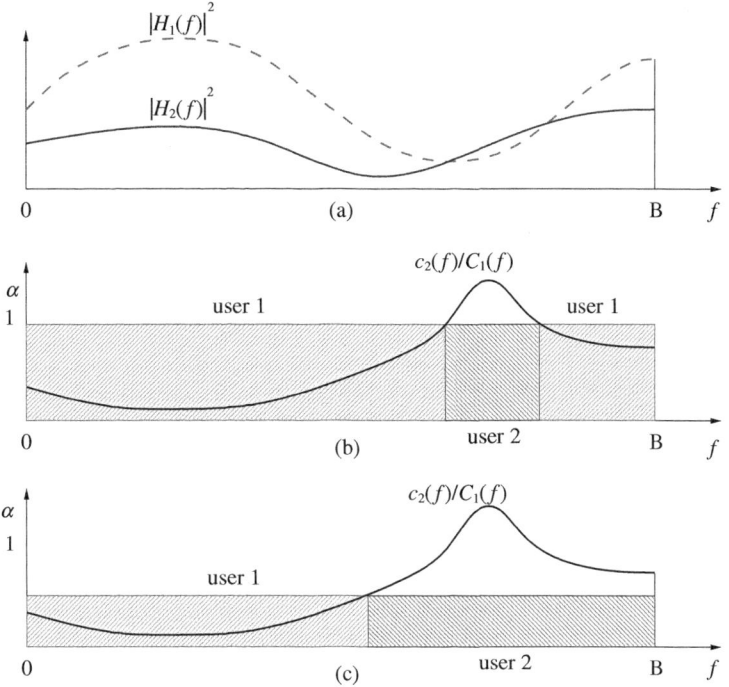

Figure 7.2 Optimal subcarrier assignment for a two-user network

(a) Frequency responses for two users; (b) Subcarrier assignment resulting from throughput-based optimization; (c) Subcarrier assignment resulting from utility-based optimization

where

$$U_i'(r) = \frac{dU_i(r)}{dr}.$$

Figure 7.2 demonstrates the difference between utility-based optimization and traditional throughput-based optimization. For traditional optimization, $U_i(r) = r$; therefore, the threshold, $\alpha^* = \frac{U_1'(r_1)}{U_2'(r_2)}$ is always 1. Consequently, a subcarrier or frequency is allocated to the user with the larger channel gain, as in Figure 7.2(b). To balance efficiency and fairness, an increasing utility curve with a decreasing slope is usually used. In this case, the threshold α^* depends on how much resource each user has already occupied. Since the channel corresponding to user 2 is not as good as that of user 1 in Figure 7.2(a), user 2 gets more frequency resource in the utility-based optimization than in the throughput-based optimization, as in Figure 7.2(c).

It should be noted that the optimal subcarrier assignment is not unique as we can see in network with flat-fading channels. However, α^*, r_1^*, and r_2^* are unique.

Network with multiple users

The results for a two-user network can be extended to the general case of more than two users, which is summarized in the following theorem:

THEOREM 7.2 *For a network with M users, if the subcarrier assignment, D_i^* for $i = 1, 2, \ldots, M$, maximizes the average utility, then for any $f \in D_i^*$, we have*

$$U_j'(r_j^*)c_j(f) \leq U_i'(r_i^*)c_i(f), \quad \text{for any } j \neq i, \tag{7.13}$$

and

$$r_i^* = \int_{D_i^*} c_i(f)df.$$

The proof of the above theorem is very similar to that of Theorem 7.1 and is omitted here.

7.2.3 Adaptive power allocation

In the previous section, in which power allocation is assumed to be fixed, we discussed using DSA to maximize network performance. In this section, we first investigate APA with fixed subcarrier assignment and then study joint DSA and APA. Since achievable throughput is a function of power allocation, it becomes

$$c_i(f) = \log_2(1 + \beta p(f)\rho_i(f)).$$

Adaptive power allocation with fixed subcarrier assignment

When a subcarrier assignment is fixed, APA optimization can be formulated as follows: given a fixed subcarrier assignment, D_i for $i = 1, 2, \ldots, M$, assign the power density, $p(f)$, to maximize

$$\frac{1}{M} \sum_{i=1}^{M} U_i(r_i) = \frac{1}{M} \sum_{i=1}^{M} U_i \left(\int_{D_i} \log_2[1 + \beta p(f)\rho_i(f)]\, df \right), \tag{7.14}$$

subject to

$$\frac{1}{B} \int_0^B p(f)df \leq 1, \quad \text{and} \quad p(f) \geq 0. \tag{7.15}$$

To achieve optimality, a utility-based multi-level water-filling is needed, which is stated in the following theorem:

THEOREM 7.3 *For a given fixed subcarrier assignment, D_i for all i, the optimal power allocation, $p^*(f)$, satisfies*

$$p^*(f) = \left[\frac{U_i'(r_i^*)}{\lambda} - \frac{1}{\beta\rho_i(f)} \right]^+ \quad \lambda > 0, \ f \in D_i, \tag{7.16}$$

where λ is a constant for the normalization of the optimal power density,

$$[x]^+ = \begin{cases} x & x \geq 0 \\ 0 & x < 0 \end{cases},$$

and λ as well as r_i^* satisfy

$$\frac{1}{B} \int_0^B p^*(f)\,df = 1,$$

$$and \; r_i^* = \int_{D_i} \log_2[1 + \beta p^*(f)\wp_i(f)]\,df,$$

where r_i^* and $p^*(f)$ are the optimal values of the rates and the power density, respectively.

It should be indicated that Theorem 7.3 only gives a necessary condition for globally optimal power allocation. The proof of the above theorem is similar to the water-filling theorem [183], which is summarized in Appendix A.2.

Similar to classical water-filling [183], optimal power allocation cannot be directly calculated from (7.16), and iterative algorithms are needed to obtain the optimal one satisfying the power constraint.

There are two major differences between the classical water-filling and the one in Theorem 7.3. First, the *water level* for each user is proportional to its current marginal utility value, $U_i'(r_i)$. In other words, the power allocation is also related to the utility functions. Since the data rates of users are unlikely to be equal, it is from (7.16) that the water levels, $\frac{U_i'(r_i^*)}{\lambda}$, are different for different users. Second, the power constraint is the total transmission power rather than the power of an individual user. As shown in Figure 7.3, utility-based multi-level water-filling (7.16) can be regarded as an extension of fixed-priority multi-level water-filling [90].

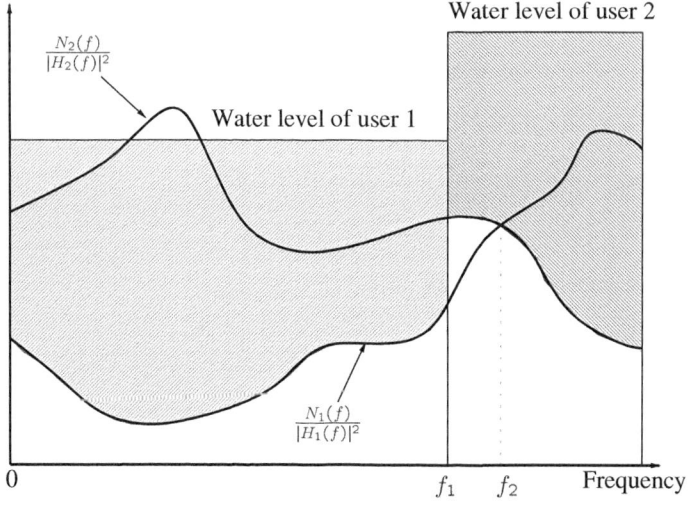

Figure 7.3 Multi-level water-filling for adaptive power allocation in a two-user network

Joint dynamic subcarrier assignment and adaptive power allocation

The DSA and APA can be used simultaneously for cross-layer optimization. Joint DSA and APA optimization can be formulated as follows: adjust the D_is and $p(f)$ to maximize

$$\frac{1}{M}\sum_{i=1}^{M} U_i(r_i) = \frac{1}{M}\sum_{i=1}^{M} U_i\left(\int_{D_i}\log_2[1 + \beta p(f)\rho_i(f)]\,df\right), \qquad (7.17)$$

subject to

$$\bigcup_{i=1}^{M} D_i \subseteq [0, B], \qquad (7.18)$$

$$D_i \bigcap D_j = \varnothing, \quad i \neq j \text{ and } i, j = 1, 2, \ldots, M, \qquad (7.19)$$

and

$$\frac{1}{B}\int_0^B p(f)\,df \leq 1 \text{ and } p(f) \geq 0. \qquad (7.20)$$

Obviously, there are two necessary conditions for the global optimum for the joint DSA and APA:

(i)   Fixing the optimal subcarrier assignment: any change of the power allocation does not increase the total utility.
(ii) Fixing the optimal power allocation: any change of the subcarrier assignment does not increase the total utility.

Therefore, an optimal frequency assignment, D_i^* for all i, and power allocation $p^*(f)$ must satisfy the conditions in both Theorems 7.2 and 7.3. Consequently, we have the following theorem:

THEOREM 7.4   *Let D_i^* for $i = 1, 2, \ldots, M$ and $p^*(f)$ be the optimal subcarrier assignment and power allocation, respectively. Then the following conditions are satisfied:*

$$\begin{cases} U_j'(r_j^*)\log_2(1 + \beta p^*(f)\rho_j(f)) \leq U_i'(r_i^*)\log_2(1 + \beta p^*(f)\rho_i(f)) & f \in D_i^*, \\ p^*(f) = \left[\dfrac{U_i'(r_i^*)}{\lambda} - \dfrac{1}{\beta\rho_i(f)}\right]^+ & \lambda > 0 \;\; f \in D_i^*, \end{cases}$$

$$(7.21)$$

where r_i^* and λ are constrained by

$$\frac{1}{B}\int_0^B p^*(f)\,df = 1,$$

$$\text{and } r_i^* = \int_{D_i^*}\log_2(1 + \beta p^*(f)\rho_i(f))\,df.$$

When the utility function is just the throughput, $U_i(r_i) = r_i$, the optimal subcarrier assignment is independent of the optimal power allocation. In this case, the optimal subcarrier assignment and power allocation have the following closed forms:

$$
\begin{cases}
D_i^* = \{f \in [0 : B] : \rho_i(f) = \max_m \rho_m(f)\} \\
p^*(f) = \left[\dfrac{1}{\lambda} - \dfrac{1}{\beta \max_m \rho_m(f)}\right]^+ \\
\dfrac{1}{B} \displaystyle\int_0^B p^*(f)df = 1,
\end{cases}
$$

which is identical to the result in [196]. This illustrates that *frequency division multiple access* (FDMA)-type systems can achieve Shannon capacity when they are optimized for the sum of throughputs.

7.2.4 Properties of cross-layer optimization

In this section, we will prove the convexity of the achievable data rate region and show that, if the utility function is concave, then a local maximum is also a global maximum. Therefore, the necessary conditions in Theorems 7.2, 7.3, and 7.4 are also sufficient ones.

Convexity of instantaneous data rate region

A data rate vector $\mathbf{r}$ is defined as

$$
\mathbf{r} = (r_1, r_2, \ldots, r_M)^T \in \mathbb{R}_+^M,
$$

where M is the number of users. The *instantaneous data rate region*, $\mathcal{C}_\pi$, is a set that consists of the total achievable data rate vectors under the constraint of a resource allocation policy π, such as DSA, APA, or joint DSA and APA. The instantaneous data rate region is obviously determined by the channel conditions and the resource allocation constraints. It is intuitive that more adaptive resource allocation techniques will result in a larger feasible region.

The objective function is

$$
U(\mathbf{r}) = \frac{1}{M} \sum_{i=1}^M U_i(r_i).
$$

Thus, the optimization problem can be regarded as

$$
\max_{\mathbf{r} \in \mathcal{C}_\pi} U(\mathbf{r}).
$$

Therefore, if $\mathcal{C}_\pi$ is convex, the optimization problem will become tractable. The convexity of the instantaneous data rate region with frequency assignment and power allocation can be described by the following theorem, which is proved in Appendix A.3:

THEOREM 7.5 *For an OFDM-based network with infinitesimal subcarrier space and with DSA, APA, or joint DSA and APA, the achievable data rate region is convex.*

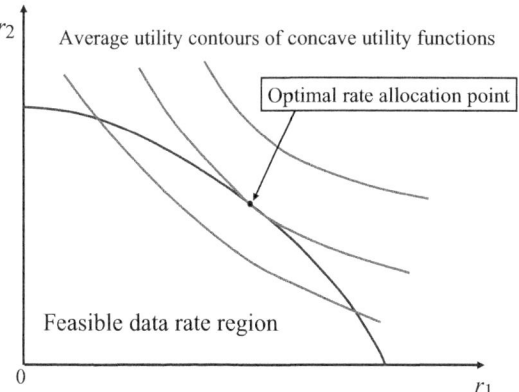

Figure 7.4 Feasible data rate region and optimal rate allocation

With the above theorem, we can obtain the following property of the cross-layer optimization:

LEMMA 7.6   *Let the boundary of the instantaneous data rate region be a subset of the data rate region with the following property: no component of any data rate vector can be increased while the other data rate components remain fixed. The data rate with respect to the maximum of the average utility must be on the boundary of the data rate region if each utility function is strictly increasing.*

Proof   Suppose that the maximum can be achieved by a data rate vector $\mathbf{r}$, which is not on the boundary of the data rate region. There must exist a vector $\mathbf{r}^*$ such that $\mathbf{r} \leqslant \mathbf{r}^*$ with $r_i < r_i^*$ for some i, then $U(\mathbf{r}) < U(\mathbf{r}^*)$.[1] The contradiction shows Lemma 7.6.   □

Lemma 7.6 implies that using a strictly increasing utility function intends to assign all resources including all power and bandwidth to users.

Global optimum

For general differentiable utility functions, the conditions (7.13), (7.16), and (7.21) are sufficient and necessary for *locally* optimal solutions of respective optimization problems; hence, they are only necessary for *global* optimality. With concave utility functions, however, the global optimality of the cross-layer optimization can be described by the following theorem:

THEOREM 7.7   *If all $U_i(r_i)$ are concave functions, then a local maximum of $U(\mathbf{r})$ is also a global maximum, and the conditions (7.13), (7.16), and (7.21) are not only necessary but also sufficient, respectively.*

Proof   The proof simply uses the following two consequences in convex analysis [169].

(i)   If all $U_i(r_i)$ are concave functions, then the objective function $U(\mathbf{r}) = \frac{1}{M}\sum_{i=1}^{M} U_i(r_i)$ is also a concave function.

---

[1] $\mathbf{r} \leqslant \mathbf{r}^*$ means $r_i \leq r_i^*$, for all i.

(ii)   If $\mathcal{C}_\pi \in \mathbb{R}^n$ is a convex set and $U : \mathcal{C}_\pi \mapsto \mathbb{R}$ is a concave function, then a local maximum of U is also a global maximum.

$\square$

The sufficiency of the conditions (7.13), (7.16), and (7.21) for a global optimum is indispensable for algorithm design. If, in addition, the $U_i(r_i)$ are all strictly concave, there is a unique global maximum solution to the optimization problems. Note that the unique global maximum implies that there is only one optimal data rate vector. However, there may be different frequency and power allocation schemes corresponding to the optimal data rate vector as we can see from a network with flat-fading channels for all users.

The relation between the feasible data rate region and concave utility functions is shown in Figure 7.4. Heuristically, Lemma 7.6 shows that the rate vector corresponding to the maximum is located on the boundary of the achievable rate region. Therefore, the optimal rate vector should be a point of tangency between the region boundary and an average utility contour.

8 Algorithm development for utility-based optimization

We have established a theoretical framework for cross-layer optimization in orthogonal frequency division multiplexing (OFDM) wireless networks in Section 7.2. In this chapter, we focus on effective and practical algorithms for efficient and fair resource allocation in OFDM wireless networks. In practical OFDM wireless networks, the number of subcarriers is finite; therefore, the optimization problem turns from continuous to discrete. This discrete optimization, together with non-linear utility functions, challenges algorithm design.

The system model is presented in Section 6.1, but the time parameter n is omitted in this chapter. Given a power vector $\mathbf{p}$, the achievable transmission efficiency at subcarrier k is denoted as $c_i^{\mathbf{p}}[k]$. To make optimization problems more tractable, we will further assume that the utility curve is continuously differentiable. We take various conditions into account and develop a variety of efficient algorithms, including sorting-search dynamic subcarrier assignment, greedy bit loading, and power allocation, as well as objective aggregation algorithms. Furthermore, we will also extend our discussion to a special type of non-concave utility function in Section 8.5.

The use of utility functions with respect to average data rates can further improve performance by exploiting time diversity. A low-pass time filter can be easily incorporated into all of the algorithms.

8.1 Dynamic subcarrier assignment (DSA) algorithms

In this section, we develop DSA algorithms by assuming a fixed power allocation. When only DSA is used, the problem can be mathematically formulated as follows: given a fixed power allocation, $\mathbf{p}$,

$$\max_{D_i, i \in \mathcal{M}} \sum_{i \in \mathcal{M}} U_i(r_i), \tag{8.1}$$

$$\text{subject to } \bigcup_{i \in \mathcal{M}} D_i \subseteq \mathcal{K}, \tag{8.2}$$

$$D_i \bigcap D_j = \varnothing, \quad i \neq j \ \forall i, j \in \mathcal{M}. \tag{8.3}$$

Unlike the scenario of infinite subcarriers, which is analyzed in Section 7.2, the instantaneous data rate region determined by (8.2) and (8.3) is not convex anymore. Therefore, it is necessary to investigate the corresponding optimality conditions.

8.1.1 Optimality conditions

In order to study the optimality, we reformulate the above discrete DSA problem as a non-linear integer (0-1) programming one. Let x_{ik} indicate whether subcarrier k is assigned to user i or not, i.e.

$$x_{ik} = \begin{cases} 1, & \text{if subcarrier } k \text{ is assigned to user } i, \\ 0, & \text{otherwise.} \end{cases}$$

Then the equivalent non-linear integer (0-1) programming problem can be described as follows:

$$\max_{\mathbf{x}} \sum_{i \in \mathcal{M}} U_i \left(\Delta f \sum_{k \in \mathcal{K}} c_i^{\mathbf{p}}[k] x_{ik} \right),$$

$$\text{subject to } \sum_{i \in \mathcal{M}} x_{ik} = 1, \ k \in \mathcal{K}, \text{ and}$$

$$x_{ik} \in \{0, 1\}, \quad i \in \mathcal{M}, \ k \in \mathcal{K},$$

where $\mathbf{x} = [x_{11}, \ldots, x_{1K}, x_{21}, \ldots, x_{2K}, \ldots, x_{M1}, \ldots, x_{MK}]^T$. Thus, there is a one-to-one correspondence between $\mathbf{x}$ and the D_i.

Let

$$U(\mathbf{x}) = \sum_{i \in \mathcal{M}} U_i \left(\Delta f \sum_{k \in \mathcal{K}} c_i^{\mathbf{p}}[k] x_{ik} \right),$$

and $\mathcal{B}$ be the feasible region of $\mathbf{x}$. If the utility functions $U_i(r)$ are concave and differentiable, from the property of the subgradient of concave functions [169], $\forall \mathbf{x} \in \mathcal{B}$,

$$U(\mathbf{x}) - U(\mathbf{y}) \geq \nabla_{\mathbf{x}} U(\mathbf{x})^T (\mathbf{x} - \mathbf{y}) \quad \forall \mathbf{y} \in \mathcal{B}. \tag{8.4}$$

where the gradient of $U(\mathbf{x})$ is defined as

$$\nabla_{\mathbf{x}} U(\mathbf{x}) = \begin{bmatrix} U_1'(r_1) c_1^{\mathbf{p}}[1] \Delta f \\ \vdots \\ U_1'(r_1) c_1^{\mathbf{p}}[K] \Delta f \\ \vdots \\ U_M'(r_M) c_M^{\mathbf{p}}[1] \Delta f \\ \vdots \\ U_M'(r_M) c_M^{\mathbf{p}}[K] \Delta f \end{bmatrix},$$

with

$$U_i'(r) = \frac{dU_i(r)}{dr}.$$

It can be directly derived from (8.4) that if $\mathbf{x}^*$ satisfies

$$\nabla_{\mathbf{x}} U(\mathbf{x}^*)^T (\mathbf{x}^* - \mathbf{x}) \geq 0 \quad \forall \mathbf{x} \in \mathcal{B}, \tag{8.5}$$

then

$$U(\mathbf{x}^*) - U(\mathbf{x}) \geq 0, \ \forall \mathbf{x} \in \mathcal{B},$$

which means that $\mathbf{x}^*$ is globally optimal. The condition (8.5) is equivalent to the following expression. $\forall k \in \mathcal{K}$, letting i be the user index with respect to k such that $x_{ik}^* = 1$,

$$U_i'(r_i^*)c_i^{\mathbf{p}}[k] \geq U_j'(r_j^*)c_j^{\mathbf{p}}[k], \ \ \forall j \neq i \in \mathcal{M}$$

$$\text{and } r_i^* = \sum_{k \in \mathcal{K}} c_i^{\mathbf{p}}[k] \triangle f x_{ik}^*.$$

The above condition can be also expressed as follows. For a fixed power allocation $\mathbf{p}$ and concave utility functions, a set of D_i^* is globally optimal if

$$U_i'(r_i^*)c_i^{\mathbf{p}}[k] \geq U_j'(r_j^*)c_j^{\mathbf{p}}[k], \ \forall k \in D_i^*, \ \forall i,j \in \mathcal{M} \tag{8.6}$$

$$\text{and } r_i^* = \sum_{k \in D_i^*} c_i^{\mathbf{p}}[k] \triangle f. \tag{8.7}$$

Therefore, the variation the optimality conditions for the continuous-frequency case developed in Theorem 7.2 also holds for the discrete frequency case. It is worth noting that the above conditions are only *sufficient* for optimality, and that its *necessity* is lost due to the non-convexity of the achievable data rate region in this case.

We consider two specific cases. First, continuous rate adaptation is used. In this scenario, since the channel fading levels, $H_i[k]$, are continuous random variables, $c_i^{\mathbf{p}}[k]$ are continuous random variables as well, which implies that $\mathbb{P}\left\{c_i^{\mathbf{p}}[k] = c_i^{\mathbf{p}}[k']\right\} = 0$ for the pair $(i,k) \neq (i',k')$. According to (8.6) and (8.7), subcarrier k should be assigned to user m according to the following rule:

$$m(k) = \arg \max_{i \in \mathcal{M}} \{U_i'(r_i^*) \cdot c_i^{\mathbf{p}}[k]\}, \tag{8.8}$$

where $m(k)$ represents that subcarrier k should be assigned to user $m(k)$, and

$$r_i^* = \sum_{k \in D_i^*} c_i^{\mathbf{p}}[k] \triangle f.$$

Another scenario is that linear utility functions are used. For a linear utility function $U_i(r_i)$, its marginal utility function $U_i'(r_i)$ is a constant, which is denoted as U_i'. With the linearity, (8.4) becomes

$$U(\mathbf{x}) - U(\mathbf{y}) = \nabla_{\mathbf{x}} U^T(\mathbf{x} - \mathbf{y}) \ \ \forall \mathbf{y} \in \mathcal{B}.$$

Using the same method, we have that with linear utility functions, a set of D_i^* is globally optimal *if and only if*

$$U_i' c_i^{\mathbf{p}}[k] \geq U_j' c_j^{\mathbf{p}}[k], \ \forall k \in D_i^*, \ \forall i,j \in \mathcal{M}. \tag{8.9}$$

Therefore, the optimal subcarrier assignment has the following closed form:

$$m(k) = \arg \max_{i \in \mathcal{M}} \{U_i' \cdot c_i^{\mathbf{p}}[k]\}. \tag{8.10}$$

8.1.2   Sorting-search algorithm of subcarrier assignment

Utility-based subcarrier assignment optimization belongs to the set of non-linear combinatorial optimization problems, in which there is no general approach to achieve optimality. In this subsection, we discuss a sorting-search algorithm to seek the optimal subcarrier assignment.

Let us first consider the two-user case, in which each subcarrier in a set of subcarrier indices $\mathcal{A}$ ($\mathcal{A} \subseteq \mathcal{K}$) will be assigned to either user 1 or user 2. For this combinatorial optimization problem, there are $2^{|\mathcal{A}|}$ choices to assign $|\mathcal{A}|$ subcarriers, where $|\mathcal{A}|$ denotes the number of elements in set $\mathcal{A}$. The key idea of the sorting-search algorithm is to assume that the conditions (8.6) and (8.7) are both sufficient and necessary. Thus, we have the following rule for an optimal subcarrier assignment: *In the two-user case, if subcarrier i is assigned to user 1, and* $\dfrac{c_2^{\mathbf{P}}[j]}{c_1^{\mathbf{P}}[j]} < \dfrac{c_2^{\mathbf{P}}[i]}{c_1^{\mathbf{P}}[i]}$*, then subcarrier j must be assigned to user 1 as well.* From the above rule, after the $\dfrac{c_2^{\mathbf{P}}[k]}{c_1^{\mathbf{P}}[k]}$ for all $k \in \mathcal{A}$ are sorted in an increasing order, there are only $|\mathcal{A}| + 1$ possible assignments that may result in the optimal point, including the two extreme cases: all subcarriers are assigned to user 1 or user 2. In other words, if the conditions (8.6) and (8.7) are both sufficient and necessary, the optimal rate vector should be located on the boundary of the convex hull of the feasible data rate vector set.

For example, if there are three subcarriers for users 1 and 2, and

$$\frac{c_2^{\mathbf{P}}[1]}{c_1^{\mathbf{P}}[1]} \leq \frac{c_2^{\mathbf{P}}[2]}{c_1^{\mathbf{P}}[2]} \leq \frac{c_2^{\mathbf{P}}[3]}{c_1^{\mathbf{P}}[3]},$$

then there are only four possible choices for D_1/D_2: $\{\varnothing\}/\{1, 2, 3\}$, $\{1\}/\{2, 3\}$, $\{1, 2\}/\{3\}$, $\{1, 2, 3\}/\{\varnothing\}$. The data rates for users 1 and 2 with those four different subcarrier assignments are shown in Figure 8.1. From the figure, we can see that the slopes of the lines ab, bc, and cd are $-\dfrac{c_2^{\mathbf{P}}[1]}{c_1^{\mathbf{P}}[1]}$, $-\dfrac{c_2^{\mathbf{P}}[2]}{c_1^{\mathbf{P}}[2]}$, and $-\dfrac{c_2^{\mathbf{P}}[3]}{c_1^{\mathbf{P}}[3]}$, respectively. Note that the data rate vectors a, b, c, and d are located on the boundary of the convex hull of the feasible data rate vector set.

The remaining problem is to find out the optimal one among $|\mathcal{A}| + 1$ choices. Let T be a threshold that subcarriers satisfying $\dfrac{c_2^{\mathbf{P}}[k]}{c_1^{\mathbf{P}}[k]} > T$ are assigned to user 2, and the rest to user 1. Therefore, T will determine a subcarrier assignment. From the previous discussion, the optimal T should be close to $\dfrac{U_1'(r_1)}{U_2'(r_2)}$. With the increase of T, r_1 increases and r_2 decreases. Because of the concavity of utility functions, $\dfrac{U_1'(r_1)}{U_2'(r_2)}$ decreases with the increase of T. Clearly, binary search is the best way to arrive at the optimal threshold.

The efficient algorithm is described in Algorithm 8.1. The average computational complexity of sorting is about $|\mathcal{A}| \log_2(|\mathcal{A}|)$, and that of the binary search is only $\log_2(|\mathcal{A}|)$. Therefore, the average computational complexity of this algorithm is less

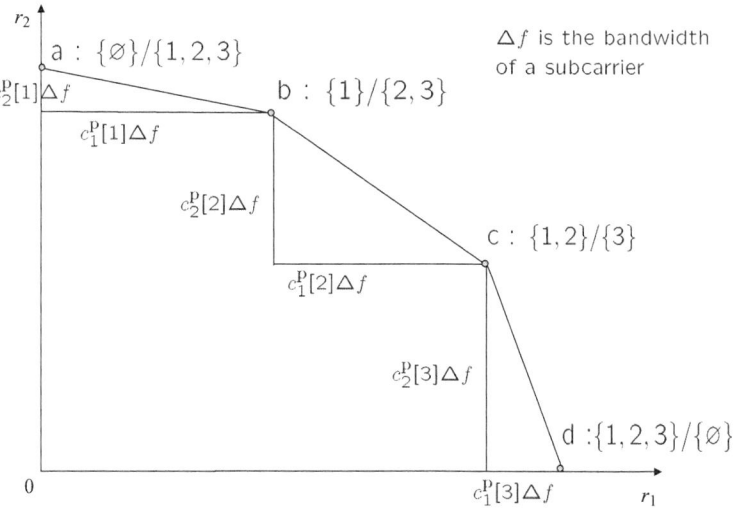

Figure 8.1 An illustration of properties of DSA

Algorithm 8.1: Sorting-search subcarrier assignment for the two-user case

sort $\frac{c_2^{\mathrm{p}}[k]}{c_1^{\mathrm{p}}[k]}$, $k \in \mathcal{A}$ in increasing order
get thresholds: $T[k]$, $k \in \{1 : |\mathcal{A}| + 1\}$ in increasing order
$low = 1$;　$high = |\mathcal{A}| + 1$
while $high - low > 0$ **do**
　$center \leftarrow \lfloor (low + high)/2 \rfloor$
　$T \leftarrow center$
　if $T - \frac{U_1'(r_1)}{U_2'(r_2)} > 0$ **then**
　　$high \leftarrow center$
　else
　　$low \leftarrow center$
　end if
end while
choose the best T between low and $high$

than $(K + 1) \log_2(K)$. An implementation in Python can be found at https://github.com/songgc/utility-based-allocation

For the M-user case, we can update the subcarrier assignment of every two users iteratively by means of the subcarrier assignment algorithm for the two-user case. Obviously, the computational complexity is nearly $(M - 1)^2(K + 1) \log_2(K)$, which is still efficient compared to the number of choices of this combinatorial optimization problem, K^M. Moreover, the algorithm is robust to both continuous and discrete rate adaptation.

It should be indicated that the above sorting-search algorithm is in general suboptimal. However, the algorithm will be optimal in each of the following cases:

(i) *The utility functions are all linear*: This is because the condition (8.9) is sufficient and necessary for optimality.

(ii) *The bandwidth of a subcarrier of the OFDM signal is infinitesimal*: In this case, $\frac{\Delta f}{B} \to 0$, the feasible data rate region becomes convex. From Section 7.2, this condition leads to the sufficient and necessary condition for optimality

$$\nabla_{\mathbf{x}} U(\mathbf{x})^T (\mathbf{x} - \mathbf{y}) \geq 0 \quad \forall \mathbf{y} \in \mathcal{B}.$$

In practical OFDM systems, $\frac{\Delta f}{B}$ is usually small, and thus the performance of the sorting-search algorithm is nearly optimal in practical situations.

8.2   Adaptive power allocation (APA) algorithms

We develop algorithms for adaptive power allocation in this section. First, we assume that subcarrier assignment is fixed, and then we extend the developed algorithms to the joint DSA and APA.

8.2.1   APA for fixed subcarrier assignment

When only APA is allowed in the system, the subcarrier assignment, D_i for all i, is fixed, and we have

$$\max_{\mathbf{p}} \sum_{i \in \mathcal{M}} U_i(r_i) \tag{8.11}$$

$$\text{subject to} \sum_{k \in \mathcal{K}} p[k] \leq \bar{P} \tag{8.12}$$

$$p[k] \geq 0. \tag{8.13}$$

When continuous rate adaptation is used, the optimal power allocation for a fixed subcarrier assignment has the following solution, which comes from Theorem 7.3:

$$p^*[k] = \left[\frac{U_i'(r_i^*)}{\lambda} - \frac{1}{\beta \rho_i[k]} \right]^+ \quad \lambda > 0, \quad k \in D_i \tag{8.14}$$

and λ and r_i^* satisfy

$$\sum_{k \in \mathcal{K}} p^*[k] = \bar{P}$$

$$r_i^* = \sum_{k \in D_i} \log_2(1 + \beta p^*[k] \rho_i[k]) \Delta f.$$

This is actually a utility-based water-filling.

Algorithm 8.2: Sequential-linear-approximation water-filling algorithm for continuous rate adaptation

Iterate until $\sum_{i \in \mathcal{M}} U_i'(r_i^{(n)})(r_i^{(n+1)} - r_i^{(n)}) \leq \epsilon$

(i)   Get the new power allocation from the linear optimization problem and the corresponding data rates.

$$p[k] \leftarrow \left[\frac{\gamma_{m(k)}^{(n)}}{\lambda} - \frac{1}{\beta \rho_{m(k)}[k]} \right]^+ \text{ for all } k$$

$$r_i^{(n+1)} \leftarrow \sum_{k \in D_i} \log_2(1 + \beta p[k]\rho_i[k])\Delta f \text{ for all } i,$$

where $m(k)$ means that subcarrier k is assigned to user $m(k)$.

(ii) Update $\gamma_i^{(n)}$ with a positive step size $\mu \in (0,1)$

$$\gamma_i^{(n+1)} \leftarrow (1-\mu)\gamma_i^{(n)} + \mu U_i'(r_i^{(n+1)}) \text{ for all } i.$$

8.2.2 Sequential-linear-approximation water-filling algorithm for continuous rate adaptation

With continuous rate adaptation, APA optimization is still a non-linear convex programming problem, and (8.14) is both sufficient and necessary for global optimality. When a subcarrier assignment is fixed, the non-linear optimization problem can be approached by a series of linear optimization problems by means of the sequential-linear-approximation algorithm (Frank–Wolfe method) [141], which can be summarized by Algorithm 8.2. Each iteration of the algorithm contains two steps. First, we solve an optimization problem with fixed marginal utilities, which is a regular water-filling problem, and then update their marginal utilities using a subgradient method. Intuitively, by solving the group of optimization problems with a linear objective $\sum_{i \in \mathcal{M}} \gamma_i r_i$ subject to the same constraints as those of the original problem, for all possible $\gamma_i \geq 0$, we can trace out the entire boundary of the data rate region.

8.2.3 Greedy power allocation algorithm based on maximizing total utility for discrete rate adaptation

In practice, continuous rate adaptation is unfeasible, and there are only several modulation levels. Thus, the optimal power level at each subcarrier for discrete rate adaptation is not continuous either. As a result, the previous water-filling algorithm cannot achieve the optimal power allocation. Therefore, we develop a greedy algorithm for discrete modulation levels.

The key idea of the greedy algorithm is to allocate bits and the corresponding power successively and maximize the utility argument per power in each step of bit loading.

Algorithm 8.3: Greedy power allocation algorithm for discrete rate adaptation

$b_k \leftarrow 0$; $\triangle p_k \leftarrow 0$ for all k
$r_i \leftarrow 0$ for all i
$p_{total} \leftarrow 0$; $\triangle p \leftarrow 0$
while $p_{total} + \triangle p < \bar{P}$ **do**
    $p_{total} \leftarrow p_{total} + \triangle p$
    $\triangle p_k \leftarrow f(b_k + \triangle b_k) - f(b_k)$ for all k,
    where b_k is the current modulation level of subcarrier k, and $\triangle b_k$ is the
    difference between the next modulation level and the current one for subcarrier k.
    if $p_{total} + \triangle p_k > \bar{P}$ **then**
        $\triangle p_k \leftarrow \infty$
    end if
    $\triangle U_k \leftarrow U_{m(k)}(r_{m(k)} + \triangle b_k) - U_{m(k)}(r_{m(k)})$
    $\hat{k} \leftarrow \arg\max_{k \in \mathcal{K}}(\frac{\triangle U_k}{\triangle p_k})$
    $\triangle p \leftarrow \triangle p_{\hat{k}}$
    if $p_{total} + \triangle p \leq \bar{P}$ **then**
        $b_{\hat{k}} \leftarrow b_{\hat{k}} + \triangle b_{\hat{k}}$
        $r_{m(\hat{k})} \leftarrow r_{m(\hat{k})} + \triangle b_{\hat{k}}$
    end if
end while

Let $f(b)$ be the required power to transmit b b/s/Hz, which is usually determined by the system design. If M-ary quadrature amplitude modulation (M-QAM) is used, according to (7.2), $f(b)$ is given by

$$f(b) = \frac{2^b - 1}{\beta \rho[k]}.$$

In initialization, zero bits are assigned to all subcarriers. During each bit loading iteration, power is allocated at some subcarrier so that the increase in utility per power is maximized. The iteration process will stop when the total transmission power constraint is reached. The greedy power allocation is summarized in Algorithm 8.3. Note that nonlinear concave utility functions do not increase the algorithm complexity compared to linear utility functions.

Using the following three steps, we can prove that the greedy algorithm results in *global* optimal bit loading and power allocation with concave utility functions. First, we show that the objective function (8.11) is also concave with respect to the power vector **p**. Then, we check that the feasible region of power allocation vector **p** constrained by (8.12) is a *polymatroid*, which satisfies the normalized, non-decreasing, and submodular properties [73]. Finally, taking advantage of the concavity of the objective function and the polymatroid structure of the feasible region, we can demonstrate the optimality of the greedy algorithm according to [73].

8.3   Joint dynamic subcarrier assignment and adaptive power allocation

As in Section 7.2, when both power allocation and subcarrier assignment can be changed, the joint DSA and APA optimization problem can be expressed as follows:

$$\max_{D_i, i \in \mathcal{M}, \mathbf{p}} \sum_{i \in \mathcal{M}} U_i(r_i), \tag{8.15}$$

$$\text{subject to} \bigcup_{i \in \mathcal{M}} D_i \subseteq \mathcal{K}, \tag{8.16}$$

$$D_i \bigcap D_j = \varnothing, \quad i \neq j \ \forall i, j \in \mathcal{M}, \tag{8.17}$$

$$\sum_{k \in \mathcal{K}} p[k] \leq \bar{P} \tag{8.18}$$

$$p[k] \geq 0. \tag{8.19}$$

Obviously, the optimal resource allocation (optimal rate vector) must simultaneously satisfy the conditions for the DSA-only and APA-only problems. Similar to the discussion in Section 8.2.2, for concave functions the algorithm for the joint DSA and APA using continuous rate adaptation is a combination of iterative subcarrier assignment, power allocation, and the update of marginal utility, which is summarized in Algorithm 8.4. For those concave utility functions, using this algorithm with an appropriate update-step μ, we can find a global maximum. The computational complexity of the subcarrier assignment is only $\mathcal{O}(MK)$. For discrete rate adaptation, we can iteratively use the sorting-search DSA and the greedy APA algorithms.

Algorithm 8.4: Joint DSA and APA with continuous rate adaptation

Iterate until $\sum_{i \in \mathcal{M}} U_i'(r_i^{(n)})(r_i^{(n+1)} - r_i^{(n)}) \leq \epsilon$.

(i)   Get the new subcarrier assignment, according to the condition (8.9), using

$$m(k) \leftarrow \arg \max_{i \in \mathcal{M}} \{ \gamma_i^{(n)} c_i^{\mathbf{p}}[k] \} \quad \text{for all } k.$$

(ii)   Get the new power allocation from the linear optimization problem and the corresponding data rates

$$p[k] \leftarrow \left[\frac{\gamma_{m(k)}^{(n)}}{\lambda} - \frac{1}{\beta \rho_{m(k)}[k]} \right]^+ \quad \text{for all } k$$

$$r_i^{(n+1)} \leftarrow \sum_{k \in D_i} \log_2(1 + \beta p[k] \rho_i[k]) \Delta f \quad \text{for all } i.$$

(iii)   Update $\gamma_i^{(n)}$ with a positive step size $\mu \in (0, 1)$

$$\gamma_i^{(n+1)} \leftarrow (1 - \mu) \gamma_i^{(n)} + \mu U_i'(r_i^{(n+1)}) \quad \text{for all } i,$$

8.4 Algorithm modification for non-concave utility functions

Utility functions depend on the type of applications and are not always concave. For instance, it is demonstrated in [108] that the utility function for best-effort applications is

$$U(r) = [0.16 + 0.8 \ln(r - 0.3)]^+, \tag{8.20}$$

where r is in units of kbps. For more general use, we express the utility function as

$$U(r) = \begin{cases} a + b \ln(r - c) & r \geq r_{thr}, \\ 0 & 0 \leq r < r_{thr}, \end{cases} \tag{8.21}$$

where $b > 0$ and $a = -b \ln(r_{thr} - c)$ is a threshold. Even though the above utility function (8.20) is not exactly concave over $[0, +\infty)$, it is strictly concave and differentiable when the data rate is above a threshold. For this utility function (8.20), the threshold, $r_{thr} = 1.119$ kbps, is very small. As a result, the non-concavity of this function may not significantly affect the solution of the optimization problem, especially in the case of high signal-to-noise ratio (SNR). However, the non-concavity sometimes does affect the solution; therefore, we will discuss an approach to deal with the non-concavity problem.

Finding the global optimum of a non-convex optimization problem is in general very difficult. An intuition can be obtained from (8.21); this utility function implies the need for admission control. r_{thr} is actually the threshold for admission control. Our solution includes the following two steps:

- Modify this utility function to $\tilde{U}(r)$ as follows:

$$\tilde{U}(r) = \begin{cases} U(r), & r \geq r_{thr}, \\ U'(r_{thr})(r - r_{thr}), & 0 \leq r < r_{thr}, \end{cases} \tag{8.22}$$

  which is concave over $[0, +\infty)$, and a global maximum for the modified utility function can be obtained by using the previous algorithms. Note that the modification is suitable for any utility function that is concave over $[r_{thr}, +\infty)$.
- Using admission control, shown in Figure 8.2, the solution obtained from the modified utility function $\tilde{U}(r)$ can be corrected to that of the original utility function $U(r)$.

8.5 Cross-layer optimization based on utility functions with respect to average data rates

All of the forementioned resource allocation algorithms underlying maximizing the aggregate utility just consider the instantaneous channel conditions, fairness, and efficiency. In reality, however, users mainly care about the average data rate during a certain period of time, not the instantaneous one. In this section, we investigate the impact of time diversity on the performance of cross-layer optimization. We start with a general case and then study the asymptotic performance.

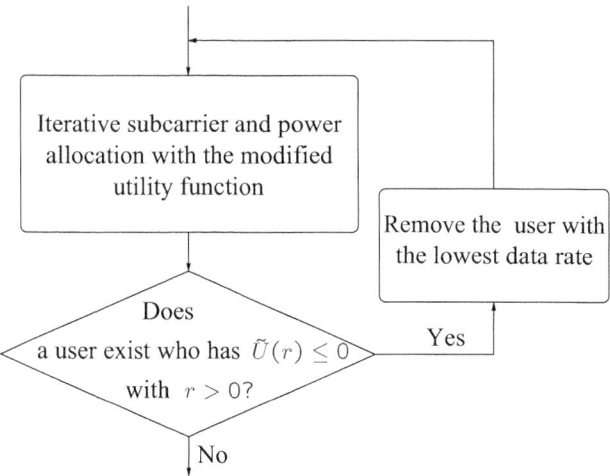

Figure 8.2 Modified dynamic resource allocation algorithm

The average data rate $\bar{r}_i[n]$ of each user at time n can be expressed by using an exponentially weighted low-pass time window as

$$\bar{r}_i[n] = (1 - \rho_w)\bar{r}_i[n-1] + \rho_w r_i[n]. \tag{8.23}$$

where $r_i[n]$ is the instantaneous data rate of user i at time n. $\rho_w = \dfrac{T_s}{T_w}$, where T_s is the slot length, and T_w is the length of the window. Therefore, the optimization problem should be expressed as maximizing the total utility with respect to the average data rates, $\bar{r}_i[n]$, i.e.

$$\max_{\mathbf{r}[n] \in \mathcal{C}_\pi(\mathbf{H})} \sum_{i \in \mathcal{M}} U_i(\bar{r}_i[n]), \tag{8.24}$$

where $\mathbf{r}[n]$ is the data rate vector $[r_1[n], r_2[n], \ldots, r_M[n]]^T$, and $\mathcal{C}_\pi(\mathbf{H})$ is the instantaneous feasible data rate region at time n, determined by the current channel states $\mathbf{H}$, which is given by

$$\mathbf{H} = (H_1[1], \ldots, H_1[K], \ldots, H_M[1], \ldots, H_M[K]),$$

and the allocation constraints of a certain resource allocation policy π, such as DSA, APA, as well as joint DSA and APA.

Since $\bar{r}_i[n]$ is a function of $r_i[n]$ in (8.23), the optimization problem (8.24) can be rewritten as

$$\max_{\mathbf{r}[n] \in \mathcal{C}_\pi(\mathbf{H})} \sum_{i \in \mathcal{M}} V_i(r_i[n]), \tag{8.25}$$

where $V_i(r_i[n]) = U_i((1 - \rho_w)\bar{r}_i[n-1] + \rho_w r_i[n])$.

The above problem can be regarded as an optimization based on utility functions, $V_i(\cdot)$, with respect to the instantaneous data rate, $r_i[n]$, as well. The marginal utility function is given by

$$\frac{\partial}{\partial r_i[n]} V_i(r_i[n]) = \rho_w U_i'(r_i)\big|_{r_i=(1-\rho_w)\bar{r}_i[n-1]+\rho_w r_i[n]}.$$

Therefore, all previous algorithms work well as long as $U_i(r_i)$ and $U_i'(r_i)$ are replaced by $U_i(\bar{r}_i[n])$ and $\rho_w U_i'(\bar{r}_i[n])$, respectively. The computational complexity of maximizing the total utility function with respect to the average data rates is the same as that of optimization with respect to the instantaneous data rates.

Without a time window, the optimization problem must guarantee fairness in each slot period. However, when the time window is used, the fairness requirement is relaxed to a time-window length. This provides more flexibility to improve spectral efficiency. The current resource allocation is related to the previous ones. If one user has a higher average data rate, his priority is set to be lower. Hence, the use of a time window may enhance fairness as well. The length of the time window should be longer than the correlation time of the channel in order to get more time diversity. But if it is too long, the utility function cannot capture the short-term preference of users.

If ρ_w is small enough, the computational complexity of the optimization problem can be reduced furthermore. With a small ρ_w, we have

$$\frac{\partial U_i(\bar{r}_i[n])}{\partial r_i[n]} \approx \rho_w U_i'(r_i)\big|_{r_i=\bar{r}_{i[n-1]}},$$

which means that the current marginal utility values are totally determined by the previous resource allocation. With the one-order Taylor formula, it follows that

$$\sum_{i\in\mathcal{M}} U_i(\bar{r}_i[n]) - \sum_{i\in\mathcal{M}} U_i(\bar{r}_i[n-1])$$
$$\approx \sum_{i\in\mathcal{M}} U_i'(\bar{r}_i[n-1])(\rho_w r_i[n] - \rho_w \bar{r}_i[n-1]). \tag{8.26}$$

Since all $\bar{r}_i[n-1]$ are fixed at time n, the optimization problem becomes the one with a linear objective function as follows:

$$\max_{\mathbf{r}[n]\in\mathcal{C}_\pi(\mathbf{H})} \sum_{i\in\mathcal{M}} U_i'(\bar{r}_i[n-1])r_i[n], \tag{8.27}$$

which maximizes the sum of weighted rates. The weights are adaptively controlled by the marginal utility with respect to the current average rates.

The linear objective function greatly simplifies the corresponding algorithms. In particular, for DSA, we have the following closed form according to (8.10):

$$m(k,n) = \arg\max_{i\in\mathcal{M}}\{U_i'(\bar{r}_i[n-1]) \cdot c_i^{\mathbf{P}}[k,n]\}, \tag{8.28}$$

where $m(k,n)$ represents that subcarrier k is assigned to user $m(k,n)$ at time n, and $c_i^{\mathbf{P}}[k,n]$ denotes the achievable transmission efficiency of subcarrier k at time n. Its complexity is only $M \cdot K$. If there is only one carrier (single-carrier system), and if $U_i(\bar{r}_i[n]) = \ln(\bar{r}_i[n])$, (8.28) is simplified as

$$m(n) = \arg \max_{i \in \mathcal{M}} \left\{ \frac{c_i^{\mathbf{p}}[n]}{\bar{r}_i[n-1]} \right\},$$

which is just the proportional fair scheduling proposed for code division multiple access (CDMA) systems in [200]. Therefore, the utility-based resource allocation we presented here is a general framework for allocating multi-user shared resources. For APA or joint DSA and APA, iteration is still needed, but the linear objective function offers fast convergence.

8.6   Efficiency and fairness

Both efficiency and fairness issues are very important for resource allocation in wireless networks. An allocation scheme is said to be *efficient* if there is no other scheme that would simultaneously benefit someone and harm nobody in terms of their utilities. Therefore, utility-based optimization is obviously efficient. Note that it differs from *spectral efficiency* that is measured in terms of the total throughput over the bandwidth. Clearly, the maximum spectral efficiency is achieved by using a utility function $U_i(r_i) = r_i$ for all i.

With the channel knowledge for each user at the base station, the DSA scheme tends to assign subcarriers to users with a better SNR at the corresponding subcarriers, thereby having high spectral efficiency. It is obvious from (7.13) that the utility-based DSA penalizes the users with poor channel conditions.

When $U_i(r_i) = r_i$, $U_i'(r_i) = 1$. In this case, each subcarrier is assigned to the user with the best channel conditions among all users; therefore, the system can obtain the largest multi-user diversity with respect to spectral efficiency. Although multi-user diversity is similar to traditional selection diversity, its diversity gain results from the number of users, rather than from the number of antennas.

Fairness requires a fair share of bandwidth among competing users and protection from aggressive connections. Two representative types of fairness are *proportional* fairness [116] and *max–min* fairness [44]. Proportional fairness provides each connection a priority inversely proportional to its data rate. A vector of rates $\mathbf{r} \in \mathcal{C}$ is said to be *proportional fair* if for any other feasible rate vector $\mathbf{r}' \in \mathcal{C}$, the aggregate of proportional changes is zero or negative:

$$\sum_{i=1}^{M} \frac{r_i' - r_i}{r_i} \leq 0. \tag{8.29}$$

For a concave utility function $U(\mathbf{r})$ and a convex set $\mathcal{C}_\pi$, $\mathbf{r}$ is optimal if and only if

$$\nabla U(\mathbf{r})^T (\mathbf{r}' - \mathbf{r}) \leq 0 \quad \text{for all } \mathbf{r}' \in \mathcal{C}_\pi, \tag{8.30}$$

where $\nabla U(\mathbf{r}) = [U_1'(r_1), U_2'(r_2), \ldots, U_M'(r_M)]^T$. When the logarithmic utility function, $U(r) = \ln(r)$, is used, (8.30) is identical to (8.29). Therefore, the logarithmic utility function is associated with the proportional fairness for utility-based optimization.

A data rate vector $\mathbf{r}$ is *max–min fair* if for each $m \in \mathcal{M}$, r_m cannot be increased without decreasing r_i for some i for which $r_i < r_m$. Obviously, max–min fairness has a strict fairness criterion since lower rates can get absolute priority.

Consider a family of utility functions expressed as

$$U(r) = -\frac{r^{-\alpha}}{\alpha}, \quad \alpha > 0. \tag{8.31}$$

Obviously, the parameter α determines the degree of fairness. As α increases, the fairness of the corresponding utility function becomes stricter and stricter. When $\alpha \to \infty$, it turns out to be the max–min fairness.

It can be also seen from (7.13) that increasing utility functions encourage the users having good channel conditions, and decreasing marginal utility functions assign a high priority to the users with a low data rate. Therefore, utility-based resource allocation can guarantee both efficiency and fairness.

8.6.1 Fairness of "extreme OFDM" using utility functions with respect to instantaneous data rates

Since the instantaneous capacity is convex in the "extreme OFDM" system in Section 7.2, in which the number of subcarriers is assumed to be infinite, utility-based optimization related to instantaneous data rates in Section 7.2 can maintain a fairness defined as (8.30) with respect to the instantaneous capacity region.

8.6.2 Fairness of "practical OFDM" using utility functions with respect to average data rates

In practical OFDM systems, in which the number of subcarriers is finite, the instantaneous capacity region is not convex anymore. Thus, we focus on the steady state of the system and the fairness related to the long-term average data rate region.

The scheduling algorithm (8.28) is assumed to be used. We consider the situation in the steady state when the window size $T_w \to \infty$. It is assumed that the channel processes $\mathbf{H}$ are ergodic. We denote by $\tilde{r}_i$ the limit data rate of user i; due to the ergodicity of $\mathbf{H}$, it follows that

$$\tilde{r}_i = \lim_{n \to \infty} \bar{r}_i[n]$$

$$= \mathbb{E}\{r_i\}.$$

Let $\tilde{\mathcal{C}}_\pi$ be the long-term *average data rate region* under the allocation constraints of the policy π, which consists of the average data rate vectors obtained by all possible *stationary* resource allocation schemes.

It is easy to prove that $\tilde{\mathcal{C}}_\pi$ is a convex set; i.e., $\forall \, \tilde{\mathbf{r}}^{(1)}, \tilde{\mathbf{r}}^{(2)} \in \tilde{\mathcal{C}}_\pi, \alpha \in [0, 1]$, we will show that $\alpha \tilde{\mathbf{r}}^{(1)} + (1 - \alpha) \tilde{\mathbf{r}}^{(2)} \in \tilde{\mathcal{C}}_\pi$. According to the definition of $\tilde{\mathcal{C}}_\pi$, there must exist such a resource allocation scheme $F^{(1)}$ such that $\tilde{\mathbf{r}}^{(1)} = \mathbb{E}\{F^{(1)}(\mathbf{H})\}$, where $F^{(1)}(\mathbf{H})$ is the data rate vector of user i under the channel states $\mathbf{H}$ when a resource allocation scheme $F^{(1)}$ is employed. Likely, $\tilde{\mathbf{r}}^{(2)}$ results from another resource allocation scheme

$F^{(2)}$ so that $\tilde{\mathbf{r}}^{(2)} = \mathbb{E}\{F^{(2)}(\mathbf{H})\}$. We can construct such a new scheme F that under the channel states $\mathbf{H}$,

$$F = \begin{cases} F^{(1)} & \xi = 1, \\ F^{(2)} & \xi = 0, \end{cases}$$

where ξ is a binary random variable with $P\{\xi = 1\} = \alpha$. It follows that

$$\begin{aligned} \tilde{\mathbf{r}} &= \mathbb{E}\{F(\mathbf{H})\} \\ &= \mathbb{E}\{\xi F^{(1)}(\mathbf{H}) + (1 - \xi)F^{(2)}(\mathbf{H})\} \\ &= P\{\xi = 1\}\mathbb{E}\{F^{(1)}(\mathbf{H})\} + (1 - P\{\xi = 1\})\mathbb{E}\{F^{(2)}(\mathbf{H})\} \\ &= \alpha\tilde{\mathbf{r}}^{(1)} + (1 - \alpha)\tilde{\mathbf{r}}^{(2)}. \end{aligned}$$

The data rate vector $\tilde{\mathbf{r}}$ with respect to the scheme F lies in $\tilde{\mathcal{C}}_\pi$; therefore, $\alpha\tilde{\mathbf{r}}^{(1)} + (1 - \alpha)\tilde{\mathbf{r}}^{(2)} \in \tilde{\mathcal{C}}_\pi$.

The optimization problem (8.27) in the steady state can be expressed as

$$\max_{\mathbf{r} \in \mathcal{C}_\pi(\mathbf{H})} \sum_{i \in \mathcal{M}} U_i'(\tilde{r}_i)r_i, \tag{8.32}$$

where $\tilde{\mathbf{r}}$ is the steady-state data rate vector. The data rate vector $\mathbf{r}^*$ under the channel conditions $\mathbf{H}$ with respect to the scheme (8.32) is given by

$$\mathbf{r}^* = \arg\max_{\mathbf{r} \in \mathcal{C}_\pi(\mathbf{H})} \sum_{i \in \mathcal{M}} U_i'(\tilde{r}_i)r_i. \tag{8.33}$$

Due to being in the steady state, $\mathbb{E}\{\mathbf{r}^*\} = \tilde{\mathbf{r}}$. Obviously, with any other scheme F, it follows that $\mathbf{r}' = F(\mathbf{H})$, and

$$\nabla U(\tilde{\mathbf{r}})^T(\mathbf{r}' - \mathbf{r}^*) \leq 0, \quad \mathbf{r}' \in \mathcal{C}_\pi(\mathbf{H}), \tag{8.34}$$

where, $\nabla U(\tilde{\mathbf{r}}) = [U_1'(\tilde{r}_1), U_2'(\tilde{r}_2), \dots, U_M'(\tilde{r}_M)]^T$. Taking expectations on both sides, we have

$$\nabla U(\tilde{\mathbf{r}})^T(\tilde{\mathbf{r}}' - \tilde{\mathbf{r}}) \leq 0, \quad \tilde{\mathbf{r}}' \in \tilde{\mathcal{C}}_\pi. \tag{8.35}$$

Due to the convexity of the feasible average rate region $\tilde{\mathcal{C}}_\pi$ and the concavity of utility functions $U_i(r)$, the condition (8.35) is sufficient and necessary for the optimality of the following problem [169]:

$$\max_{\tilde{\mathbf{r}} \in \tilde{\mathcal{C}}_\pi} \sum_{i \in \mathcal{M}} U_i(\tilde{r}_i). \tag{8.36}$$

Therefore, when $\rho_w \to 0$, the optimization problem in the instantaneous rate region (8.27) can achieve the optimality of the optimization problem with respect to the long-term average data rates in the average rate region (8.36). In this scenario, the properties of efficiency and fairness that utility functions offer are all concerned with long-term average data rates. For instance, if $U_i(\tilde{r}_i) = \ln(\tilde{r}_i)$, then $U_i'(\tilde{r}_i) = 1/\tilde{r}_i$. It follows from (8.35) that

$$\sum_{i \in \mathcal{M}} \frac{\tilde{r}_i' - \tilde{r}_i}{\tilde{r}_i} \le 0, \quad \text{for all } \tilde{\mathbf{r}}' \in \tilde{\mathcal{C}}_\pi,$$

in which the long-term average data rate vector $\tilde{\mathbf{r}}$ is proportional fair.

8.7   Simulation results

In this section, we present simulation results to illustrate the performance of the various resource allocation approaches developed in this chapter. In our simulation, the channel is assumed to have a *bad-urban* (BU) delay profile [189] and suffer from shadowing with a standard deviation of 80dB. Let the acceptable bit error rate (BER) be 10^{-6} for rate adaptation. The bandwidth of each subcarrier is 10 kHz, and the utility function in (8.20) is used. To be able to compare those results properly, we set the average bandwidth per user, B/M, to be 80 kHz and show the average total utility per 80 kHz in simulation results.

At first, we assume the distances from the base station to all users are identical and compare various resource allocation schemes without a time window. Figure 8.3 shows some numerical results for different resource allocation schemes. Here continuous rate adaptation is used for all schemes. The *fixed subcarrier assignment* (FSA) results in the same performance when the number of users changes, while DSA offers significant multi-user diversity, which increases with the number of users. However, in the continuous rate adaptation case, the joint DSA and APA only leads to a very small improvement compared to DSA. The contribution of the APA is limited in this case as well.

Figure 8.4 shows the performance of different adaptive resource allocation policies with discrete rate adaptation. The variable M-QAM with modulation levels $\{0, 2, 4, 6, \cdots\}$ is assumed to be employed. The improvement from the DSA is similar to that in the continuous rate adaptation case. However, the contribution of the APA is significant in sharp contrast to that of continuous rate adaptation. Besides, the DSA in conjunction with the APA is able to substantially improve network performance even in the two-user case. For example, to achieve an average utility of 3, the gain from the joint DSA and APA is about 8 dB for the two-user case, and it increases to around 11 dB for the 16-user case.

Next, we evaluate fairness and spectral efficiency in the scenario when the distances between users and the base station are different. The path loss is modeled by

$$PL(d) = 128.1 + 37.6 \log_{10} d \quad [dB],$$

where d (km) is the distance between a user and the base station. Each user is assumed to be stationary or slowly moving so that the maximum Doppler shift is 5 Hz; as a result, their path loss and shadowing values are fixed during the simulation. In this simulation, the number of users is 8. We sort the eight users according to their distances to the base station. The path loss difference between the users closest to and farthest from the base station is about 18 dB in the simulation.

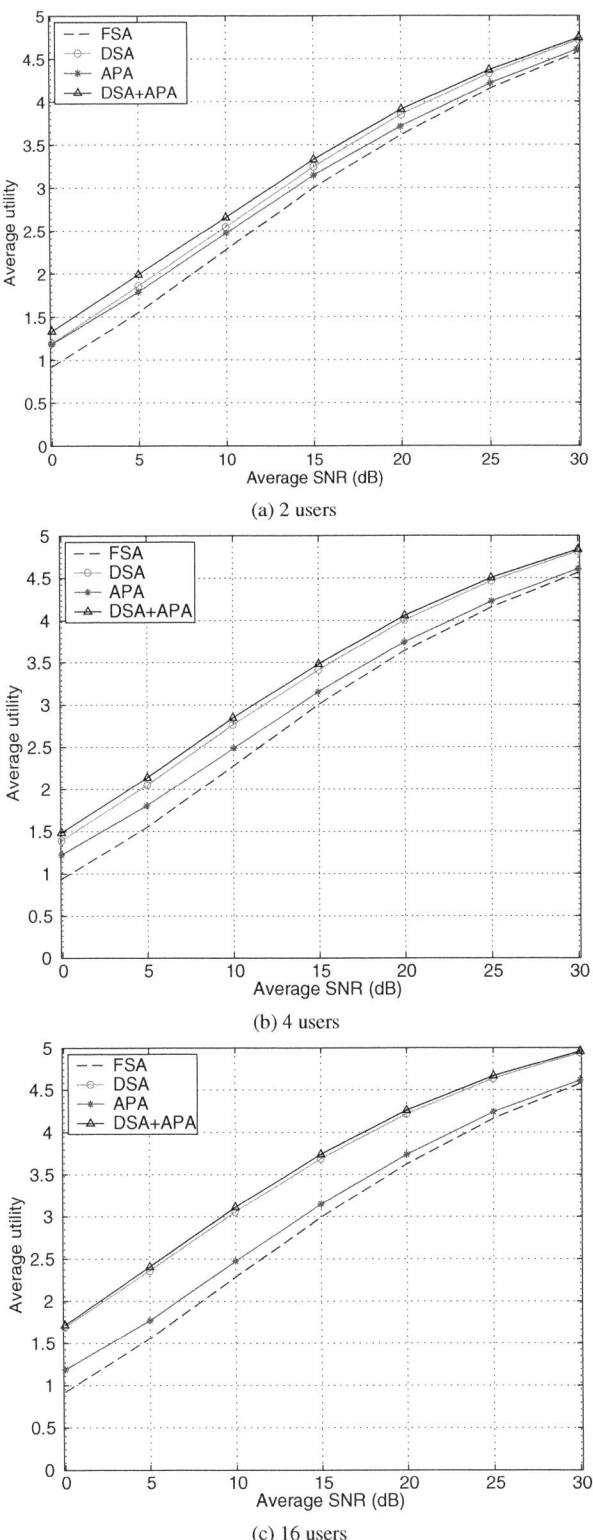

Figure 8.3 Average user utility versus SNR for OFDM wireless network with different resource allocation schemes

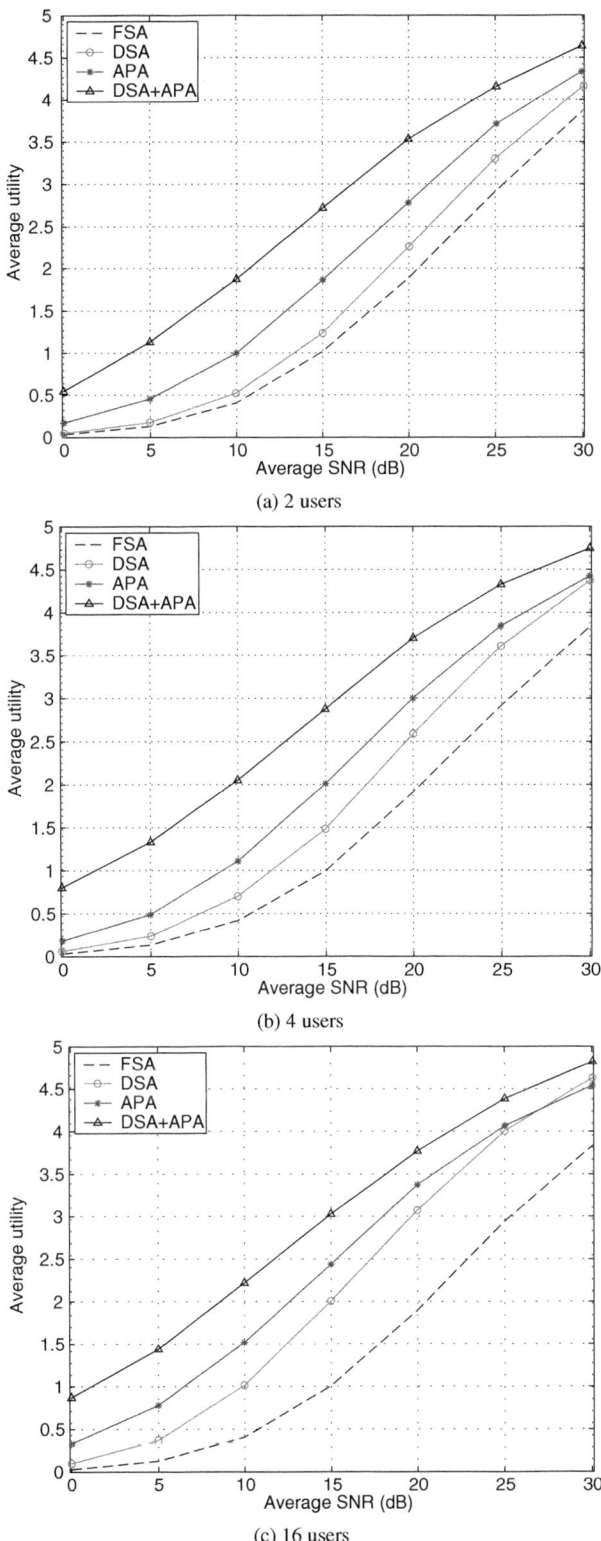

Figure 8.4 Average user utility versus SNR by using discrete rate adaptation and different resource allocation schemes

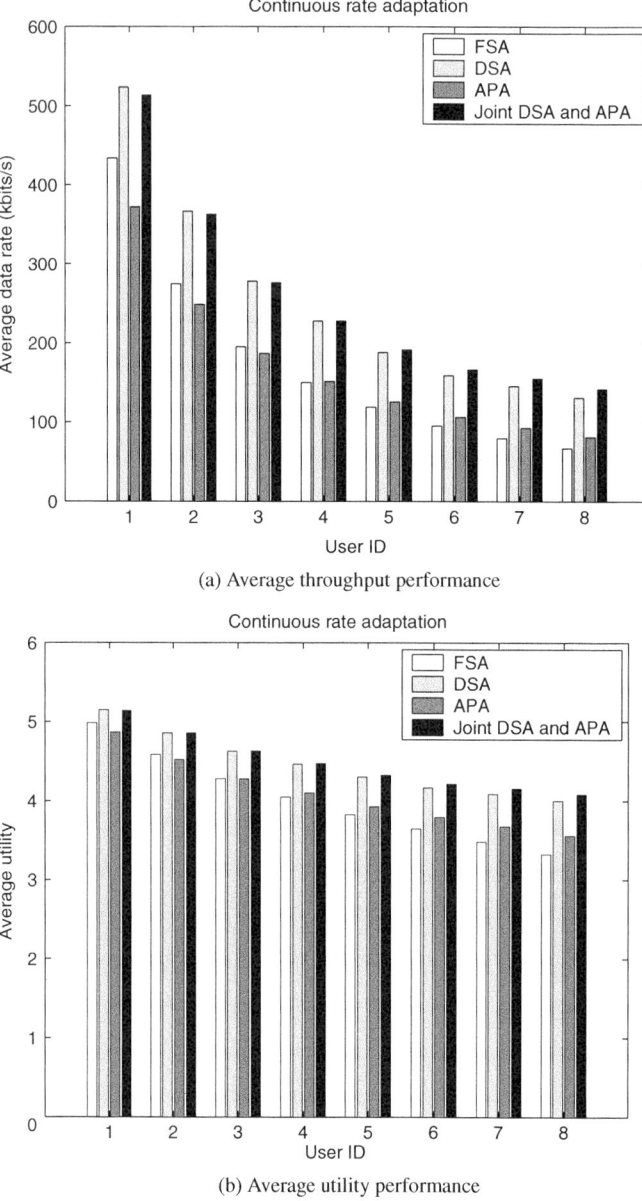

(a) Average throughput performance

(b) Average utility performance

Figure 8.5 Average performance of various resource allocation schemes with continuous rate adaptation

Figures 8.5 and 8.6 show the average throughput and average utility performance of each user with various resource allocation policies when continuous and discrete rate adaptation techniques are deployed, respectively. It is clear that although each user gets the same bandwidth and power, different path loss values result in different data rates. It can be seen from both figures that all utility-based resource allocation

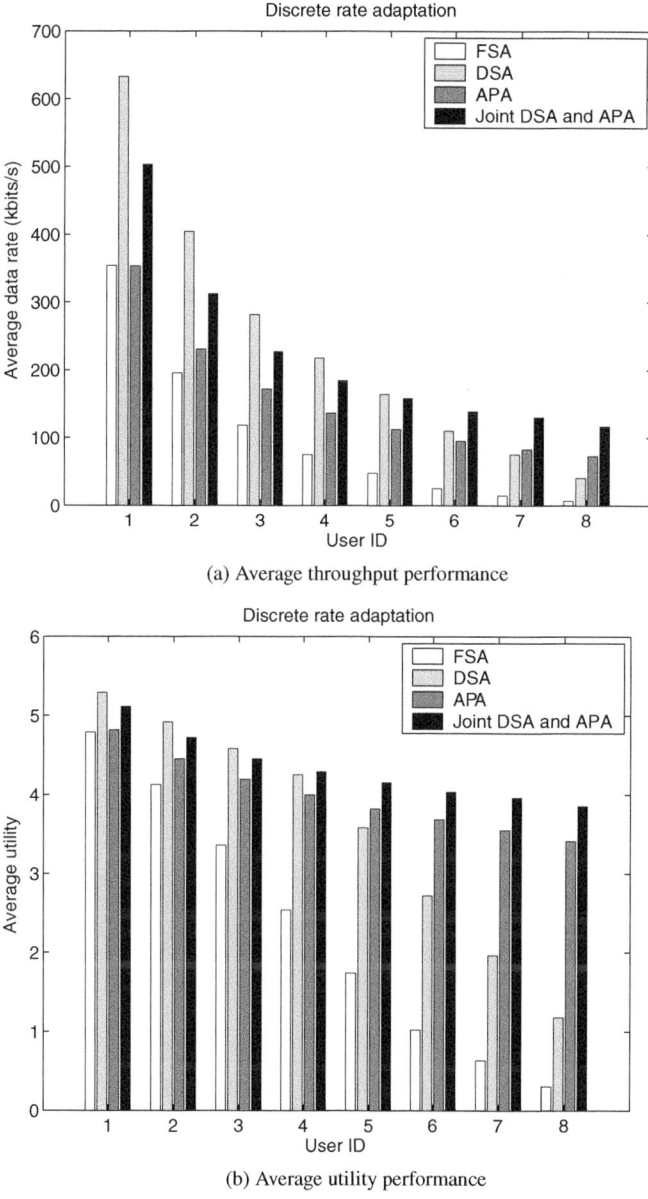

(a) Average throughput performance

(b) Average utility performance

Figure 8.6 Average performance of various resource allocation schemes with discrete rate adaptation

policies can increase the total throughput. Furthermore, the poorer the channel conditions, the greater the improvement in throughput; hence all utility-based resource allocation schemes can provide fairer services than the FSA. Figure 8.5 also confirms that using the DSA can provide similar performance as the joint DSA and APA for

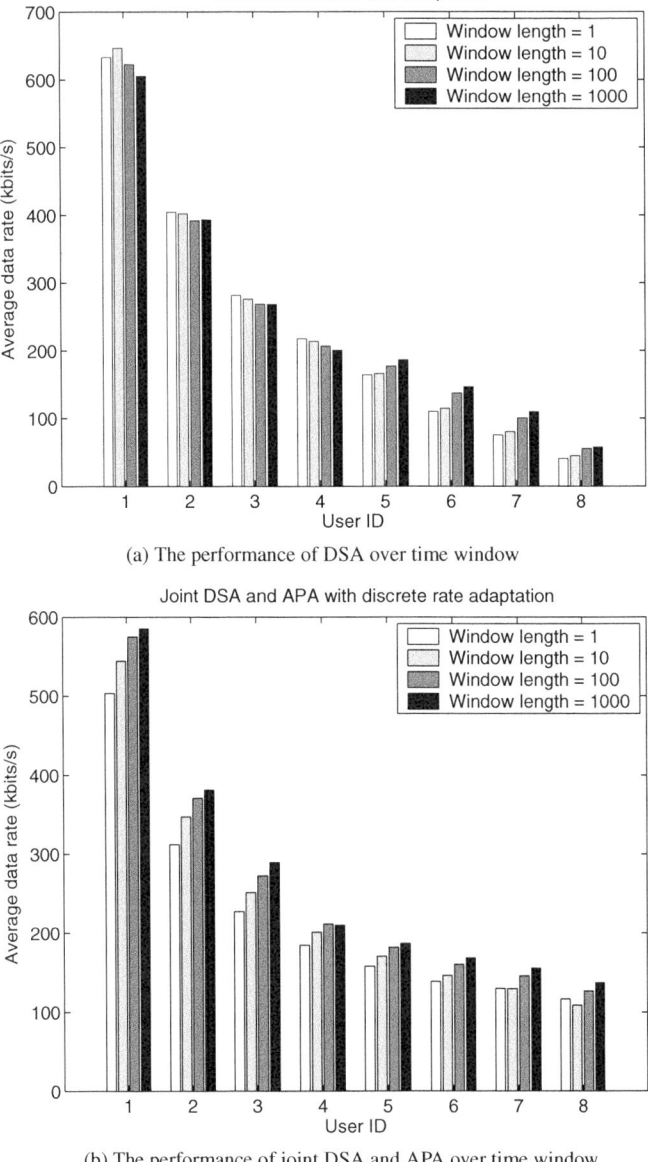

(a) The performance of DSA over time window

(b) The performance of joint DSA and APA over time window

Figure 8.7 Performance of addition of time window

continuous rate adaptation. For discrete rate adaptation, however, Figure 8.6 shows the significant improvement the joint DSA and APA in offering efficient and fair allocation.

Figure 8.7 demonstrates the performance of the DSA and the joint DSA and APA over time windows with different window lengths when discrete rate adaptation is employed. For the DSA, the time window helps to enhance the fairness of resource allocation. On the other hand, for the joint DSA and APA, a time window can further improve the average throughput of each user. However, the complexity of implementing the time window is not negligible.

8.8   Summary

In this chapter, we have presented utility-based cross-layer optimization for OFDM-based wireless networks. The utility is used here to build a bridge between the physical and medium access control (MAC) layers and to balance the efficiency and fairness of resource allocation. In particular, we have investigated the necessary and sufficient conditions for finding an optimum for the DSA, APA, and joint DSA and APA schemes when instantaneous-rate-based utility functions are used and the number of subcarriers is assumed to be infinite. Based on the theoretical framework, we have developed a variety of efficient algorithms, including the sorting-search DSA, the greedy bit-loading and power allocation, and the objective aggregation algorithms for practical OFDM systems. We have also modified the algorithms for non-concave utility functions. The use of average-rate-based utility functions is very suitable to best-effort traffic. A low-pass time filter resulting from average-rate-based utility functions can easily be incorporated into all algorithms to exploit time diversity. The extensive computer simulation results demonstrate the significant performance gain for the developed algorithms. In the next chapter, we will focus on using utility-based optimization for delay-sensitive traffic.

9 Joint channel- and queue-aware multi-carrier scheduling using delay-based utility functions

The relationship between rate-based utility functions and fairness in wireless networks has been shown in Chapter 8. Rate-based scheduling schemes, which apply the channel state information (CSI) and rate-based utility functions, do not take traffic burstiness into account. In this chapter, utility functions with respect to average delays is used for designing channel- and queue-aware scheduling, which is highly advantageous to data transmission with a low latency requirement.

This chapter is organized as follows. In Section 9.1, we introduce the background and motivations of this work. In Section 9.2, we briefly introduce how to extend scheduling schemes existing in single-carrier systems into the corresponding multi-channel scheduling schemes. In Section 9.3, we develop maximum delay utility (MDU) scheduling based on maximizing the total utility in terms of average waiting time. In Section 9.4, we state the maximum stability region and develop the results regarding stability. In Section 9.6, we use delay transmit diversity and adaptive power allocation to further improve the system performance. Finally, in Section 9.7, we compare several multi-carrier scheduling schemes using simulation.

9.1   Introduction

It is increasingly clear that most information traffic would be delivered based on IP networks because of the efficient bandwidth use and the low-cost infrastructure construction. Thus, the queue state information, such as queue length and packet delay, which is a reflection of traffic burstiness, should be utilized in scheduling packets. On the other hand, since the queue state information is tightly connected with quality of service (QoS), wisely controlling queues is one of the most effective ways for QoS provisioning. As compared to channel-aware scheduling, joint channel- and queue-aware scheduling would be more beneficial to wireless resource allocation and QoS provisioning. *Modified largest weighted delay first* (M-LWDF) and *exponential* (EXP) scheduling rules have been proposed for code division multiple access (CDMA) downlink transmission in [37, 182], respectively. Neither rules require statistical information about arrival traffic and wireless channels. The stability properties of the M-LWDF and the EXP scheduling rules over time-varying channels have been also studied by using the fluid limit technique in [37, 182], respectively. Other work on packet scheduling with emphasis on queueing system stability can be found in [192, 193, 139, 146, 69].

94

In this chapter, we investigate joint channel- and queue-aware scheduling in orthogonal frequency division multiplexing (OFDM)-based networks with emphasis on designing joint channel- and queue-aware scheduling schemes for multi-carrier networks. It should be indicated that the scheduling design for multi-carrier networks is not just a simple extension of existing scheduling approaches in single-carrier networks. First, multicarrier networks have nice granularity for resource allocation since the whole bandwidth is divided into many subchannels. Second, multi-carrier scheduling actually works in a parallel fashion. Unlike in single-carrier networks, multiple users can be served simultaneously in multi-carrier networks; thus, from a queueing point of view, there are multiple servers in multi-carrier scheduling.

9.2 Extending scheduling rules in single-carrier networks into OFDMA networks

In this section, dynamic subcarrier assignment (DSA) is used, but power allocation is fixed. Some existing scheduling schemes exploiting multi-user diversity in single-carrier networks can be directly extended to multi-carrier networks. In dynamically assigning subcarriers, we usually need to solve the optimization problem expressed as follows:

$$\max_{D_i^{(n)}, i \in \mathcal{A}^n} \sum_{i \in \mathcal{A}^n} w_i[n] r_i[n], \tag{9.1}$$

$$\text{subject to } \bigcup_{i \in \mathcal{A}^n} D_i^{(n)} \subseteq \mathcal{K}, \tag{9.2}$$

$$D_i^{(n)} \bigcap D_j^{(n)} = \varnothing, \ i \neq j \ \forall i, j \in \mathcal{A}^n, \tag{9.3}$$

where $\mathcal{A}^n = \{i : Q_i[n] > 0\}$ is the set in which each queue is not empty at time slot n, and the optimization objective is to maximize the sum weighted data rate with the weights $w_1[n], w_2[n], \ldots, w_M[n]$. In Section 8.1, the optimal assignment for the above problem is derived as

$$m(k, n) = \arg\max_{i \in \mathcal{A}^n} \{w_i[n] c_i[k, n]\}, \tag{9.4}$$

where $m(k, n)$ $(m(k, n) \in \mathcal{A}^n)$ represents subcarrier k to be assigned to user $m(k, n)$ at time n. This result is very useful to design scheduling approaches or to extend some scheduling rules in the single-carrier case to the OFDM scenario.

9.2.1 Max-sum-capacity (MSC) rule

The MSC rule is a channel-aware scheduling scheme that maximizes the total throughput in the system. Thus, the optimization problem can be expressed as (9.1)–(9.3) with $w_i[n] = 1$, for all i. Clearly, the MSC rule is given by

$$m(k, n) = \arg\max_{i \in \mathcal{A}^n} \{c_i[k, n]\}. \tag{9.5}$$

Although the MSC rule makes the most efficient use of the bandwidth, it can lead to unfairness and instability, especially for non-symmetrical channel conditions and non-uniform traffic patterns.

9.2.2   Proportional fair (PF) scheduling

The PF scheduling is a channel-aware scheduling rule aiming to maximize $\sum_i \ln(\bar{r}_i[n])$, where $\bar{r}_i[n]$ is the average data rate for user i. The scheduling rule in multi-carrier networks is obtained in [200, 186] also in Chapter 8 as

$$m(k, n) = \arg\max_{i \in \mathcal{A}^n} \left\{ \frac{c_i[k, n]}{\bar{r}_i[n]} \right\}. \tag{9.6}$$

Since ρ_w is very small, $\bar{r}_i[n] \approx \bar{r}_i[n-1]$. Although this DSA algorithm guarantees the proportional fairness [186], it is not throughput-optimal [36].[1] The PF scheduling is suitable to best-effort traffic, which has no specific QoS requirements.

9.2.3   Modified largest weighted delay first (M-LWDF) rule

In [37], the M-LWDF scheme is proposed for single-carrier CDMA networks with a shared downlink channel. From an optimization point of view, the M-LWDF intends to maximize $\sum_i \frac{T_{\mathrm{HOL},i}[n]}{\bar{r}_i[n]} r_i[n]$, where $T_{\mathrm{HOL},i}$ is the delay of the *head-of-line* (HOL) packet of user i. Using the result (9.4), we have the multichannel version of M-LWDF as

$$m(k, n) = \arg\max_{i \in \mathcal{A}^n} \left\{ \frac{c_i[k, n]}{\bar{r}_i[n]} T_{\mathrm{HOL},i[n]} \right\}.$$

9.2.4   Exponential (EXP) rule

The EXP scheduling rule is also designed for single-carrier CDMA networks with a shared downlink channel [182]. The structure of the EXP rule is very similar to the M-LWDF, but with different weights. The multi-channel version of EXP rule can be expressed as

$$m(k, n) = \arg\max_{i \in \mathcal{A}^n} \left\{ \frac{c_i[k, n]}{\bar{r}_i[n]} \exp\left(\frac{T_{\mathrm{HOL},i}[n]}{1 + \sqrt{\overline{T}_{\mathrm{HOL}}[n]}} \right) \right\},$$

where $\overline{T}_{\mathrm{HOL}}[n] = \frac{1}{|\mathcal{A}^n|} \sum_{i \in \mathcal{A}^n} T_{\mathrm{HOL},i}[n]$.

The M-LWDF and EXP rules have been proven to be throughput-optimal in single-carrier networks [37, 182]. With a few modifications, the proofs are valid in OFDM networks. Both scheduling rules can be used for delay-sensitive traffic.

---

[1] A scheduling algorithm is called *throughput-optimal* if it stabilizes a queueing system in which stability is feasible at all to do with any algorithms [37] .

9.3 Max-delay-utility (MDU) scheduling

Designing channel-aware-only scheduling is usually tractable. It is shown in Chapter 8 that most channel-aware-only scheduling schemes can be derived by maximizing the sum of specific utility functions with respect to data rates. However, there are two difficulties in designing joint channel- and queue-aware scheduling. First, it is hard to formulate the desired optimization goals related to the QoS requirements such as average waiting time, delay violation probability, etc. Second, the optimal solutions to those optimization problems usually require dynamic programming with exponential computational complexity, which makes them impossible in practice. In this section, we use the utility functions with respect to average waiting times in designing joint channel- and queue-aware scheduling, which was first reported in [187].

9.3.1 Utility functions

Assume that user i is associated with an average waiting time W_i and the corresponding utility is $U_i(W_i)$. Obviously, with a long delay, the user has a low level of satisfaction (utility). It is reasonable to assume that $U_i(W_i)$ is decreasing. There are usually two approaches to obtaining utility functions. For a specific type of application, the utility function may be obtained by sophisticated subjective surveys. Another method is to design utility functions based on the habits of the traffic and appropriate fairness in the network.

9.3.2 Optimization objective

Assume the average arrival bit rate of user i as λ_i, defined as

$$\lambda_i = \frac{1}{T_s} \lim_{n \to \infty} \frac{A_i[n]}{n},$$

where $A_i[n]$ is the total number of bits arriving during $(0, nT_s]$. Assuming that $Q_i[n]$ is ergodic, with Little's law, the average waiting time for user i, W_i, is

$$W_i = \frac{Q_i}{\lambda_i},$$

where $Q_i = \lim_{N \to \infty} \frac{\sum_{n=0}^{N-1} Q_i[n]}{N}$.

Let the base station control service bit rates so that

$$r_i[n]T_s \leq Q_i[n]. \tag{9.7}$$

Then, the queue evolution equation (6.3) becomes

$$Q_i[n+1] = Q_i[n] - r_i[n]T_s + a_i[n]. \tag{9.8}$$

By exploiting an exponentially weighted low-pass filter, the average queue length, $\bar{Q}_i[n]$, can be updated as

$$\bar{Q}_i[n] = (1 - \rho_w)\bar{Q}_i[n-1] + \rho_w Q_i[n], \tag{9.9}$$

where $0 < \rho_w < 1$.

Define the average waiting time over the time window at time nT_s as

$$W_i[n] = \frac{\bar{Q}_i[n]}{\lambda_i}. \tag{9.10}$$

At time nT_s (the beginning of time slot n), given the service rate $r_i[n]$, the predicted average waiting time at the end of time slot n, $(n + 1)T_s$, is obtained by

$$\hat{W}_i[n + 1] = \frac{\mathbb{E}_{a_i[n]}\{\bar{Q}_i[n + 1]\}}{\lambda_i},$$

where $\mathbb{E}_{a_i[n]}\{\cdot\}$ denotes expectation with respect to $a_i[n]$. According to (6.3) and (9.9), we have

$$\mathbb{E}_{a_i[n]}\{\bar{Q}_i[n + 1]\} = (1 - \rho_w)\bar{Q}_i[n] + \rho_w(Q_i[n] - r_i[n]T_s + \mathbb{E}\{a_i[n]\}).$$

Using $\mathbb{E}\{a_i[n]\} = \lambda_i T_s$, $\hat{W}_i[n + 1]$, is obtained by

$$\hat{W}_i[n + 1] = (1 - \rho_w)\frac{\bar{Q}_i[n]}{\lambda_i} + \rho_w\frac{Q_i[n]}{\lambda_i} + \rho_w T_s - \frac{\rho_w}{\lambda_i}T_s r_i[n]$$

$$= (1 - \rho_w)W_i[n] + \rho_w\frac{Q_i[n]}{\lambda_i} + \rho_w T_s - \frac{\rho_w}{\lambda_i}T_s r_i[n].$$

Therefore, the predicted average waiting time at time $(n + 1)T_s$ is a function of the service rate during time slot n, $r_i[n]$.

The optimization objective is to maximize the total utility with respect to the predicted average waiting times at each time slot in the network, i.e.

$$\max_{r_i[n],i\in\mathcal{M}} \sum_{i=1}^{M} U_i(\hat{W}_i[n + 1]).$$

Given the arrival processes, the average waiting time is actually determined by the service rate. It is obvious that

$$\frac{\partial U_i}{\partial r_i} = -\frac{\partial U_i}{\partial W_i}\frac{\rho_w}{\lambda_i}T_s.$$

If ρ_w is small enough, and using the properties of $U_i(W_i)$, we have

$$\sum_{i=1}^{M} U_i(\hat{W}_i[n + 1]) - \sum_{i=1}^{M} U_i(\hat{W}_i[n]),$$

$$\approx \sum_{i=1}^{M} \frac{\partial U_i}{\partial r_i}\bigg|_{r_i=r_i[n-1]} (r_i[n] - r_i[n - 1]),$$

$$\approx \sum_{i=1}^{M} -\frac{\partial U_i}{\partial W_i}\bigg|_{W_i=W_i[n]} \rho_w T_s \left(\frac{r_i[n]}{\lambda_i} - \frac{r_i[n - 1]}{\lambda_i}\right),$$

$$= \sum_{i=1}^{M} \left|\frac{\partial U_i}{\partial W_i}\right|\bigg|_{W_i=W_i[n]} \rho_w T_s \left(\frac{r_i[n]}{\lambda_i} - \frac{r_i[n - 1]}{\lambda_i}\right).$$

Since $r_i[n-1]$ is fixed at time slot n, the optimization objective turns out to be a linear function of $r_i[n]$,

$$\max \sum_{i=1}^{M} \frac{\left|U_i'(W_i[n])\right|}{\lambda_i} r_i[n], \tag{9.11}$$

where $U_i'(W_i[n]) = \frac{\partial U_i(W_i)}{\partial W_i}\Big|_{W_i=W_i[n]}$, and $W_i[n]$ can be obtained from (9.10).

9.3.3   Problem formulation in OFDMA

If the subcarriers can be dynamically assigned, with objective (9.11), we formulate this problem in the orthogonal frequency division multiple access (OFDMA) system as

$$\max_{D_i^{(n)}, i \in \mathcal{A}^n} \sum_{i \in \mathcal{A}^n} \frac{\left|U_i'(W_i[n])\right|}{\hat{\lambda}_i} r_i[n], \tag{9.12}$$

$$\text{subject to} \bigcup_{i \in \mathcal{A}^n} D_i^{(n)} \subseteq \mathcal{K}, \tag{9.13}$$

$$D_i^{(n)} \bigcap D_j^{(n)} = \varnothing, \quad i \neq j \ \forall i,j \in \mathcal{A}^n, \tag{9.14}$$

$$r_i[n] \leq \frac{Q_i[n]}{T_s}, \quad i \in \mathcal{A}^n, \tag{9.15}$$

where the constraint (9.15) comes from the queue control rule (9.7), which means that the scheduler does not waste service rate. We refer to (9.15) as the *frugality constraint* (FC). Note that in the optimization objective (9.12) the estimated arrival rate $\hat{\lambda}_i$ replaces the expected (exact) value λ_i. This is because the base station does not know the arrival rates and λ_i must be estimated. They can also be estimated through an exponentially weighted low-pass window. Besides, there is another way to estimate the arrival rates. Since the FC is applied, the scheduler does not serve any empty queue and waste service rate; therefore, λ_i equals the long-term average of the service rate of user i, $\mathbb{E}\{r_i\}$, in this scenario. In practice, we let

$$\hat{\lambda}_i = \bar{r}_i[n].$$

Letting

$$h(r; r_{max}) = \begin{cases} r & \text{if } r < r_{max}, \\ r_{max} & \text{if } r \geq r_{max}, \end{cases}$$

we can rewrite the optimization problem defined in (9.12)–(9.15) as

$$\max_{D_i^{(n)}, i \in \mathcal{A}^n} \sum_{i \in \mathcal{A}^n} \frac{\left|U_i'\left(\frac{\bar{Q}_i[n]}{\hat{\lambda}_i}\right)\right|}{\hat{\lambda}_i} h\left(r_i[n]; \frac{Q_i[n]}{T_s}\right), \tag{9.16}$$

$$\text{subject to} \bigcup_{i \in \mathcal{A}^n} D_i^{(n)} \subseteq \mathcal{K}, \tag{9.17}$$

$$D_i^{(n)} \bigcap D_j^{(n)} = \varnothing \ i \neq j \ \forall i,j \in \mathcal{A}^n. \tag{9.18}$$

9.3.4   Algorithms

The integer optimization problem (9.16)–(9.18) is NP-hard. In Section 8.1, an efficient and fast suboptimal DSA algorithm, a sorting-search algorithm, is introduced for the subcarrier assignment problem with the concave objective function. Note that the function $h(r; r_{max})$ is concave with respect to r. Therefore, the sorting-search algorithm can work well to solve the problem described in (9.16)–(9.18).

The FC is not necessary for MDU scheduling. Without the FC, the MDU scheme can be implemented according to (9.4). To avoid ambiguity, we use the MDU-FC to indicate the MDU working with the FC in this chapter. On the other hand, the FC can be applied in other scheduling schemes, such as those schemes mentioned in Section 9.2. Certainly, the sorting-search algorithm is needed in the case in which the FC is used.

9.4   Stability

It is shown in Chapter 8 that a utility function with respect to the data rate is directly associated with a kind of fairness. The tradeoff between spectral efficiency and fairness is a core problem of resource allocation, especially for best-effort traffic. In addition, Chapter 8 demonstrated that concave utility functions can provide clear, tractable efficiency–fairness relations. Fortunately, a logarithmic function, which is concave, is usually used to describe best-effort traffic [116, 184].

Unlike best-effort traffic, the necessary condition for guaranteeing the QoS requirements of a delay-sensitive stream is that the service rate must be larger than the incoming rate of the stream. Therefore, the study of the stability issue is the key to analyzing scheduling algorithms for delay-sensitive traffic. In this chapter, we show the relationship between utility functions and stability. In fact, utility functions with very loose conditions (e.g. convexity/concavity is not required) are able to stabilize the system using MDU scheduling.

9.4.1   Background and definition of stability

The interaction between queueing and time-varying wireless channels is not well understood in a multi-user environment since multiple interacting queues result in difficulty in analysis. Currently, the stability property of scheduling is becoming more and more important [192, 69, 146]. First, the stability issue is essential for QoS provisioning and admission control. Moreover, the stability issue is mathematically tractable in many cases. There are two important methods to deal with the stability issue: Foster–Lyapunov drift [140] and fluid limit [60]. The Foster–Lyapunov method is classical for stability and harmonic analysis, but it may be very intricate in complicated scenarios. The fluid limit technique establishes the equivalency on stability between the original network and the associated fluid model with deterministic and continuous arrival streams. However, the above equivalent relationship for stability is usually built on the Markovian property of the system, and it is still unknown if the Markov assumption can be relaxed to just a stationary condition for the fluid limit technique in a general case. Moreover, both methods applied in most previous work such as [192, 139, 146, 69] (using the Foster–Lyapunov method) and [37, 182] (using the fluid limit technique) are challenged by the fact that the weights used in MDU scheduling are functions of the current and previous queue states. We incorporate the concept and the properties of limits into the Foster–Lyapunov method to deal with the above difficulty and to make the proofs concise.

For a queueing system, the system is stable if each queue length reaches a steady state and does not go to infinity. Mathematically, we define stability as follows. The system is stable if there exists $p > 0$ such that

$$\limsup_{N\to\infty} \frac{1}{N} \sum_{n=0}^{N-1} \mathbb{E}\left\{ \left| (\mathbf{Q}[n])^p \right| \right\} < \infty, \tag{9.19}$$

where $\mathbf{Q}[n] = (Q_1[n], Q_2[n], \ldots, Q_M[n])^T$, and for a vector $\mathbf{x} = [x_1, x_2, \ldots, x_M]^T$, $|\mathbf{x}| = \sum_{i=1}^{M} x_i$. To investigate the stability issue, we will first discuss the capacity region of the downlink system.

9.4.2 Capacity region

Define a data rate vector $\mathbf{r}$ as

$$\mathbf{r} = (r_1, r_2, \ldots, r_M)^T \in \mathbb{R}_+^M,$$

where M is the number of users. The instantaneous capacity region for service data rates, $\mathcal{C}(\mathbf{H})$, is a set that consists of the total achievable data rate vectors in the current channel state $\mathbf{H}$. For instance, if DSA is allowed in the system, then the instantaneous data rate region is given by

$$\mathcal{C}_{DSA}(\mathbf{H}) = \left\{ \mathbf{r}(\mathbf{D}) : D_i \bigcap D_j = \varnothing, \ \forall i \neq j, \ \bigcup_{i\in\mathcal{M}} D_i \subseteq \mathcal{K} \right\},$$

where $\mathbf{D} = \{D_1, D_2, \ldots, D_M\}$. Usually, in practical systems, $\mathcal{C}(\mathbf{H})$ is a non-convex set since real systems can only provide finite modulation and coding schemes.

A resource allocation policy $\mathcal{R}(\mathbf{H})$ is said to be channel-stationary if the rate allocation depends only on the channel state $\mathbf{H}$. Note that channel-stationary policies can exploit time-sharing for the achievable data rate vectors in $\mathcal{C}(\mathbf{H})$. Hence, all available channel-stationary resource allocation policies can construct the convex hull of $\mathcal{C}(\mathbf{H})$; i.e.,

$$cov(\mathcal{C}(\mathbf{H})) = \{\mathcal{R}(\mathbf{H}) : \text{for all } \mathcal{R}\}. \tag{9.20}$$

Hence, $cov(\mathcal{C}(\mathbf{H})$ can be seen as the capacity region that can be achieved by time-averaging two or more feasible rate vectors in the instantaneous capacity region in the channel state $\mathbf{H}$.

Assume the channel state process $\mathbf{H}(t)$ to be ergodic. Let $\tilde{\mathcal{C}}$ be the ergodic capacity region under the allocation constraints, which consists of the average data rate vectors obtained by all possible channel-stationary resource allocation schemes. Thus,

$$\tilde{\mathcal{C}} = \{\mathbb{E}\{\mathcal{R}(\mathbf{H})\} : \text{ for all } \mathcal{R}\}.$$

Explicitly, the ergodic capacity region is a closed, convex, and compact set. Nevertheless, we do not consider non-channel-stationary resource allocation policies, for which the average service data rate vector is defined as

$$\liminf_{t\to\infty} \frac{\int_{\tau=0}^{t} \mathbf{r}(\tau)\,d\tau}{t}.$$

However, we have the following lemma:

LEMMA 9.1   *With ergodic channel state processes, any average service data rate vector under any non-channel-stationary policy still lies in the ergodic capacity region $\tilde{\mathcal{C}}$.*

The proof is shown in Appendix A.4.

The lemma claims that the long-term average service rate vector under any resource allocation policy lies in the ergodic capacity region, which is determined by the physical layer techniques and the channel distributions.

9.4.3   Maximum stability region

Assume that the input streams are stationary and ergodic with rate vector $\lambda = [\lambda_1, \lambda_2, \ldots, \lambda_M]^T$, and that the channel processes are stationary and ergodic as well. Then, with a similar proof to that of Lemma 1b in [146], we have the following lemma:

LEMMA 9.2   *The necessary condition for stability is $\lambda \in \tilde{\mathcal{C}}$.*

However, in the case $\lambda \in \tilde{\mathcal{C}}$, not all scheduling policies can stabilize the system. The *stability region* of a policy is defined to be the set of all possible arrival rate vectors for which the system is stable under the policy [192]. Note that the capacity region is concerned with the service data rates, whereas the stability region is concerned with the arrival rates. The *maximum stability region* is defined as the largest stability region that can be achieved by some scheduling schemes. Similarly, a policy is called a maximum-stability-region policy if the stability region of the policy covers all stability regions under all other policies. Thus, the concept of maximum-stability-region policy is interchangeable with the concept of throughput-optimal policy in [37].

Naturally, we are interested in the following questions. First, does the maximum stability region always exist? Second, how large can the maximum stability region be? Finally, how do we identify and design maximum-stability-region policies without the statistical information about the arrivals and the wireless channels? The answer to the first question is yes. Mathematically, the maximum stability region is the superset of stability regions of all possible policies. Let $\mathcal{S}_1$ and $\mathcal{S}_2$ be the stability regions of policy

$\mathcal{R}_1$ and $\mathcal{R}_2$, respectively. Then, we can construct a policy $\mathcal{R}$ such that

$$\mathcal{R} = \begin{cases} \mathcal{R}_1 & \text{if } \lambda \in \mathcal{S}_1 - \mathcal{S}_2, \\ \mathcal{R}_2 & \text{if } \lambda \in \mathcal{S}_2 - \mathcal{S}_1, \\ \mathcal{R}_1 \text{ or } \mathcal{R}_2 & \text{if } \lambda \in \mathcal{S}_1 \bigcap \mathcal{S}_2. \end{cases}$$

Thus, the policy $\mathcal{R}$ has the stability region $\mathcal{S}_1 \bigcup \mathcal{S}_2$. With the same method, we can always construct a policy with the superset of stability regions of all possible policies – the maximum stability region. For the second question, it is easy to show from Lemma 9.2 that the maximum stability region must be a subset of $\tilde{\mathcal{C}}$. We will explore the remaining questions by investigating a more general scheduling policy that allocates data rate vectors such that

$$\max_{\mathbf{r}[n] \in \mathcal{C}(\mathbf{H}[n])} \underline{g}^T (\mathbf{V}[n]) \mathbf{r}[n], \tag{9.21}$$

where the vector function $\underline{g}(\cdot)$ and the vector $\mathbf{V}[n]$ are described as follows:

- Let $\underline{g}(\mathbf{x}) = [g_1(x_1), g_2(x_2), \dots, g_M(x_M)]^T$ and assume that $g_i(\cdot)$ are non-negative and non-decreasing functions such that

$$\text{for } x < \infty, \ g_i(x) < \infty, \tag{9.22}$$

$$\lim_{x \to \infty} g_i(x) = \infty, \tag{9.23}$$

and given any constant $|A| < \infty$,

$$\lim_{x \to \infty} \frac{g_i(x + A)}{g_i(x)} = 1. \tag{9.24}$$

- Let $\mathbf{V}[n] = (V_1[n], V_2[n], \dots, V_M[n])^T$, where $V_i[n] = f(Q_i[n], Q_i[n-1], \dots)$. The function f is non-negative and non-decreasing with respect to $Q_i[n]$ for all n. Furthermore,

$$\mathbb{E}\{|Q_i[n] - V_i[n]|\} < \infty \text{ for all } i. \tag{9.25}$$

- In addition to the ergodicity of the channel processes and arrival streams, we assume that

$$\mathbb{E}\{g_i(a_i[n])a_i[n]\} < \infty \text{ for all } i, \tag{9.26}$$

where $a_i[n]$ is the arrival bits during a time slot for user i. From a practical point of view, any achievable instantaneous data rates are bounded, which can also simplify the proof of the following theorem:

We first consider the optimization problem given by

$$\max_{\mathbf{r} \in \mathcal{C}(\mathbf{H}[n])} \mathbf{w}^T \mathbf{r}, \tag{9.27}$$

where $\mathbf{w} = [w_1, w_2, \dots, w_M]^T$. Clearly, the scheduling policy based on (9.27) is stationary. Furthermore, the following lemma shows that the scheduling policy also leads to optimality in the long-term sense.

LEMMA 9.3   *For a given weight vector* $\mathbf{w}$, *assume that* $\mathbf{r}^*(\mathbf{H})$ *is the optimization problem (9.27) in the instantaneous capacity region* $\mathcal{C}(\mathbf{H})$. *Let* $\tilde{\mathbf{r}}^* = \mathbb{E}\{\mathbf{r}^*(\mathbf{H})\}$, *then* $\tilde{\mathbf{r}}^*$ *is the optimal solution to the following optimization problem in the ergodic capacity region* $\tilde{\mathcal{C}}$:

$$\max_{\tilde{\mathbf{r}} \in \tilde{\mathcal{C}}} \mathbf{w}^T \tilde{\mathbf{r}}. \tag{9.28}$$

The proof is shown in Appendix A.5. Then, we present the major results for the stability issue in the following theorem:

THEOREM 9.4   *If the average arrival rate vector is within the interior of the ergodic capacity region,* $Int(\tilde{\mathcal{C}})$, *where* $Int(\tilde{\mathcal{C}}) = \tilde{\mathcal{C}} - -$*the boundary of* $\tilde{\mathcal{C}}$, *then the scheduling (9.21) satisfying the conditions (9.22)–(9.26) stabilizes the queues in the following sense:*

$$\limsup_{N \to \infty} \frac{1}{N} \sum_{n=0}^{N-1} \mathbb{E}\left\{ \left| \underline{g}(\mathbf{V}[n]) \right| \right\} < \infty. \tag{9.29}$$

In other words, the scheduling has the maximum stability region, which is $Int(\tilde{\mathcal{C}})$.

The proof is shown in Section 9.5.

Note that the performance of the scheduling (9.21) is worse than that of the scheduling (9.21) with the FC since the scheduling (9.21) may waste some subcarriers in empty queues. The system without the use of the FC is called the *dominant system*. Therefore, the scheduling (9.21) with the FC has the maximum stability region as well.

To study MDU scheduling, we have to prove the relation (9.25) first. We have the following lemma:

LEMMA 9.5   *Let* $\bar{Q}_i[0] = Q_i[0]$ *for all i. Then*

$$\mathbb{E}\{|Q_i[n] - \bar{Q}_i[n]|\} < \infty \text{ for all } i.$$

The proof is shown in Appendix A.6. The following corollary states the stability property of MDU scheduling:

COROLLARY 9.6   *Express the weights* $\frac{|U_i'(\bar{Q}_i[n]/\hat{\lambda}_i)|}{\hat{\lambda}_i}$ *in MDU scheduling as* $g_i(\bar{Q}_i[n])$. *Then MDU scheduling with functions* $g_i(\cdot)$ *satisfying conditions (9.22)–(9.26) has the maximum stability region,* $Int(\tilde{\mathcal{C}})$. *If the average arrival rate vector is within* $Int(\tilde{\mathcal{C}})$, *then the MDU scheduling policy is stable, i.e.*

$$\limsup_{N \to \infty} \frac{1}{N} \sum_{n=0}^{N-1} \mathbb{E}\left\{ \left| \underline{g}(\rho_w \mathbf{Q}[n]) \right| \right\} < \infty,$$

where $\rho_w \mathbf{Q}[n] = (\rho_w Q_1[n], \rho_w Q_2[n], \ldots, \rho_w Q_M[n])^T$.

Proof   The weights of MDU scheduling are $g_i(V_i[n])$, where $V_i[n] = \bar{Q}_i[n]$. Lemma 9.5 shows the validity of (9.25) for MDU scheduling. It follows from (9.9) that

$$\rho_w \mathbf{Q}[n] \le \bar{\mathbf{Q}}[n].$$

Since $\underline{g}(\cdot)$ is non-decreasing, then

$$\underline{g}(\rho_w \mathbf{Q}[n]) \le \underline{g}(\bar{\mathbf{Q}}[n]).$$

Therefore, we obtain

$$\limsup_{N\to\infty} \frac{1}{N} \sum_{n=0}^{N-1} \mathbb{E}\left\{ \left| \underline{g}(\rho_w \mathbf{Q}[n]) \right| \right\} \le \limsup_{N\to\infty} \frac{1}{N} \sum_{n=0}^{N-1} \mathbb{E}\left\{ \left| \underline{g}(\bar{\mathbf{Q}}[n]) \right| \right\}$$

$$< \infty.$$

Remarks □

- The general scheduling rule (9.21) that is able to achieve the maximum stability region does not require statistical information about the arrivals and the wireless channels.
- If $g_i(x)$ is continuously differentiable, the condition (9.24) can be replaced by

$$\lim_{x\to\infty} \frac{g_i'(x)}{g_i(x)} = 0.$$

  Intuitively, any non-negative and increasing function whose increasing order is higher than or equal to the logarithm function and lower than the exponential function satisfies both conditions (9.23) and (9.24). Therefore, there are many degrees of freedom for designing scheduling policies with the maximum stability region. Similarly, there is enough room to choose the $V_i[n]$. For instance, given a finite positive integer J, $V_i[n] = Q_i[n-J]$, $V_i[n] = \sum_{j=0}^{J-1} Q_i[n-j]/J$, $V_i[n] = (Q_i[n] \cdot Q_i[n-2] \cdot Q_i[n-J+1])^{\frac{1}{J}}$, etc., are all able to stabilize the system.

- The maximum stability region is shown to be the interior of the ergodic capacity region that is determined by the physical layer techniques. Figure 9.1 illustrates the stability regions of some scheduling schemes for the two-user case. According to Lemma 9.3, the rate allocation of the MSC scheduling is the solution to the following problem

$$\max_{[\tilde{r}_1, \tilde{r}_2]^T \in \tilde{\mathcal{C}}} \tilde{r}_1 + \tilde{r}_2.$$

Thus, the optimal solution $[\tilde{r}_1^*, \tilde{r}_2^*]^T$ should be the tangent point between the boundary of $\tilde{\mathcal{C}}$ and a line $\tilde{r}_1 + \tilde{r}_2 = b$ for an appropriate b; the stability region is, therefore, $\lambda_1 < \tilde{r}_1^*$ and $\lambda_2 < \tilde{r}_2^*$. The PF scheduling with a very small ρ_w leads to the optimization problem:

$$\max_{[\tilde{r}_1, \tilde{r}_2]^T \in \tilde{\mathcal{C}}} \frac{1}{\tilde{r}_1^\dagger} \tilde{r}_1 + \frac{1}{\tilde{r}_2^\dagger} \tilde{r}_2.$$

Similar to the MSC, the optimal rate vector for the PF $[\tilde{r}_1^\dagger, \tilde{r}_2^\dagger]^T$ should be the point of tangency between the boundary of $\tilde{\mathcal{C}}$ and a line $\frac{1}{\tilde{r}_1^\dagger}\tilde{r}_1 + \frac{1}{\tilde{r}_2^\dagger}\tilde{r}_2 = b'$ with an appropriate b'. Since the MSC and PF scheduling schemes have small stability

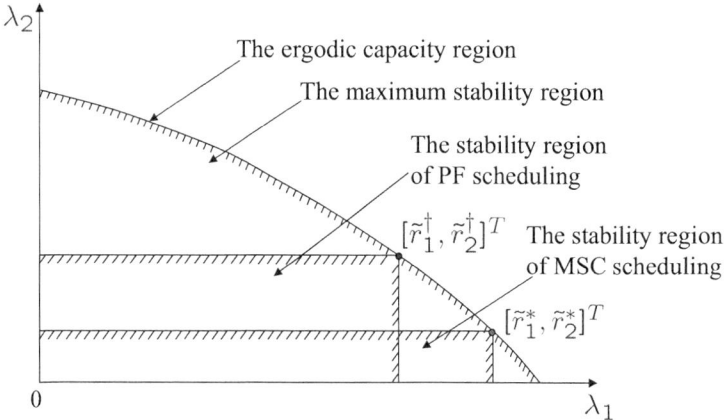

Figure 9.1 Stability regions for different scheduling schemes in the two-user case

regions, they cannot stabilize all arrival vectors inside the ergodic capacity region but outside their stability regions.

- In the proof of Theorem 9.4, we see that the FC cannot stabilize scheduling approaches without the maximum stability region, and that the effect that FC may become marginal with a heavy traffic load. The impact of the FC on different scheduling policies will be discussed in Section 9.7.

- To obtain more system gains, we should jointly design and optimize techniques in multiple layers, but cross-layer design usually seems complicated and not transparent. However, the above result gives us a guideline for cross-layer optimization. First, use advanced physical layer techniques to enlarge the ergodic capacity region. Second, design a scheduling scheme with the maximum stability region to fully exploit the ergodic capacity region.

9.5   Proof of Theorem 9.4

The primary method used in the proof is the Foster–Lyapunov method. A new tip is to apply Fatou's lemma and the definition of the upper limit. The proof does not require the Markovian property on the channel states and/or the arrival traffic.

Let the Lyapunov function be

$$L(\mathbf{Q}[n]) = \sum_{i \in \mathcal{M}} L_i(Q_i[n]),$$

where $\frac{dL_i(x)}{dx} = g_i(x)$. Define

$$\mathbf{Q}'[n+1] = \mathbf{Q}[n] - \mathbf{r}[n]T_s + \mathbf{a}[n]$$
$$\text{and } \boldsymbol{\xi} = -\mathbf{r}[n]T_s + \mathbf{a}[n]. \tag{9.30}$$

Using the mean value theorem [170], we obtain

$$
L(\mathbf{Q}'[n+1]) - L(\mathbf{Q}[n]) = \nabla L^T (\mathbf{Q}'[n] + \boldsymbol{v} \odot \boldsymbol{\xi}) \boldsymbol{\xi}
$$
$$
= \underline{g}^T (\mathbf{Q}'[n] + \boldsymbol{v} \odot \boldsymbol{\xi}) \boldsymbol{\xi},
$$

where $\mathbf{a} \odot \mathbf{b} = [a_1 b_1, a_2 b_2, \dots, a_M b_M]^T, 0 < v_i < 1$ for all i. Clearly,

$$
L(\mathbf{Q}'[n+1]) - L(\mathbf{Q}[n])
$$
$$
= \underline{g}^T (\mathbf{V}[n]) \boldsymbol{\xi} + \left[\underline{g}(\mathbf{Q}[n] + \boldsymbol{v} \odot \boldsymbol{\xi}) - \underline{g}(\mathbf{V}[n]) \right]^T \boldsymbol{\xi}. \tag{9.31}
$$

Define the state pair $\mathbf{Y}[n] = (\mathbf{Q}[n], \mathbf{V}[n])$. Conditioning on $\mathbf{Y}[n]$ and taking expectation, we obtain

$$
\mathbb{E}\left\{ L(\mathbf{Q}'[n+1]) - L(\mathbf{Q}[n]) \big| \mathbf{Y}[n] \right\}
$$
$$
= \underline{g}^T (\mathbf{V}[n]) \mathbb{E}\left\{ \boldsymbol{\xi} \big| \mathbf{Y}[n] \right\}, \tag{9.32}
$$
$$
+ \mathbb{E}\left\{ \left[\underline{g}(\mathbf{Q}[n] + \boldsymbol{v} \odot \boldsymbol{\xi}) - \underline{g}(\mathbf{V}[n]) \right]^T \boldsymbol{\xi} \big| \mathbf{Y}[n] \right\}. \tag{9.33}
$$

We will study parts (9.32) and (9.33) separately. Equation (9.32) then becomes

$$
\underline{g}^T (\mathbf{V}[n]) \mathbb{E}\left\{ \boldsymbol{\xi} \big| \mathbf{Y}[n] \right\} = T_s \underline{g}^T (\mathbf{V}[n]) \mathbb{E}\left\{ \frac{\mathbf{a}[n]}{T_s} - \mathbf{r}[n] \big| \mathbf{Y}[n] \right\}
$$
$$
= T_s \underline{g}^T (\mathbf{V}[n]) \left(\boldsymbol{\lambda} - \mathbb{E}\left\{ \mathbf{r}[n] \big| \mathbf{Y}[n] \right\} \right). \tag{9.34}
$$

According to Lemma 9.3, the scheduling policy (9.21) results in

$$
\mathbb{E}\left\{ \mathbf{r}[n] \big| \mathbf{Y}[n] \right\} = \arg \max_{\tilde{\mathbf{r}}[n] \in \tilde{\mathcal{C}}} \underline{g}^T (\mathbf{V}[n]) \tilde{\mathbf{r}}[n],
$$

which minimizes (9.34). Since $\boldsymbol{\lambda}$ is located within the interior ergodic capacity region $\tilde{\mathcal{C}}$, there exists a rate vector $\mathbf{r}' \in \tilde{\mathcal{C}}$ such that $r_i' > \lambda_i$ for all i. Let $\delta = \min_i (r_i' - \lambda_i)$. Thus, under the scheduling policy,

$$
\underline{g}^T (\mathbf{V}[n]) \mathbb{E}\left\{ \boldsymbol{\xi} \big| \mathbf{Y}[n] \right\} \leq T_s \underline{g}^T (\mathbf{V}[n]) \left(\boldsymbol{\lambda} - \mathbf{r}'[n] \right)
$$
$$
< -T_s |\underline{g}(\mathbf{V}[n])| \delta. \tag{9.35}
$$

To explore the property of (9.33), we consider

$$
\limsup_{\mathbf{V}[n] \to \infty} \frac{\mathbb{E}\left\{ \left[\underline{g}(\mathbf{Q}[n] + \boldsymbol{v} \odot \boldsymbol{\xi}) - \underline{g}(\mathbf{V}[n]) \right]^T \boldsymbol{\xi} \big| \mathbf{Y}[n] \right\}}{|\underline{g}(\mathbf{V}[n])|}. \tag{9.36}
$$

Using the dual of Fatou's lemma [89], we obtain

$$(9.36) = \limsup_{\mathbf{V}[n] \to \infty} \mathbb{E} \left\{ \frac{\left[\underline{g}(\mathbf{Q}[n] + \boldsymbol{v} \odot \boldsymbol{\xi}) - \underline{g}(\mathbf{V}[n]) \right]^T \boldsymbol{\xi}}{|\underline{g}(\mathbf{V}[n])|} \right\}$$

$$\leq \mathbb{E} \left\{ \limsup_{\mathbf{V}[n] \to \infty} \frac{\left[\underline{g}(\mathbf{Q}[n] + \boldsymbol{v} \odot \boldsymbol{\xi}) - \underline{g}(\mathbf{V}[n]) \right]^T \boldsymbol{\xi}}{|\underline{g}(\mathbf{V}[n])|} \right\}$$

$$\leq \mathbb{E} \left\{ \limsup_{\mathbf{V}[n] \to \infty} \sum_{i \in \mathcal{M}} \frac{|g_i(Q_i[n] + v_i \xi_i) - g_i(V_i[n])| \, |\xi_i|}{g_i(V_i[n])} \right\}$$

$$= \mathbb{E} \left\{ \sum_{i \in \mathcal{M}} \limsup_{V_i[n] \to \infty} \left| \frac{g_i(Q_i[n] + v_i \xi_i)}{g_i(V_i[n])} - 1 \right| |\xi_i| \right\}.$$

Let $\boldsymbol{\zeta} = \mathbf{Q}[n] - \mathbf{V}[n]$, then $\mathbb{E}\{|\zeta_i|\} < \infty$ for all i according to the condition (9.25). Equation (9.30) and the condition (9.26) lead to

$$\mathbb{E}\{|\xi_i|\} \leq \mathbb{E}\{r_i[n]\}T_s + \mathbb{E}\{a_i[n]\}$$
$$< \infty \text{ for all } i.$$

Hence, it follows from the properties of the $g_i(\cdot)$s given by (9.23) and (9.24) that

$$\lim_{V_i[n] \to \infty} \frac{g_i(Q_i[n] + v_i \xi_i)}{g_i(V_i[n])}$$

$$= \lim_{V_i[n] \to \infty} \frac{g_i(V_i[n] + \zeta_i + v_i \xi_i)}{g_i(V_i[n])}$$

$$= 1 \text{ with probability 1.}$$

Therefore, we have

$$\limsup_{V_i[n] \to \infty} \left| \frac{g_i(Q_i[n] + v_i \xi_i)}{g_i(V_i[n])} - 1 \right| |\xi_i| = 0 \text{ with probability 1,}$$

and

$$\limsup_{\mathbf{V}[n] \to \infty} \frac{\mathbb{E} \left\{ \left[\underline{g}(\mathbf{Q}[n] + \boldsymbol{v} \odot \boldsymbol{\xi}) - \underline{g}(\mathbf{V}[n]) \right]^T \boldsymbol{\xi} \Big| \mathbf{Y}[n] \right\}}{|\underline{g}(\mathbf{V}[n])|}$$

$$\leq \mathbb{E} \left\{ \limsup_{\mathbf{V}[n] \to \infty} \sum_{i \in \mathcal{M}} \frac{[g_i(Q_i[n] + v_i|\xi_i|) - g_i(V_i[n])] \, |\xi_i|}{g_i(V_i[n])} \right\}$$

$$= 0,$$

which means that

$$\limsup_{\mathbf{V}[n] \to \infty} \frac{\mathbb{E} \left\{ \left[\underline{g}(\mathbf{Q}[n] + \boldsymbol{v} \odot \boldsymbol{\xi}) - \underline{g}(\mathbf{V}[n]) \right]^T \boldsymbol{\xi} \Big| \mathbf{Y}[n] \right\}}{|\underline{g}(\mathbf{V}[n])|} = -\Omega_0,$$

where $\Omega_0 \geq 0$.

The definition of the upper limit[2] implies that for any $\epsilon > 0$, there exists $\mathbf{V}^* > 0$ such that for $\mathbf{V}[n] > \mathbf{V}^*$,

$$\frac{\mathbb{E}\left\{\left[\underline{g}(\mathbf{Q}[n] + \boldsymbol{v} \odot \boldsymbol{\xi}) - \underline{g}(\mathbf{V}[n])\right]^T \boldsymbol{\xi} \middle| \mathbf{Y}[n]\right\}}{|\underline{g}(\mathbf{V}[n])|} < -\Omega_0 + \epsilon \qquad (9.37)$$

$$< \epsilon. \qquad (9.38)$$

It follows from (9.37) and the fact that $|\underline{g}(\mathbf{V}[n])| > 0$ for $\mathbf{V}[n] > \mathbf{V}^*$ that

$$\mathbb{E}\left\{\left[\underline{g}(\mathbf{Q}[n] + \boldsymbol{v} \odot \boldsymbol{\xi}) - \underline{g}(\mathbf{V}[n])\right]^T \boldsymbol{\xi} \middle| \mathbf{Y}[n]\right\} < \epsilon |\underline{g}(\mathbf{V}[n])|. \qquad (9.39)$$

Due to the assumptions (9.22) and (9.26), there must exist a positive number $\Omega_1 < \infty$ such that

$$\sup_{\mathbf{V}[n] \leq \mathbf{V}^*} \mathbb{E}\left\{\left[\underline{g}(\mathbf{Q}[n] + \boldsymbol{v} \odot \boldsymbol{\xi}) - \underline{g}(\mathbf{V}[n])\right]^T \boldsymbol{\xi} \middle| \mathbf{Y}[n]\right\} < \Omega_1. \qquad (9.40)$$

Part (9.33) can be obtained from (9.39) and (9.40) as

$$\mathbb{E}\left\{\left[\underline{g}(\mathbf{Q}[n] + \boldsymbol{v} \odot \boldsymbol{\xi}) - \underline{g}(\mathbf{V}[n])\right]^T \boldsymbol{\xi} \middle| \mathbf{Y}[n]\right\} < \begin{cases} \Omega_1 & \mathbf{V}[n] \leq \mathbf{V}^* \\ \epsilon |\underline{g}(\mathbf{V}[n])| & \mathbf{V}[n] > \mathbf{V}^*, \end{cases}$$

which can be combined into

$$\mathbb{E}\left\{\left[\underline{g}(\mathbf{Q}[n] + \boldsymbol{v} \odot \boldsymbol{\xi}) - \underline{g}(\mathbf{V}[n])\right]^T \boldsymbol{\xi} \middle| \mathbf{Y}[n]\right\} < \epsilon |\underline{g}(\mathbf{V}[n])| + \Omega_1. \qquad (9.41)$$

Therefore, it follows from (9.35) and (9.41) that

$$\mathbb{E}\left\{L(\mathbf{Q}'[n+1]) - L(\mathbf{Q}[n]) \middle| \mathbf{Y}[n]\right\} < -(T_s\delta - \epsilon)|\underline{g}(\mathbf{V}[n])| + \Omega_1.$$

Since ϵ is an arbitrary positive, we let ϵ be small enough so that $\epsilon < T_s\delta$.

On the other hand,

$$\mathbf{Q}[n+1] - \mathbf{Q}'[n+1] = \begin{cases} 0 & \mathbf{r}[n]T_s \leq \mathbf{Q}[n] \\ \mathbf{r}[n]T_s - \mathbf{Q}[n] & \mathbf{r}[n]T_s > \mathbf{Q}[n], \end{cases}$$

and

$$\mathbb{E}\left\{L(\mathbf{Q}[n+1]) - L(\mathbf{Q}'[n+1]) \middle| \mathbf{Y}[n]\right\}$$

$$= \mathbb{E}\left\{\underline{g}\left(\mathbf{Q}'[n+1] + \boldsymbol{v}' \odot (\mathbf{r}[n]T_s - \mathbf{Q}[n])\right)(\mathbf{r}[n]T_s - \mathbf{Q}[n]) \cdot \mathbf{1}\{\mathbf{r}[n]T_s > \mathbf{Q}[n]\} \middle| \mathbf{Y}[n]\right\},$$

where $\mathbf{1}\{event\}$ equals 1 if the event is true, whereas it equals 0. Therefore, $\mathbb{E}\left\{L(\mathbf{Q}[n+1]) - L(\mathbf{Q}'[n+1]) \middle| \mathbf{Y}[n]\right\}$ is bounded by a positive number, Ω_2.

---

[2] $\limsup_{x \to \infty} f(x) = \lim_{y \to \infty} \sup\{f(x) : x > y\} = \inf_y \sup\{f(x) : x > y\}$

Consequently, it follows that

$$\mathbb{E}\left\{L(\mathbf{Q}[n+1]) - L(\mathbf{Q}[n]) \big| \mathbf{Y}[n]\right\}$$
$$= \mathbb{E}\left\{L(\mathbf{Q}[n+1]) - L(\mathbf{Q}'[n+1]) \big| \mathbf{Y}[n]\right\} + \mathbb{E}\left\{L(\mathbf{Q}'[n+1]) - L(\mathbf{Q}[n]) \big| \mathbf{Y}[n]\right\}$$
$$< -(T_s\delta - \epsilon)\big|\underline{g}(\mathbf{V}[n])\big| + \Omega_1 + \Omega_2.$$

Taking expectation with respect to $\mathbf{Y}[n]$, we obtain

$$\mathbb{E}\{L(\mathbf{Q}[n+1]) - L(\mathbf{Q}[n])\} < -(T_s\delta - \epsilon)\mathbb{E}\left\{\big|\underline{g}(\mathbf{V}[n])\big|\right\} + \Omega_1 + \Omega_2.$$

Taking summation, we have

$$\mathbb{E}\{L(\mathbf{Q}[N])\} - L(\mathbf{Q}[0]) < -(T_s\delta - \epsilon)\sum_{n=0}^{N-1}\mathbb{E}\left\{\big|\underline{g}(\mathbf{V}[n])\big|\right\} + N(\Omega_1 + \Omega_2),$$

and

$$\frac{1}{N}\sum_{n=0}^{N-1}\mathbb{E}\left\{\big|\underline{g}(\mathbf{V}[n])\big|\right\} < \frac{\Omega_1 + \Omega_2}{T_s\delta - \epsilon} + \frac{1}{N}L(\mathbf{Q}[0]).$$

Let $N \to \infty$, $\frac{1}{N}L(\mathbf{Q}[0]) \to 0$, and then the theorem is proved.

9.6   Further improvement through delay transmit diversity and adaptive power allocation

The studies on the stability issue in Section 9.4 can directly guide us to techniques for improving the performance of multi-carrier scheduling. According to the properties of the maximum stability region, we use joint stabilizing scheduling and power allocation to extend the maximum stability region so as to enhance the throughput-delay performance. Moreover, we use delay transmit diversity to increase the fluctuations in the frequency domain, by which we can obtain more frequency diversity.

9.6.1   Joint dynamic subcarrier assignment and adaptive power allocation

In Section 9.4, we showed that the maximum stability region is the interior of the ergodic capacity region at the physical layer. Thus, any techniques that are able to enlarge the ergodic capacity region can definitely improve the system performance. Adaptive power allocation lets the transmit power at each subcarrier be adjustable and only constrained by the total power limit $\bar{P}$; let $p[k]$ be the power at subcarrier k, then $\sum_{k\in\mathcal{K}} p[k] \le \bar{P}$.

Then, the instantaneous data rate region for the joint DSA and adaptive power allocation (APA) is given by

$$\mathcal{C}_{DSA+APA}(\mathbf{H}) = \left\{\mathbf{r}(\mathbf{D}, \mathbf{p}) : D_i \bigcap D_j = \varnothing, \ \forall i \neq j, \ \bigcup_{i\in\mathcal{M}} D_i \subseteq \mathcal{K}, \sum_{k\in\mathcal{K}} p[k] \le \bar{P}\right\},$$

where $\mathbf{p} = \{p[1], p[2], \ldots, p[K]\}$. Since the joint DSA and APA has looser constraints on resource allocation than the DSA, $\mathcal{C}_{DSA}(\mathbf{H}) \subseteq \mathcal{C}_{DSA+APA}(\mathbf{H})$ for each channel state $\mathbf{H}$;

therefore, the ergodic capacity region of joint DSA and APA is larger than that of DSA as well. As long as a scheduling scheme with the maximum stability region is applied in the system, the enlarged ergodic capacity region can be fully exploited. Obviously, the scheduling rule (9.21) on $C_{DSA+APA}(\mathbf{H})$ still has the maximum stability region.

The effect of APA is influenced by the rate adaptation used in the system. With continuous rate adaptation, the improvement in APA is trivial; however, the improvement in APA becomes substantial when there are only a small number of modulation levels, which is shown in Chapter 8.

For discrete rate adaptation, water-filling is not optimal for power allocation. In Section 8.2.3, a greedy power allocation algorithm is introduced to achieve the optimality of optimization problems with a concave objective function. To solve the joint DSA and APA problem, the sorting-search DSA and the greedy APA algorithms can be used iteratively. Mathematically, MDU with the FC can be expressed as the above optimization problem; thus, the sorting-search DSA and the greedy APA algorithms discussed in Section 8 can be implemented with no change.

9.6.2   Delay transmit diversity

In a single-carrier network, the multi-user diversity gain is limited in environments with little scattering or slow-fading. Opportunistic beamforming is proposed in [200] to induce fast and large fluctuations so as to amplify the multi-user diversity gain. The main idea of opportunistic beamforming is to change the magnitudes and phases of antenna weights in a pseudorandom fashion.

In a multi-carrier network, the multi-user diversity gain is diminished in environments with flat-fading, which is usually caused by a line-of-sight path and/or little scattering. Therefore, we can use a simpler multiple transmit antenna scheme, *delay transmit diversity*, to increase the randomness in the frequency domain compared with the opportunistic beamforming in the time domain.

Delay transmit diversity was first proposed in single-carrier systems [206]. Delay transmit diversity actually converts spatial diversity into frequency diversity by inducing multiple paths. Thus, the Viterbi algorithm is needed in the receivers in single-carrier systems. An OFDM system with delay transmit diversity is shown in Figure 9.2. The signals from additional antennas are the same as the signal from the first antenna but with different delays. Note that delay transmit diversity is implemented after *inverse fast Fourier transform* (IFFT) processing. For traditional OFDM systems, although the delay transmit diversity does not require additional processing in the receivers, some channel coding across subcarriers is needed to obtain the frequency diversity amplified by the delay transmit diversity [126]. However, in OFDM systems using DSA, the delay transmit diversity becomes totally transparent since any modulation, coding, and scheduling schemes used in the single-antenna case remain unchanged. Moreover, the frequency diversity induced by delay transmit diversity is transformed into multi-user diversity, which can be absorbed through DSA. In brief, delay transmit diversity and opportunistic beamforming are duals of each other.

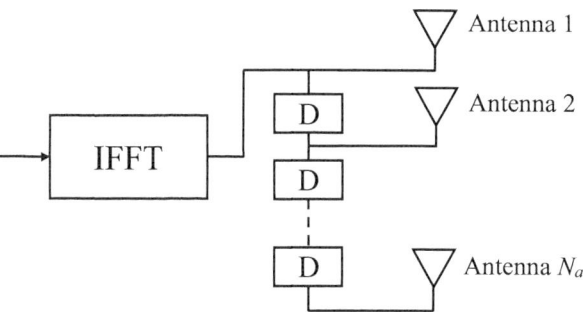

Figure 9.2 Delay transmit diversity in an OFDM system

9.7 Simulation results and performance comparison

In this section, we compare the performance of different scheduling schemes in an OFDM network. In the simulation, each user's channel suffers multi-path Rayleigh fading with the delay profile of Channel B for outdoor to indoor and pedestrian environments in [167], and each user is assumed to be stationary or slowly moving so that the maximum Doppler shift is 10 Hz. In the OFDM network, there are 128 subcarriers in a total channel bandwidth of 1.920 MHz. We assume that there are 20 users in the system. These 20 users have different distances from the base station; consequently, their average achievable transmission rates are different due to path loss.

Let the acceptable bit error rate (BER) be 10^{-6} for rate adaptation since data transmission is sensitive to error. Assume that a set of achievable transmission rates in bits/sec per Hz is $\{0, 1/2, 1, 2, 3, 4\}$. The transmission rate is chosen to be the largest available rate whose required signal-to-noise ratio (SNR) determined by (7.2) is larger than or equal to the current SNR. In practice, we can use 1/2-rate channel coding and a series of modulation schemes, including BPSK, QPSK, 16-QAM, 64-QAM, as well as 256-QAM to achieve the above feasible rates.

The packet length is assumed to be independently and exponentially distributed with an average length of 1024 bits. The packet arrival is modeled as a Poisson random variable for the following two reasons. First, delay-sensitive traffic is usually generated smoothly. Second, delays in the system are determined by two factors: the burst of arrivals and the fluctuation of scheduled service rates. Since in this chapter we are more interested in the second factor, Poisson arrival is assumed.

The length of a time slot T_s is 4 ms. All simulations were run for 300 000 slots, which correspond to 20 minutes in reality.

9.7.1 Performance comparison

The simulation results are shown in Figure 9.3 in terms of traffic load versus mean delay. In each simulation, all users have the same arrival rate. Due to the asymmetry among users' channel conditions, Figure 9.3 represents the mean delays for the worst user

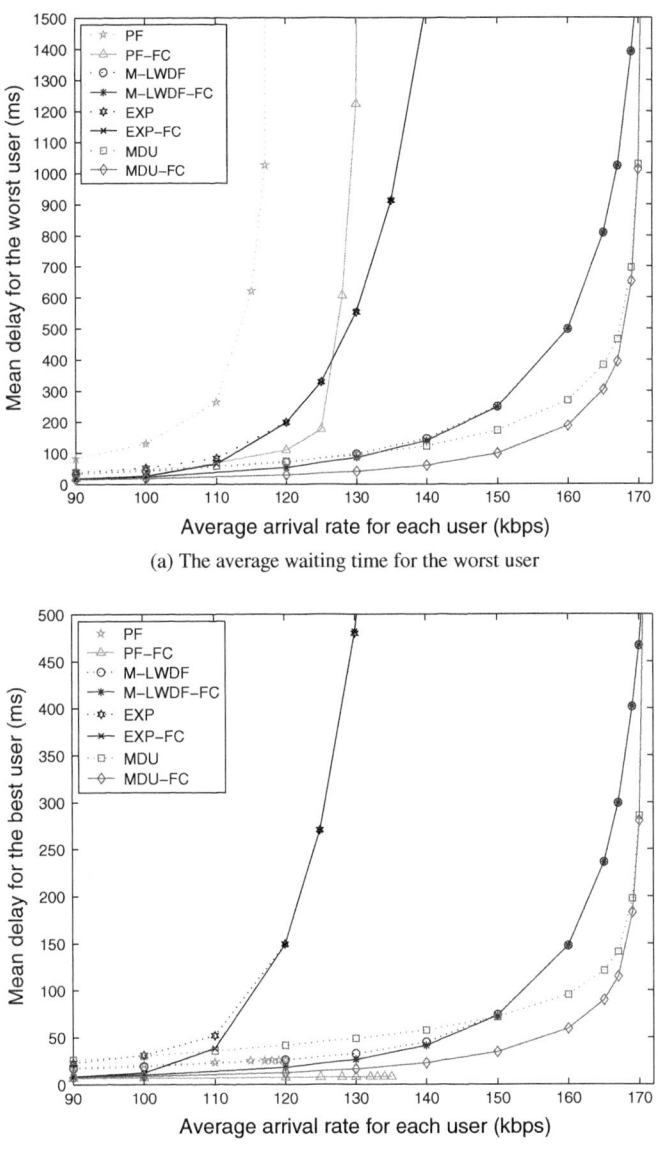

(a) The average waiting time for the worst user

(b) The average waiting time for the best user

Figure 9.3 Delay performance of different scheduling policies

who has the smallest average SNR, for the best user who has the largest average SNR, and averaging for all users in the system, respectively. We compare the performance of PF, M-LWDF, EXP, and MDU scheduling rules in an adaptive OFDM network. For comparison, the MDU uses $|U_i'(W)| = W$ for all i. Since the FC can be deployed with all scheduling rules, we run two schemes for each scheduling rule: with and without FC, respectively.

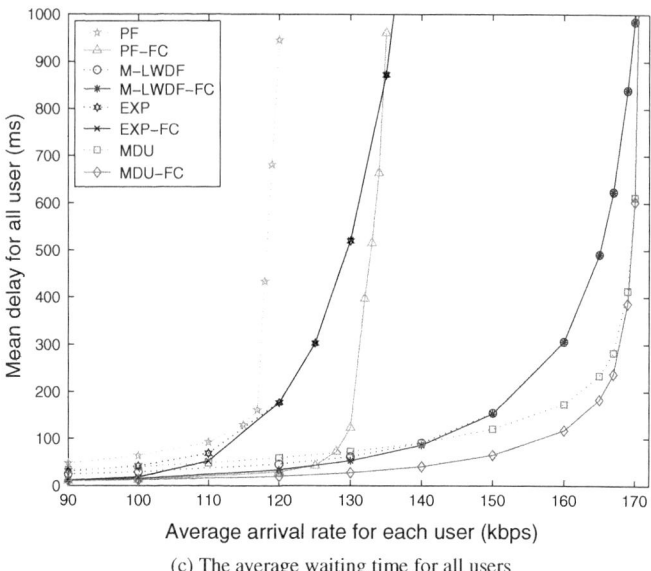

(c) The average waiting time for all users

Figure 9.3 *(cont.)*

Figure 9.3 shows the advantages of maximum-stability-region scheduling schemes. Since the PF scheduling has a small stability region, the maximum throughput the PF is able to support for the worst user is around 117 kbps, whereas the saturated throughput of the worst user with MDU is 170 kbps. Furthermore, the PF scheduling cannot guarantee delay fairness, which is concluded from the fact that with a heavy traffic load, the worst user suffers from an extremely long delay while the mean delay of the best user is very short. However, M-LWDF, EXP, and MDU, all of which have the maximum stability region, can maintain fairness in terms of delay performance.

Figure 9.3 also depicts the effect of FC. In Section 9.4, we concluded that the FC cannot stabilize the system. It is shown in Figure 9.3 that although the FC enhances the performance of the PF scheduling, the PF-FC does not have the maximum stability region. The effect of FC on the M-LWDF and EXP rules is very small, particularly with a heavy traffic load. However, the FC can boost the performance of MDU scheduling. When the traffic load is light or moderate, the MDU-FC can reduce the mean delay to half that with the MDU when not using the FC.

Finally, Figure 9.3 demonstrates that a scheduling scheme with the maximum stability region cannot provide sufficient good performance. The EXP scheduling has the maximum stability region, but its performance is still poor compared to the M-LWDF and MDU approaches. This is due to the mechanism of the EXP rule. If one user has a larger delay than others, the weight of this user becomes very large because of the exponential function used in the weight, and then this user may occupy all of the subcarriers with high probability. Because frequency-selective fading is present, assigning the whole bandwidth to one user is less efficient. Therefore, unlike single-carrier networks, aggressive weight assignments hurt the efficiency in OFDM networks.

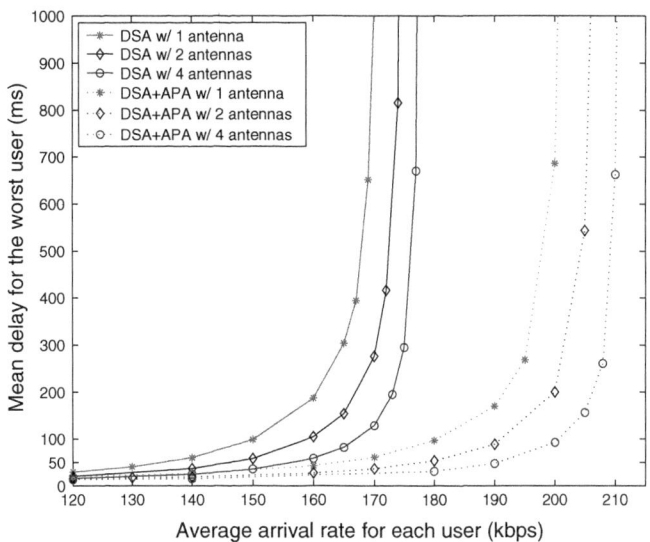

(a) The average waiting time for the worst user

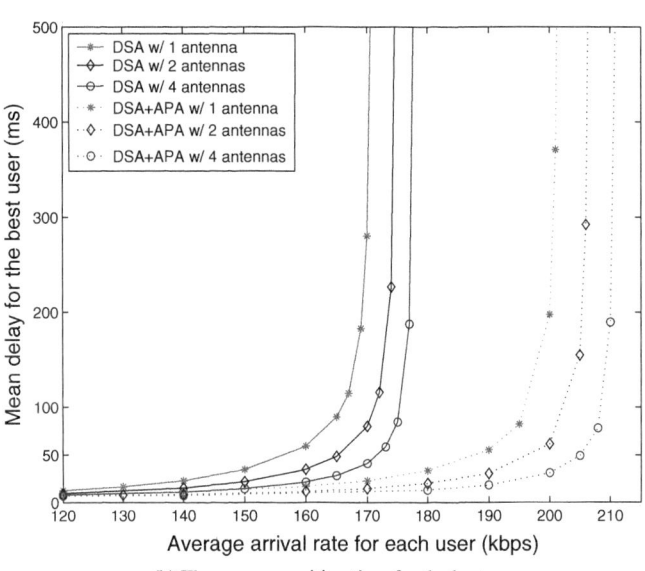

(b) The average waiting time for the best user

Figure 9.4 Delay performance of MDU-FC with delay transmit diversity and adaptive power allocation

This is a big difference in designing scheduling between single and multiple carrier networks. It is shown that the multi-channel version of M-LWDF works well. However, since the MDU policy uses the average queue lengths (delays) as the weights, which is a more moderate way, the MDU policy can allocate resources more efficiently. Figure 9.3 shows that the MDU-FC outperforms the other schemes.

9.7.2   Improvement in delay transmit diversity and adaptive power allocation

In this subsection, we present the simulation results when delay transmit diversity and power allocation are used. MDU scheduling with the FC is applied in the simulation. We let one delay tap in the delay transmit diversity be 1 μs. Figure 9.4 shows that the joint DSA and APA scheme exhibits a substantial improvement on the throughput-delay performance. Taking advantage of transparency and simplicity, the delay transmit diversity can boost the performance of the scheduling schemes based on DSA as well as those having joint DSA and APA.

9.8   Summary

We have investigated joint channel- and queue-aware multi-channel scheduling in OFDM networks from several important aspects. Based on utility functions with respect to average waiting times, we introduced MDU scheduling, which can be implemented by an on-line algorithm without knowledge of the statistical information about the channels and arrival traffic. Since the stability issue of scheduling is essential for QoS provisioning, we characterized the maximum stability region, which can reach the interior of the ergodic physical layer capacity region. Through concise proofs, we showed that with very few conditions on the scheduling schemes, the scheduling schemes can achieve the maximum stability region. To deal with environments with insufficient scattering or strong light-of-sight components, we used delay transmit diversity to induce the randomness in the frequency domain. In the simulation, we compared several scheduling schemes, and showed that MDU-FC scheduling has better throughput-delay performance than other scheduling schemes, and that the combination of scheduling and power allocation can significantly improve the performance further.

10 Utility-based generalized QoS scheduling for heterogeneous traffic

We developed maximum delay utility (MDU) scheduling with the help of channel and queue state information to enhance spectral efficiency and guarantee quality of service (QoS) in Chapter 9, in which, however, we emphasized its theoretical framework, such as queueing system stability. In this chapter, we apply MDU scheduling to allocate resources for QoS differentiation for different applications. We also present comprehensive simulation results that consider multiple traffic types, including packet-switched voice, streaming, and best-effort traffic. The simulation results demonstrate that MDU scheduling is a generalized QoS scheduling algorithm that is able to efficiently allocate resources for heterogeneous traffic with diverse QoS requirements. It substantially outperforms the multi-channel version of a combination of modified largest weighted delay first (M-LWDF) [37] and proportional fair (PF) scheduling [200, 186], called M-LWDF-PF scheduling.

10.1   Introduction

Guaranteeing QoS for multiple types of traffic is challenging to resource allocation and scheduling, especially for wireless data networks [32]. Traditionally, the main idea of QoS provisioning is to reserve resources so as to ensure that certain subjective or objective performance measures are met. In terms of scheduling, *generalized processor sharing* [153] (GPS)-based scheduling schemes, such as *weighted fair queueing* (WFQ) [43], and priority queueing are usually used for worst-case throughput and delay guarantee [49, 138, 204]. Obviously, their major drawback is that they cannot improve capacity since no channel site information (CSI) is used.

Currently, channel-aware or opportunistic scheduling has received much attention since it can exploit the variations of wireless fading channels to improve the spectral efficiency [130, 45]. Although proper fairness can be maintained by it, channel-aware scheduling is mainly suitable to best-effort applications but not efficient for delay-sensitive applications.

M-LWDF [37, 35], which makes scheduling decisions based on the current channel conditions and the states of the queues, is used for delay-sensitive applications with a QoS requirement that is defined as follows:

$$\mathbb{P}\{W_i > T_i\} \leq \delta_i, \tag{10.1}$$

where W_i is a packet delay for user i, and parameters T_i and δ_i are the delay threshold and the maximum probability of exceeding it, respectively. Its multi-channel version was used in Section 9.2. To meet QoS differentiation on delay performance, the M-LWDF maps the QoS requirement (10.1) to a scheduling weight

$$a_i = -\frac{\log \delta_i}{T_i}, \tag{10.2}$$

which is based on the results of large deviations. The M-LWDF scheduling is widely used in 1xEV-DO/DV for scheduling delay-sensitive traffic. In addition, it has the maximum stability region. However, M-LWDF scheduling cannot handle the more complicated QoS requirements and heterogeneous traffic well.

In MDU scheduling, the QoS requirements of each user are described by its utility function. From an application view, MDU scheduling captures the essence of QoS levels with a detail sufficient to predict subjective quality of users. From a network view, it provides the simplicity to enable monitoring and control mechanisms for guaranteeing QoS. By maximizing the total utility within the network, MDU scheduling establishes a simple, automatic mechanism that can simultaneously improve spectral efficiency and provide the right incentives to ensure that all applications can receive their required QoS.

10.2   MDU scheduling for heterogeneous traffic

In this section, we show how to employ MDU scheduling for a mixture of delay-sensitive and best-effort traffic by designing utility functions according to the QoS requirements. Note that MDU scheduling implements the fugality constraint (FC) in this chapter.

10.2.1   Mechanisms of MDU scheduling for diverse QoS requirements

To apply MDU scheduling, we need to design utility functions with respect to average waiting time W for the corresponding QoS requirements. Since the marginal utility functions are proportional to the scheduling weights, the marginal utility functions, $U_i'(\cdot)$, play a crucial role in scheduling. Therefore, we directly design the marginal utility functions rather than the utility functions in this section.

We design the marginal utility functions based on both certain objective and subjective performance criteria. The objective consideration is the system stability, which is studied in Section 9.4. One of results is that conditions (9.22)–(9.24) can make MDU scheduling stabilize the queueing system. From a system perspective, a significant difference between delay-sensitive and best-effort applications is that the incoming rate of a delay-sensitive stream is usually determined by its source, but the data rate of a best-effort connection is controlled by its transport layer according to the level of network congestion [107]. From a subject perspective, best-effort applications have no specific QoS requirements. Based on these two reasons, *the core idea of designing marginal utility functions is to let the marginal utility functions of delay-sensitive traffic satisfy conditions (9.22)–(9.24), but make the marginal utility functions of best-effort traffic bounded.* Assume that connections 1 to M_1 are delay-sensitive, connections $M_1 + 1$ to M ($M_1 < M$) are best-effort. Their corresponding incoming rate is λ_i. It follows from the design that

$$\lim_{W \to \infty} \frac{U_i'(W)}{U_j'(W)} = 0, \ i \in \{M_1 + 1, M_1 + 2, \dots, M\} \text{ and } j \in \{1, 2, \dots, M_1\}. \quad (10.3)$$

The above equation means that MDU scheduling can sense the level of network congestion. If the network is congested, best-effort connections hardly obtain resources to transmit packets according to (10.3). If the rate vector

$$[\lambda_1, \lambda_2, \dots, \lambda_{M_1}, 0, 0, \dots, 0]^T$$

is located within the ergodic capacity region $\tilde{\mathcal{C}}$, the above design makes all of the delay-sensitive connections stable, which comes directly from the results of Corollary 9.6. Therefore, MDU scheduling does not allow those best-effort connections to affect the stability of delay-sensitive connections. If the network load is low, the scheduler can automatically assign more resources to those best-effort connections. The more specific design of the marginal utility functions is based on the subjective performance criteria of certain applications. Section 10.2.2 shows the details.

10.2.2 Marginal utility functions for MDU scheduling

In this section, we design the marginal utility functions based on the corresponding required QoS for packet-switched voice, streaming, and best-effort traffic.

Delay-sensitive applications

For a delay-sensitive application, we set a threshold for the marginal utility function that depends on the characteristics of the application. When the average waiting time is less than the threshold, the marginal utility increases with a small order. When the average waiting time is beyond the threshold, the marginal utility increases with a relatively high order.

For packet-switched voice or *voice over IP* (VoIP), the end-to-end delay is usually required less than 100 ms [1]. Since there are other delay factors besides the delay resulting from wireless scheduling, we set the marginal utility function for voice as follows:

$$|U_V'(W)| = \begin{cases} W & W \leq 25\text{ms}, \\ W^{1.5} - 25^{1.5} + 25 & W > 25\text{ms}, \end{cases} \quad (10.4)$$

where the threshold, 25 ms, comes from one-forth of 100 ms.

Good-quality streaming transmission needs end-to-end delay between 150–400 ms. We choose the following marginal utility function for streaming traffic:

$$|U_S'(W)| = \begin{cases} W^{0.6} & W \leq 100\text{ms}, \\ W - 100 + 100^{0.6} & W > 100\text{ms}, \end{cases} \quad (10.5)$$

where the threshold, 100 ms, comes from one-forth of 400 ms. Obviously, marginal utility functions (10.4) and (10.5) both satisfy conditions (9.22)–(9.24).

Best-effort applications

Since best-effort traffic is not delay-sensitive, the utility function with respect to average waiting time is not sufficient enough to describe the performance of this traffic. From the point of view of scheduling weights, however, we can still give the marginal utility function in terms of average waiting time. For example, the marginal utility function is given by

$$|U'_D(W)| = \begin{cases} W^{0.5} & W \leq 100\text{ms}, \\ 100^{0.5} & W > 100\text{ms}. \end{cases} \tag{10.6}$$

Please note a long average waiting time means network congestion. Compared to (10.4) and (10.5), the marginal utility function for best-effort traffic (10.6) lets the scheduling weights be bounded, whereas delay-sensitive applications set higher scheduling weights according to (10.4) and (10.5). In this simulation, we intend to know the maximum throughput that best-effort traffic can obtain. Thus, we fix $|U'_D(W)|$ to the maximum value, $100^{0.5}$. Actually, MDU scheduling for best-effort traffic becomes PF scheduling, which is known to be applicable to best-effort traffic, see Chapter 8.

10.3   Simulation

In this section, we design appropriate simulations that take into account the impact of different traffic types and average signal-to-noise ratio (SNR) values on scheduling performance.

10.3.1   Simulation conditions

For comparison, we assume that the number of each traffic type is an even integer. For each type of traffic, half of the users have the same average SNR of 15 dB, and we call them good users; the rest have the same average SNR of 8 dB, and we call them bad users. In the simulation, each bad user's channel suffers multi-path Rayleigh fading with the delay profile of Channel B for outdoor to indoor and pedestrian environments in [167], and each user is assumed to be stationary or slowly moving so that the maximum Doppler shift is 10 Hz. Each good user experiences Rician fading and the delay profile and Doppler shift are the same as those of bad users' channels. The Rician factor is 0.5. In the OFDM network, there are 256 subcarriers in a total channel bandwidth of 2.048 MHz. These 256 subcarriers are grouped into 32 clusters, each of which can be dynamically assigned to a user during a time slot. Let the acceptable bit error rate (BER) be 10^{-5} for rate adaptation since data transmission is sensitive to error. Assume that a set of achievable transmission rates in bits/sec per Hz is {0,1/2,1,2,3,4}. In practice, we can use 1/2-rate channel coding and a series of modulation schemes including BPSK, QPSK, 16-QAM, 64-QAM, as well as 256-QAM to achieve the above feasible rates.

Table 10.1 Scheduling weights for M-LWDF-PF

	Voice	Streaming	Best-effort
T_i (ms)	100	400	–
δ_i	5%	5%	–
Weight	13	3.25	0.26

We consider three types of traffic: packet-switched voice, streaming, and best-effort traffic. The traffic model for voice traffic is the on–off voice activity model with exponentially distributed duration of voice spurts and gaps [180]. The average talk spurt is 1.00 s, and the average silent interval is 1.35 s. Within each talk spurt interval, a 32 kbps digital voice coding is assumed. The streaming traffic is simulated according to the model in [62]. The duration of each state is exponentially distributed with mean 160 ms. The data rate in each state is generated according to a truncated exponential distribution in which the minimum, maximum, and average data rates are 64, 256, and 180 kbps, respectively. As mentioned before, we only care about the maximum throughput of best-effort traffic in this simulation and fix its scheduling weights. Therefore, we apply a full-buffer model to best-effort traffic. In the full-buffer model, there are infinite data packets in the queues. Although this model may not be realistic, it can obtain the maximum achievable throughput for best-effort traffic.

For M-LWDF-PF scheduling, the weights for delay-sensitive applications can be calculated by (10.2). The weights for the simulation are listed in Table 10.1.

10.3.2 Simulation results

We design three experiments in the simulation and compare the performance of MDU scheduling and that of M-LWDF-PF scheduling. The performance of delay-sensitive traffic is evaluated in terms of the 95th percentile delay, and that of best-effort traffic is measured in terms of average throughput. We focus on discussing the properties of MDU scheduling first.

Increase of voice users

In this experiment, we fix the numbers of streaming and best-effort users to be 14 and 20, respectively, and increase the number of voice users. It is seen from Figure 10.1 that as the number of voice users increases, the throughput of best-effort traffic decreases apparently; the delay for streaming users increases slightly. However, there is only a very small rise in the delay for voice and streaming users in the system employing MDU scheduling.

Increase of streaming users

In this experiment, we fix the numbers of voice and best-effort users both to be 20 and increase the number of streaming users. Since the average data rate of a streaming link is as large as 180 kbps, we can clearly see the performance in both less-congested and congested situations in Figure 10.2. When the network is lesscongested (the number of

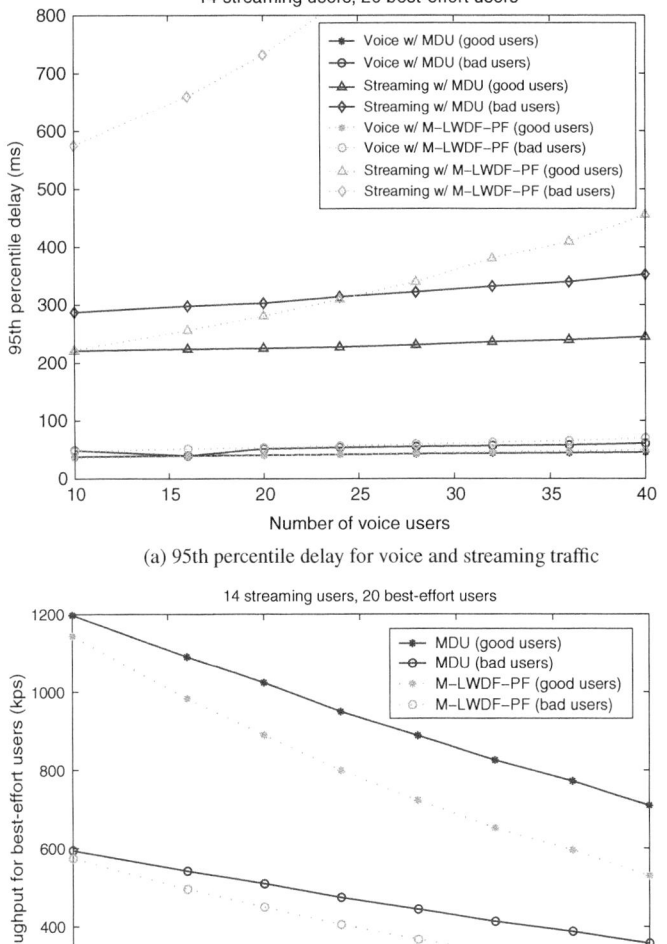

(a) 95th percentile delay for voice and streaming traffic

(b) Average total throughput for best-effort traffic

Figure 10.1 Heterogeneous traffic performance versus the number of voice users

streaming users does not exceed 16), MDU scheduling can maintain high-quality delay performance for those delay-sensitive applications and provide a high data rate for the best-effort users. When the network is congested, e.g. in the 20-streaming-user case, the throughput for best-effort users becomes extremely small, and the delay for streaming users has a dramatic increase. However, the performance of voice users is still very good.

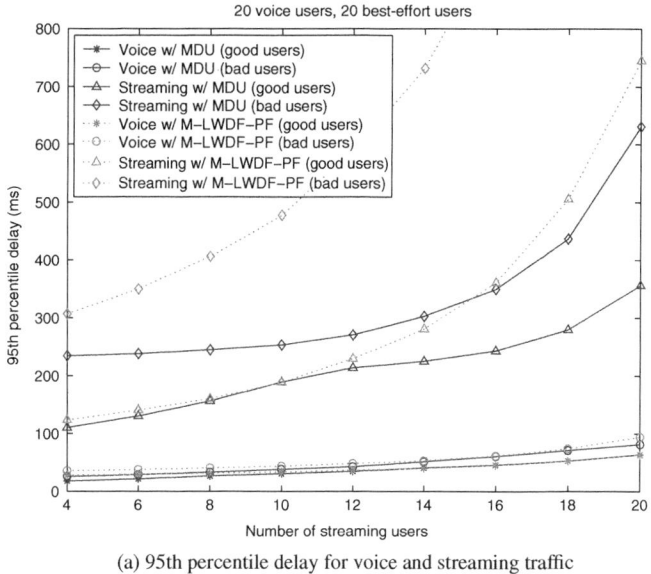

(a) 95th percentile delay for voice and streaming traffic

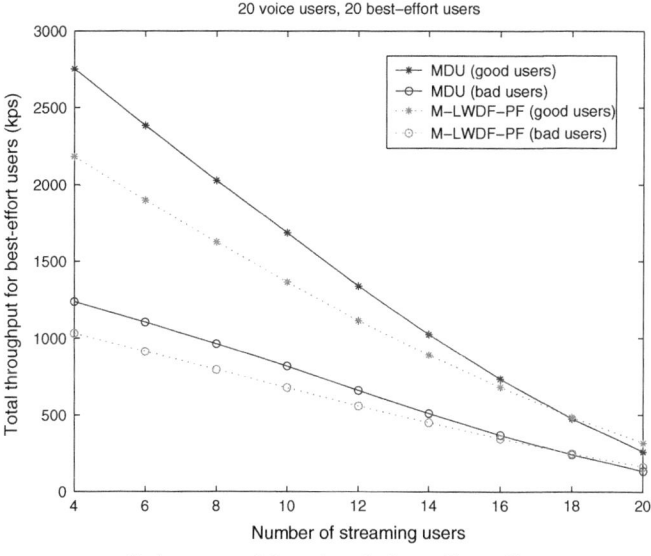

(b) Average total throughput for best-effort traffic

Figure 10.2 Heterogeneous traffic performance versus the number of streaming users

Increase of best-effort users

In the last experiment, we fix the numbers of voice and streaming users to be 20 and 10, respectively, and increase the number of best-effort users. It is seen from Figure 10.3 that as the number of best-effort users increases, the performance of voice and streaming users is maintained very well with MDU scheduling, and the throughput for best-effort increases, which results from multi-user diversity.

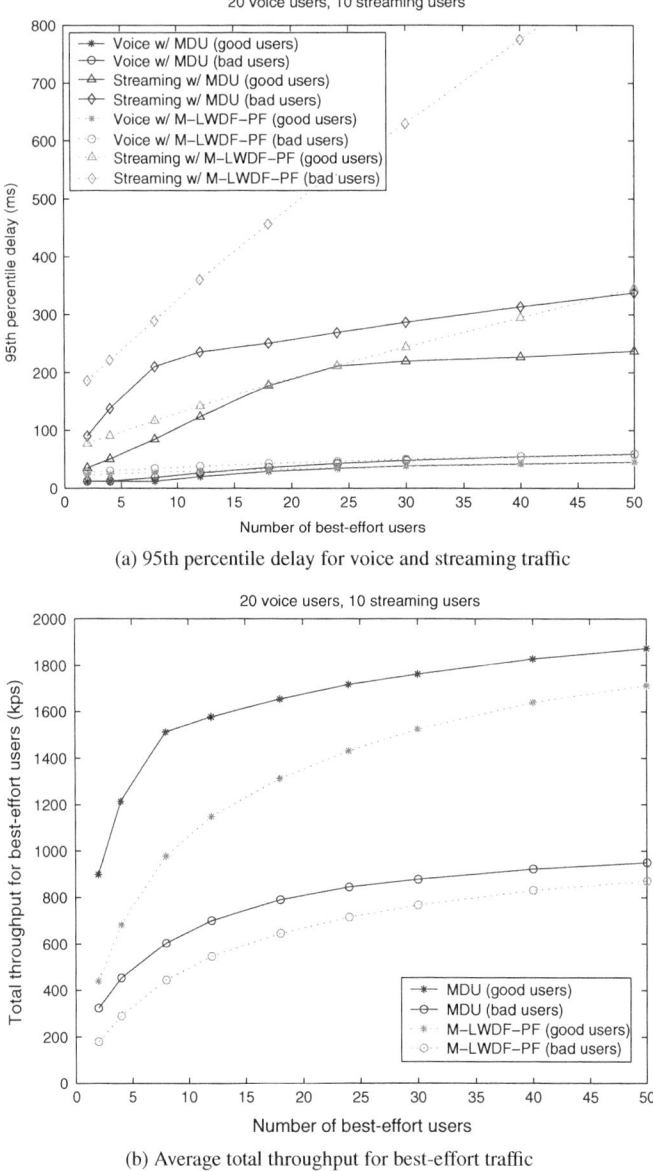

(a) 95th percentile delay for voice and streaming traffic

(b) Average total throughput for best-effort traffic

Figure 10.3 Heterogeneous traffic performance versus the number of best-effort users

Therefore, we can in these three experiments see the excellent mechanisms of MDU scheduling: high spectral efficiency by taking advantage of knowledge of CSI and good diverse QoS provisioning by exploiting utility functions. We also compare MDU with the M-LWDF-PF scheduling in Figures 10.1–10.3. Note that M-LWDF scheduling is also a scheduling scheme that can adjust resource allocation according to users' channel and queue state information and have the maximum stability region. All experiments show that both scheduling schemes offer similar delay performance for voice users, and that in most of the cases, MDU scheduling provides considerably smaller delays for streaming traffic than does M-LWDF-PF scheduling while MDU scheduling allows best-effort users to achieve higher throughput than M-LWDF-PF scheduling at the same time. This is mainly because MDU scheduling can more appropriately capture the required QoS compared to other scheduling schemes.

10.4 Summary

By simulation, we have demonstrated that MDU scheduling can effectively handle multiple traffic types with diverse QoS requirements and substantially outperforms the multi-channel version of M-LWDF-PF scheduling. MDU scheduling benefits from the awareness of channel quality and queue information, traffic multiplexing, and resource regulation through utility functions, which appropriately capture the QoS requirements of specific traffic. In addition, MDU scheduling has a very simple QoS architecture. It does not need statistical information about incoming traffic, and its implementation complexity is also low. Therefore, MDU scheduling is an attractive solution for IP-based wireless data networks.

11 Asymptotic performance analysis for channel-aware scheduling

To obtain multi-user diversity gain, adaptive modulation and channel-aware scheduling must be used. However, the channel variance and the opportunistic nature of channel-aware scheduling make throughput analysis very difficult. An asymptotic analysis of signal-to-noise ratio (SNR) for multi-user diversity is presented in [200]. Capacity analyses for Rayleigh and Nakagami fading channels are addressed in [217] and [51], respectively. However, the results of those capacity analyses are too complicated to obtain insights.

In this chapter, we provide an asymptotic performance analysis of channel-aware packet scheduling based on extreme value theory, including throughput and delay analysis for both single carrier and multi-carrier networks. In Section 11.1, we briefly describe the main results of extreme value theory used in this chapter. In Section 11.2, we use an asymptotic analysis of throughput of single-carrier systems with channel-aware scheduling. We first address the average throughput of systems with an homogeneous average SNR and obtain its asymptotic expression. Compared to the exact throughput expression, the asymptotic one, which is applicable to a broader range of channel fading distributions, is more concise and easier to obtain insights. Furthermore, we confirm the accuracy of the asymptotic results by numerical simulation. For a system with heterogeneous SNRs, normalized-SNR-based scheduling needs to be used for fairness. We also investigate the asymptotic average throughput of normalized-SNR-based scheduling and prove that the average throughput in this case is less than that in the homogeneous case with a power constraint. In Section 11.3, we provide a closed-form asymptotic average packet delay analysis for single-carrier networks exploiting multi-user diversity. In Section 11.4, asymptotic analysis of throughput and delay is extended into multi-carrier networks. The asymptotic analysis for mean packet delay demonstrates that the multi-user diversity gain in multi-carrier networks is not limited by slow-fading as in single-carrier networks.

11.1 Extreme value theory

Extreme value theory deals with asymptotic distributions of extreme values, such as maxima or minima. It can be used to analyze the performance of channel-aware scheduling approaches. In this section, we will briefly introduce the major results of extreme value theory [82, 61] that are used in the analysis.

Let $\xi_1, \xi_2, \ldots, \xi_M$ be *independently identically distributed* (i.i.d.) random variables with the common distribution function $F(x)$. We are interested in the distribution of the maximum, $Z_M = \max_{i \in M} \xi_i$ as $M \to \infty$. The *cumulative distribution function* (CDF) of the maximum, $H_M(x)$, is given by

$$H_M(x) = \mathbb{P}\{Z_M \le x\} = F^M(x).$$

When $M \to \infty$, we have

$$F^M(x) \to \begin{cases} 1 & \text{if } F(x) = 1, \\ 0 & \text{if } F(x) = 0, \end{cases}$$

which means that the limiting distribution is degenerate at either 0 or 1. In order to avoid this degeneration, we look for such normalizing constants as a_M and b_M depending on M that

$$\lim_{M \to \infty} H_M(a_M + b_M x) = \lim_{M \to \infty} \mathbb{P}\left\{ \frac{Z_M - a_M}{b_M} \le x \right\}$$

$$= \lim_{M \to \infty} F^M(a_M + b_M x)$$

$$= H(x),$$

where $H(x)$ is a limiting *non-degenerate* distribution function. We also say that $\frac{Z_M - a_M}{b_M}$ converges in the sense of distribution in this case. An important result about limiting distribution is described as follows.

Let $\xi_1, \xi_2, \ldots, \xi_M$ be i.i.d. random variables with the distribution function $F(x)$, and $Z_M = \max_{i \in M} \xi_i$. If there exist constants $a_M \in \mathbb{R}$, and $b_M > 0$, and some non-degenerate distribution function H such that the distribution of $\frac{Z_M - a_M}{b_M}$ converges to H, then H belongs to one of the three standard extreme value distributions: Frechet, Webull, and Gumbel distributions.

It is very interesting that there are only three possible non-degenerate limiting distributions for maxima. The distribution function $F(x)$ determines the exact limiting distribution. Thus, if a distribution function $F(x)$ results in one limiting distribution for extremes, we say that $F(x)$ belongs to the domain of attraction of this limiting distribution. Next, we will introduce a sufficient condition for a distribution function $F(x)$ to belong to the domain of attraction of the Gumbel distribution.

LEMMA 11.1   *Let $\xi_1, \xi_2, \ldots, \xi_M$ be i.i.d. random variables with the distribution function $F(x)$. Define $\omega(F) = \sup\{x : F(x) < 1\}$. Assume that there is a real number x_1 such that, for all $x_1 \le x < \omega(F)$, $f(x) = F'(x)$ and $F''(x)$ exist and $f(x) \ne 0$. If*

$$\lim_{x \to \omega(F)} \frac{d}{dx} \left[\frac{1 - F(x)}{f(x)} \right] = 0, \tag{11.1}$$

then there exist sequences a_M and $b_M > 0$ such that, as $M \to \infty$, $\frac{Z_M - a_M}{b_M}$ uniformly converges in distribution to a normalized Gumbel (maxima) random variable.

The normalizing constants a_M and b_M can be chosen as

$$a_M = F^{-1}\left(1 - \frac{1}{M}\right),$$

$$b_M = F^{-1}\left(1 - \frac{1}{Me}\right) - F^{-1}\left(1 - \frac{1}{M}\right),$$

where $F^{-1}(x) = \inf\{y : F(y) \geq x\}$.

For a random variable Z with the normalized Gumbel distribution for maxima, $\exp[-\exp(-x)]$, $-\infty < x < \infty$, it follows that

$$\mathbb{E}\{Z\} = E_0,$$

$$\mathbb{V}ar\{Z\} = \frac{\pi^2}{6},$$

where $E_0 = 0.5772 \cdots$ is the Euler constant [82].

In this chapter, we intend to calculate the average throughput; thus, mean convergence is used extensively. However, convergence in distribution cannot generally guarantee mean convergence. [154] established the relation between convergence in distribution and moment convergence, which is stated in the following lemma:

LEMMA 11.2   *If $\dfrac{Z_M - a_M}{b_M}$ converges in distribution to a random variable Z that has a non-degenerate distribution function, and if $\mathbb{E}\{[(Z_M)^-]^p\} < \infty$ for any positive real number p, where $(x)^- = -x$, $x < 0$, $= 0$, otherwise, then*

$$\lim_{M \to \infty} \mathbb{E}\left(\frac{Z_M - a_M}{b_M}\right)^p = \mathbb{E}\{Z^p\},$$

provided $\mathbb{E}|Z|^p < \infty$.

Obviously, convergence in distribution for the maximum of *non-negative* random variables results in moment convergence.

Lemmas 11.1 and 11.2 can be restated for minima by considering $(-\xi_i)$ instead of ξ_i. We give below the result for the asymptotic distribution of the minimum of i.i.d. random variables. Let $W_M = \min\limits_{i \in \mathcal{M}} \xi_i$.

LEMMA 11.3   *Let $\xi_1, \xi_2, \ldots, \xi_M$ be i.i.d. random variables with distribution function $F(x)$. Define $\alpha(F) = \inf\{x : F(x) > 0\}$. Assume that there is a real number x_1 such that, for all $\alpha(F) \leq x < x_1$, $f(x) = F'(x)$ and $F''(x)$ exist and $f(x) \neq 0$. If*

$$\lim_{x \to \alpha(F)} \frac{d}{dx}\left[\frac{F(x)}{f(x)}\right] = 0, \tag{11.2}$$

then there exist sequences c_M and $d_M > 0$ such that, as $M \to \infty$, $\dfrac{W_M - c_M}{d_M}$ uniformly converges in distribution to a normalized Gumbel (minima) random variable. The *normalizing constants c_M and d_M can be chosen as*

$$c_M = F^{-1}\left(\frac{1}{M}\right),$$

$$d_M = F^{-1}\left(\frac{1}{M}\right) - F^{-1}\left(\frac{1}{Me}\right).$$

For a random variable W with the normalized Gumbel distribution for minima, $1 - \exp[-\exp(-x)]$, $-\infty < x < \infty$, it follows that

$$\mathbb{E}\{W\} = -E_0,$$

$$\mathbb{V}ar\{W\} = \frac{\pi^2}{6}.$$

With extreme value theory, we can study the asymptotic performance of channel-aware scheduling.

11.2   Asymptotic throughput analysis of single-carrier networks

In this section, we focus on asymptotic throughput analysis for single-carrier networks with channel-aware scheduling.

11.2.1   System model

Consider a shared downlink channel of a single-carrier system with a bandwidth B and M users. The downlink channel is time-slotted, and each time slot can adaptively be assigned to a user. It is assumed that the base station knows the channel state information (CSI) of each user, and that continuous rate adaptation is applied in the downlink channel. Therefore, the transmission data rate, R, depends on the current SNR, Γ. The relationship can be written as [83]

$$R = B\log_2(1 + \beta\Gamma), \tag{11.3}$$

where β is a constant related to the targeted bit error rate (BER) and the used modulation and coding techniques.

First, we assume that all users experience statistically independent identical fading processes. The *max-sum-capacity* (MSC) scheduling rule [200, 185] is used in the system. The MSC rule is a channel-aware scheduling scheme that maximizes the total throughput in the system and works well in the homogeneous system. It assigns the channel to the user with the best channel condition on each time slot, which is described as

$$m = \arg\max_{i\in\mathcal{M}}\{\Gamma_i\}, \tag{11.4}$$

where $\mathcal{M} = \{1, 2, \ldots, M\}$, and Γ_i is the SNR of user i.

We also consider the heterogeneous case in which different users have different average SNR values due to various path losses. For the purpose of fairness, normalized-SNR-based scheduling [217] is used. This scheduling rule makes decisions based on the normalized SNR, rather than the absolute SNR, values, which is expressed as

$$m = \arg \max_{i \in \mathcal{M}} \left\{ \frac{\Gamma_i}{\gamma_i} \right\}, \tag{11.5}$$

where γ_i is the average SNR of user i; i.e., $\mathbb{E}\{\Gamma_i\} = \gamma_i$. It is obvious that normalized-SNR-based scheduling is equivalent to MSC scheduling in the homogeneous system.

11.2.2   Throughput analysis for Rayleigh fading

In this section, we will analyze the throughput performance of multi-user diversity for Rayleigh fading channels with the same average SNR γ_0. The CDF of the SNR for Rayleigh fading can be expressed as

$$F_\Gamma(\gamma) = 1 - \exp\left(-\frac{\gamma}{\gamma_0}\right). \tag{11.6}$$

Exact analysis

According to MSC scheduling, the base station schedules the user with the strongest channel condition. Therefore, the effective SNR at the transmitter, Γ_{eff}, is given by

$$\Gamma_{\text{eff}} = \max_{i \in \mathcal{M}} \Gamma_i, \tag{11.7}$$

and its distribution is

$$F_{\Gamma_{\text{eff}}}(\gamma) = \mathbb{P}\{\Gamma_1 < \gamma, \Gamma_2 < \gamma, \ldots, \Gamma_M < \gamma\}$$
$$= \left(1 - e^{-\gamma/\gamma_0}\right)^M.$$

By taking the derivative, the probability distribution function (PDF) of Γ_{eff} can be obtained as

$$f_{\Gamma_{\text{eff}}}(\gamma) = \frac{d}{d\gamma} F_{\Gamma_{\text{eff}}}(\gamma)$$
$$= M(1 - e^{-\gamma/\gamma_0})^{M-1} \frac{e^{-\gamma/\gamma_0}}{\gamma_0}. \tag{11.8}$$

Using (11.8), we calculate the average SNR when MSC scheduling is used as

$$\mathbb{E}\{\Gamma_{\text{eff}}\} = \int_0^\infty \gamma \cdot f_{\Gamma_{\text{eff}}}(\gamma) \, d\gamma$$
$$= \gamma_0 \sum_{i=1}^M \frac{1}{i}. \tag{11.9}$$

The throughput of MSC scheduling is expressed as

$$R_{\text{total}} = B \log_2(1 + \beta \Gamma_{\text{eff}});$$

hence, the average throughput is

$$\mathbb{E}\{R_{\text{total}}\} = B \int_0^\infty \log_2(1 + \beta\gamma) f_{\Gamma_{\text{eff}}}(\gamma) \, d\gamma. \tag{11.10}$$

To obtain a closed-form result of $\mathbb{E}\{R_{\text{total}}\}$, rewrite (11.8) using binomial expansion

$$f_{\Gamma_{\text{eff}}}(\gamma) = \frac{M}{\gamma_0} \sum_{i=0}^{M-1} (-1)^i \binom{M-1}{i} e^{-\frac{(1+i)\gamma}{\gamma_0}}, \tag{11.11}$$

where

$$\binom{M-1}{i} = \frac{(M-1)!}{(M-i-1)! \, i!}.$$

Substituting (11.11) into (11.10), we obtain

$$\mathbb{E}\{R_{\text{total}}\} = \frac{M}{\ln 2} \sum_{i=0}^{M-1} (-1)^{i+1} \binom{M-1}{i} \frac{e^{\frac{1+i}{\gamma_0}}}{i+1} E_i\left(-\frac{1+i}{\gamma_0}\right) \tag{11.12}$$

with

$$E_i(-x) = E_0 + \ln(x) + \sum_{i=1}^\infty \frac{(-1)^i x^i}{i! \ i}.$$

As seen above, the exact analysis of throughput analysis is very complicated, and the exact result lacks insights. Therefore, in the rest of the chapter, we will provide the simple results through asymptotic analysis.

Asymptotic analysis

First, we study the asymptotic distribution for the effective SNR Γ_{eff} in (11.7). The exponential distribution leads to

$$\frac{1 - F_\Gamma(\gamma)}{f_\Gamma(\gamma)} = \gamma_0.$$

As a result, it follows that

$$\frac{d}{d\gamma}\left[\frac{1 - F_\Gamma(\gamma)}{f_\Gamma(\gamma)}\right] = 0, \text{ for } \gamma > 0.$$

According to the results of extreme value theory in Section 11.1, the exponential distribution is in the domain of attraction of the Gumbel distribution, and

$$a_M = \gamma_0 \ln M,$$
$$b_M = \gamma_0.$$

According to Lemma 11.2, as $M \to \infty$,

$$\frac{\mathbb{E}\{\Gamma_{\text{eff}}\} - \gamma_0 \ln M}{\gamma_0} \to E_0.$$

With a large M, therefore,

$$\mathbb{E}\{\Gamma_{\text{eff}}\} \approx \gamma_0(\ln M + E_0). \tag{11.13}$$

It is shown in [218] that

$$\frac{1}{2(M+1)} < \sum_{i=1}^{M} \frac{1}{i} - (\ln M + E_0) < \frac{1}{2M},$$

which implies that the difference between the exact value (11.9) and the asymptotic value (11.13) is very small even for a small M.

We can also use the results of the extreme value theory in Section 11.1 to obtain an asymptotic analysis for throughput $R_{\text{total}} = \max_{i \in \mathcal{M}} R_i$. Let

$$R = T(\Gamma) \triangleq B \log_2(1 + \beta\Gamma).$$

Since $T(\Gamma)$ is a monotonic increasing function of Γ, the distribution of data rate R is given by

$$F_R(r) = F_\Gamma(T^{-1}(r)),$$

where $T^{-1}(r) = \dfrac{2^{\frac{r}{B}} - 1}{\beta}$. For the Rayleigh fading channel, we have

$$\frac{1 - F_R(r)}{f_R(r)} = \frac{1 - F_\Gamma(T^{-1}(r))}{f_\Gamma(T^{-1}(r))\left(T^{-1}\right)'(r)}$$

$$= \frac{\beta\gamma_0}{2^{\frac{r}{B}} - 1}. \tag{11.14}$$

Equation (11.14) results in

$$\lim_{r \to \infty} \frac{d}{dr}\left[\frac{1 - F_R(r)}{f_R(r)}\right] = 0. \tag{11.15}$$

According to the results of extreme value theory in Section 11.1, therefore, the maximum throughput, $R_{\text{total}} = \max_{i \in \mathcal{M}} R_i$, asymptotically behaves as a Gumbel random variable, $a_M + b_M Z$, where Z is a normalized Gumbel random variable, and

$$a_M = B \log_2(1 + \beta\gamma_0 \ln M),$$

$$b_M = B \log_2\left(\frac{1 + \beta\gamma_0(1 + \ln M)}{1 + \beta\gamma_0 \ln M}\right).$$

Moreover, as $M \to \infty$,

$$\frac{E\{R_{\text{total}}\} - a_M}{b_M} \to \mathbb{E}\{Z\} = E_0.$$

Thus, when M is large, the average throughput is given by

$$\mathbb{E}\{R_{\text{total}}\} \approx a_M + E_0 b_M$$

$$= B \log_2(1 + \beta\gamma_0 \ln M) + E_0 \cdot B \log_2\left(\frac{1 + \beta\gamma_0(1 + \ln M)}{1 + \beta\gamma_0 \ln M}\right), \tag{11.16}$$

where $\ln M$ is called the multi-user diversity gain [200]. In contrast to (11.12), (11.16) provides a very simple approximation for average throughput. The numerical results in Section 11.2.5 will show that this approximation is very accurate.

Note that as $M \to \infty$, $a_M \to \infty$, and $b_M \to 0$. Therefore, with a large M,

$$\mathbb{E}\{R_{\text{total}}\} \approx B \log_2(1 + \beta\gamma_0 \ln M)$$

is a rougher but simpler estimation for the average throughput. For the Rayleigh fading, it is easy to prove (11.15). However, proving (11.15) may be difficult for other fading distributions. We will provide a simple way to do it in the next section.

11.2.3   Throughput analysis for general channel distributions

In previous sections, we have seen that finding the limiting distribution of the maximum throughput is crucial to obtain the asymptotic throughput. In this section, we consider more general cases beyond Rayleigh fading. Mathematically, we study the limiting distribution of the throughput

$$R = T(\Gamma) = B \log_2(1 + \beta\Gamma),$$

given a SNR distribution, $F_\Gamma(\gamma)$. The major result is stated in the following theorem for the *limiting throughput distribution* (LTD):

THEOREM 11.4   (LTD Theorem): *Assume that all users' SNRs, $\{\Gamma_1, \Gamma_2, \ldots, \Gamma_M\}$, are i.i.d. random variables with a distribution $F_\Gamma(\gamma)$ such that $\omega(F_\Gamma) = \infty$, and $f_\Gamma(\gamma) = F'_\Gamma(\gamma)$ as well as $F''_\Gamma(\gamma)$ exist and $f_\Gamma(\gamma) \neq 0$ for all $x_1 \leq x < \infty$, where x_1 is some real number. If*

$$\lim_{\gamma \to \infty} \frac{d}{d\gamma}\left[\frac{1 - F_\Gamma(\gamma)}{f_\Gamma(\gamma)}\right] = 0, \tag{11.17}$$

then the distribution of throughput, $F_R(r) = F_\Gamma(T^{-1}(r))$, belongs to the domain of attraction of the Gumbel distribution (maxima). In addition,

$$a_M = B \log_2\left(1 + \beta F_\Gamma^{-1}\left(1 - \frac{1}{M}\right)\right), \tag{11.18}$$

$$b_M = B \log_2\left(\frac{1 + \beta F_\Gamma^{-1}\left(1 - \frac{1}{Me}\right)}{1 + \beta F_\Gamma^{-1}\left(1 - \frac{1}{M}\right)}\right). \tag{11.19}$$

The proof is shown in Appendix A.7. the LTD theorem tells us that we do not have to check $F_R(r)$ directly, which is usually very complicated to find its limiting distribution. In addition, Lemma 11.2 leads to

$$\frac{\mathbb{E}\{R_{\text{total}}^{\text{hom}}\} - a_M}{b_M} \to E_0,$$

as $M \to \infty$, where $R_{\text{total}}^{\text{hom}}$ is the total throughput for the homogeneous scenario. For a large M, the average total throughput can be evaluated by using the following expression:

$$\mathbb{E}\{R_{\text{total}}^{\text{hom}}\} \approx a_M + E_0 b_M. \tag{11.20}$$

Example

The Nakagami distribution is frequently used to characterize the fading statistics of wireless channels in certain environments. Then, the CDF of the received SNR is given by

$$F_\Gamma(\gamma) = \Gamma_{(m, \frac{m}{\gamma_0})}(\gamma) = \int_0^\gamma \left(\frac{m}{\gamma_0}\right)^m \frac{t^{m-1}}{\Gamma(m)} e^{-\frac{m}{\gamma_0}t} dt, \tag{11.21}$$

where m is called the fading figure, which is defined as the ratio of the total power to the power of the fading components, and $\Gamma(m)$ is the gamma function. In this subsection, we use the results of the LTD theorem to study the impact of Nakagami fading on throughput in the system with the MSC scheduling. Applying the results of extreme value theory in Section 11.1 and letting $u = \frac{m}{\gamma_0}$, we have

$$\lim_{\gamma \to \infty} \frac{d}{d\gamma} \left[\frac{1 - F_\Gamma(\gamma)}{f_\Gamma(\gamma)} \right]$$

$$= \lim_{\gamma \to \infty} -\frac{[1 - F_\Gamma(\gamma)]}{f_\Gamma^2(\gamma)/f_\Gamma'(\gamma)} - 1$$

$$= \lim_{\gamma \to \infty} \frac{1 - \int_0^\gamma t^{m-1} e^{-ut} dt}{\dfrac{\gamma^m e^{-u\gamma}}{u\gamma - m + 1}} - 1$$

$$= 0 \quad \text{(by L'Hospital's rule)}.$$

According to the results of extreme value theory in Section 11.1 and the LTD theorem, both $F_\Gamma(\gamma)$ and $F_R(r)$ belong to the domain of attraction of the Gumbel distribution. Therefore, the average total throughput for Nakagami fading can be given by

$$\mathbb{E}\{R_{\text{total}}^{\text{hom}}\} \approx B \log_2 \left(1 + \beta F_\Gamma^{-1}\left(1 - \frac{1}{M}\right)\right) + E_0 B \log_2 \left(\frac{1 + \beta F_\Gamma^{-1}\left(1 - \frac{1}{Me}\right)}{1 + \beta F_\Gamma^{-1}\left(1 - \frac{1}{M}\right)}\right)$$

$$= B \log_2 \left(1 + \beta \Gamma_{(m, \frac{m}{\gamma_0})}^{-1}\left(1 - \frac{1}{M}\right)\right) + E_0 B \log_2 \left(\frac{1 + \beta \Gamma_{(m, \frac{m}{\gamma_0})}^{-1}\left(1 - \frac{1}{Me}\right)}{1 + \beta \Gamma_{(m, \frac{m}{\gamma_0})}^{-1}\left(1 - \frac{1}{M}\right)}\right), \tag{11.22}$$

where $\Gamma_{(m, \frac{m}{\gamma_0})}^{-1}(\gamma)$ is the inverse incomplete gamma function. Despite no closed form for it, the inverse incomplete gamma function is usually provided in common software, such as Matlab and Mathematica.

Actually, besides the Rayleigh and Nakagami distributions, the normal, Rician, and log-normal distributions, which are often used to describe the statistics of wireless channels, belong to the domain of attraction of the Gumbel distribution [82].

Further properties of asymptotic throughput

Note that as $M \to \infty$, $a_M \to \infty$ since $F_\Gamma^{-1}(\gamma) \to \infty$ as $\gamma \to \infty$. In addition, in Appendix A.8, we prove that

$$\lim_{M \to \infty} \frac{b_M}{a_M} = 0. \tag{11.23}$$

Applying (11.23) to $F_\Gamma(\gamma)$ (a_M and b_M here are related to Γ_{eff} in (11.7)), we have

$$\lim_{M \to \infty} \frac{F_\Gamma^{-1}\left(1 - \frac{1}{Me}\right) - F_\Gamma^{-1}\left(1 - \frac{1}{M}\right)}{F_\Gamma^{-1}\left(1 - \frac{1}{M}\right)} = 0. \tag{11.24}$$

From (11.24), we have the limit of b_M that corresponds to the throughput as follows:

$$\lim_{M \to \infty} b_M = \lim_{M \to \infty} B \log_2 \left(\frac{1 + \beta F_\Gamma^{-1}\left(1 - \frac{1}{Me}\right)}{1 + \beta F_\Gamma^{-1}\left(1 - \frac{1}{M}\right)} \right)$$

$$= \lim_{M \to \infty} B \log_2 \left(\frac{F_\Gamma^{-1}\left(1 - \frac{1}{Me}\right)}{F_\Gamma^{-1}\left(1 - \frac{1}{M}\right)} \right)$$

$$= \lim_{M \to \infty} B \log_2 \left(\frac{F_\Gamma^{-1}\left(1 - \frac{1}{Me}\right) - F_\Gamma^{-1}\left(1 - \frac{1}{M}\right)}{F_\Gamma^{-1}\left(1 - \frac{1}{M}\right)} + 1 \right)$$

$$= 0. \tag{11.25}$$

Therefore, when the number of users M is very large, we have

$$\mathbb{E}\{R_{\text{total}}^{\text{hom}}\} \approx a_M = B \log_2 \left(1 + \beta F_\Gamma^{-1}\left(1 - \frac{1}{M}\right)\right), \tag{11.26}$$

which is a rough estimation for the average total throughput with a large M. According to (11.24), we have

$$F_\Gamma^{-1}\left(1 - \frac{1}{M}\right) = \mathbb{E}\{\Gamma_{\text{eff}}\} + o(\mathbb{E}\{\Gamma_{\text{eff}}\}).$$

Thus, (11.26) can also be rewritten as

$$\mathbb{E}\{R_{\text{total}}^{\text{hom}}\} \approx B \log_2 \left(1 + \beta \left[\mathbb{E}\{\Gamma_{\text{eff}}\} + o(\mathbb{E}\{\Gamma_{\text{eff}}\})\right]\right),$$

$$\approx B \log_2 \left(1 + \beta \mathbb{E}\{\Gamma_{\text{eff}}\}\right). \tag{11.27}$$

The above equation means that the average throughput is approximately a function of the average effective SNR.

Lemma 11.2 also shows that for any positive real number p,

$$\lim_{M \to \infty} \mathbb{E} \left(\frac{R_{\text{total}}^{\text{hom}} - a_M}{b_M} \right)^p = \mathbb{E}\{Z^p\}, \tag{11.28}$$

where Z is a normalized Gumbel random variable. We consider $p = 2$, and we have

$$\lim_{M \to \infty} \mathbb{E} \left(\frac{R_{\text{total}}^{\text{hom}} - a_M}{b_M} \right)^2 = E_0^2 + \frac{\pi^2}{6}.$$

Thus, as $M \to \infty$,

$$\mathbb{V}ar\{R^{\text{hom}}_{\text{total}}\} \to \frac{\pi^2}{6} b_M^2.$$

Because of (11.25), $\mathbb{V}ar\{R^{\text{hom}}_{\text{total}}\} \to 0$, which indicates that this asymptotic analysis of average throughput is quite accurate. In addition, it follows that

$$\lim_{M \to \infty} \mathbb{E}\left(R^{\text{hom}}_{\text{total}} - a_M\right)^p = \lim_{M \to \infty} b_M \mathbb{E}\{Z^p\}, \tag{11.29}$$

$$= 0. \tag{11.30}$$

According to [154], (11.29) guarantees that $R^{\text{hom}}_{\text{total}} - a_M$ converges in probability[1] to 0.

Channel access probability and average throughput per user

The channel access probability P_i is the probability that user i obtains the channel to transmit data. In the homogeneous fading case, due to the symmetry, each user has the same channel access probability; i.e.,

$$P_i = \frac{1}{M}.$$

Therefore, the average throughput of user i with the scheduling, $\mathbb{E}\{R_i^s\}$, is given by

$$\mathbb{E}\{R_i^s\} = \frac{1}{M}\mathbb{E}\{R^{\text{hom}}_{\text{total}}\}.$$

11.2.4   Throughput analysis for normalized-SNR-based scheduling

In previous sections, we presented the asymptotic throughput analysis for the homogeneous fading case. In reality, the values of the average SNR of users vary according to their path losses. Denote the average SNR of user i as γ_i. We consider a scenario in which different users have the same normalized SNR distribution $F(\gamma)$ but with different average SNR, γ_i. We assume that $F(\gamma)$ satisfies $\omega(F) = \infty$ and (11.17).

Obviously, the MSC scheduling results in unfair channel access probabilities. When normalized-SNR-based scheduling is used, the base station schedules the user with the largest normalized SNR to obtain the channel, which is mathematically expressed in (11.5). Define the effective normalized SNR at the transmitter as

$$\Gamma_{\text{eff}} = \max_{i \in \mathcal{M}} \frac{\Gamma_i}{\gamma_i}.$$

Because of the identical distribution of the normalized SNR, the previous results based on extreme value theory is still applicable to the effective normalized SNR, and all users have the same channel access probability as well; i.e.,

$$P_i = \frac{1}{M}.$$

---

[1] Assume X_n and X to be a random variable sequence and a random variable, if $\lim_{n \to \infty} \mathbb{P}\{|X_n - X| > \epsilon\} = 0$ for any $\epsilon > 0$, then we say that X_n converges in probability to X.

Thus, the average throughput of user i can be expressed as

$$\mathbb{E}\{R_i^s\} = \frac{1}{M} \int_0^\infty B \log_2(1 + \beta \gamma_i \gamma) f_{\Gamma_{\text{eff}}}(\gamma) d\gamma. \tag{11.31}$$

Recalling (11.10) and the LTD theorem, we know that in the i.i.d. fading case if the distribution of SNR $F_\Gamma(\gamma)$ satisfies (11.17), then, with a large M,

$$\int_0^\infty \log_2(1 + \beta \gamma) f_{\Gamma_{\text{eff}}}(\gamma) d\gamma \approx \log_2\left(1 + \beta F_\Gamma^{-1}\left(1 - \frac{1}{M}\right)\right)$$

$$+ E_0 \log_2\left(\frac{1 + \beta F_\Gamma^{-1}\left(1 - \frac{1}{Me}\right)}{1 + \beta F_\Gamma^{-1}\left(1 - \frac{1}{M}\right)}\right). \tag{11.32}$$

Comparing (11.31) and (11.32), we obtain the average throughput for user i as follows:

$$\mathbb{E}\{R_i^s\} \approx \frac{B}{M} \left\{ \log_2\left(1 + \beta \gamma_i F^{-1}\left(1 - \frac{1}{M}\right)\right) + E_0 \log_2\left(\frac{1 + \beta \gamma_i F^{-1}\left(1 - \frac{1}{Me}\right)}{1 + \beta \gamma_i F^{-1}\left(1 - \frac{1}{M}\right)}\right) \right\},$$

with large M. Therefore, with normalized SNR-based scheduling, each user obtains the same multi-user diversity gain as that in the homogeneous scenario and has the same channel access probability, but its own average throughput depends on its average SNR.

Furthermore, we will compare the total throughput in the heterogeneous and homogeneous scenarios. We assume that

$$\gamma_0 = \frac{1}{M} \sum_{i=1}^M \gamma_i, \tag{11.33}$$

and define

$$\sigma_\gamma^2 = \frac{1}{M} \sum_{i=1}^M (\gamma_i - \gamma_0)^2.$$

When the number of users M is large, we only consider the first term, a_M, to evaluate the average throughput. Thus, the average total throughput in the heterogeneous scenario is given by

$$\mathbb{E}\{R_{\text{total}}^{\text{het}}\} = \sum_{i=1}^M E\{R_i^s\}$$

$$\approx \frac{B}{M} \sum_{i=1}^M \log_2\left(1 + \beta \gamma_i F^{-1}\left(1 - \frac{1}{M}\right)\right),$$

and the average total throughput in the homogeneous scenario is

$$\mathbb{E}\{R_{\text{total}}^{\text{hom}}\} \approx B \log_2\left(1 + \beta \gamma_0 F^{-1}\left(1 - \frac{1}{M}\right)\right),$$

We obtain

$$\mathbb{E}\{R_{\text{total}}^{\text{het}}\} - \mathbb{E}\{R_{\text{total}}^{\text{hom}}\} \approx \frac{B}{M} \sum_{i=1}^{M} \log_2 \left(\frac{1 + \beta \gamma_i F^{-1}\left(1 - \frac{1}{M}\right)}{1 + \beta \gamma_0 F^{-1}\left(1 - \frac{1}{M}\right)} \right)$$

$$\rightarrow \frac{B}{M} \sum_{i=1}^{M} \log_2 \left(\frac{\gamma_i}{\gamma_0} \right), \quad \text{as } M \rightarrow \infty. \tag{11.34}$$

(11.34) is valid since $F^{-1}(1 - \frac{1}{M}) \rightarrow \infty$ as $M \rightarrow \infty$.

With the following inequality:

$$x - \frac{1}{2}x^2 \leq \ln(1 + x) \leq x, \quad \text{for } x \geq 0, \tag{11.35}$$

we will consider the upper and lower bounds, respectively. For the upper bound, it follows from (11.34) and (11.35) that

$$\mathbb{E}\{R_{\text{total}}^{\text{het}}\} - \mathbb{E}\{R_{\text{total}}^{\text{hom}}\} \leq \frac{B}{\ln(2)M} \sum_{i=1}^{M} \left(\frac{\gamma_i}{\gamma_0} - 1 \right).$$

$$= 0.$$

Similarly, the lower bound is given by

$$\mathbb{E}\{R_{\text{total}}^{\text{het}}\} - \mathbb{E}\{R_{\text{total}}^{\text{hom}}\} > \frac{B}{\ln(2)M} \sum_{i=1}^{M} \left(\frac{\gamma_i}{\gamma_0} - 1 \right) - \frac{1}{2 \ln 2} \frac{B}{M} \sum_{i=1}^{M} \left(\frac{\gamma_i}{\gamma_0} - 1 \right)^2$$

$$= 0 - \frac{B}{2 \ln 2} \left[\frac{1}{M} \sum_{i=1}^{M} \left(\frac{\gamma_i}{\gamma_0} \right)^2 - 1 \right]$$

$$= -\frac{B}{2 \ln 2} \frac{\sigma_\gamma^2}{\gamma_0^2}.$$

Therefore, the main result is stated as follows: when the number of users M is large,

$$-\frac{B}{2 \ln 2} \frac{\sigma_\gamma^2}{\gamma_0^2} \leq \mathbb{E}\{R_{\text{total}}^{\text{het}}\} - \mathbb{E}\{R_{\text{total}}^{\text{hom}}\} \leq 0. \tag{11.36}$$

This means that the homogeneous case leads to the maximum total throughput when (11.33) holds.

11.2.5   Numerical results

We assume that all users experience i.i.d. Nakagami fading. Let $\beta \gamma_0 = 1$. Figure 11.1 shows the average total throughput in the Nakagami fading channels with different values of m. For comparison, we also plot the average throughput in the *additive white Gaussian noise* (AWGN) channel with the same average SNR in Figure 11.1.

It is shown in Figure 11.1 that the asymptotic results are still accurate even if the number of users is small. The figure shows that the throughput increases with the number of users in the fading scenario with dynamic scheduling. As m increases, the fading fluctuation of the channel reduces, and the multi-user diversity gain is also diminished.

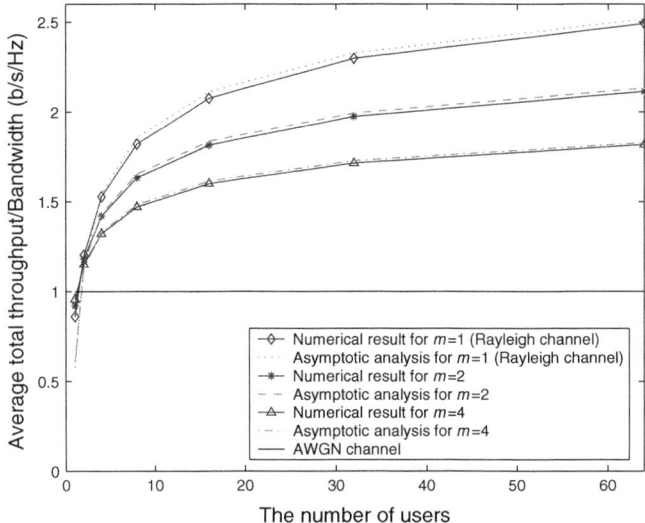

Figure 11.1 Average throughput for different environments. $\beta\gamma_0 = 1$

11.3   Asymptotic delay analysis of single-carrier networks

Besides throughput, delay is another crucial factor for wireless data services, particularly for time-sensitive applications. In [128], an analysis for mean delays is presented; however, only a rough relationship between the average waiting time and the number of users in the system is provided. In [45], the system performance in the mean sense is studied by using the multi-class processor-sharing model. Since the dynamic user configuration is considered in [45], the results are applicable for more general cases, but the complicated model has difficulty in capturing the explicit relationship between system performance and multi-user diversity. In this section, we use an asymptotically analytical result to reveal the impact of dynamic scheduling on average waiting times in single-carrier networks. Compared to [45], our analysis has two major differences. First, we just consider the static user scenario to reveal the impact of multi-user diversity on the mean delay. Although our analysis overestimates the mean delay with a light traffic load, they would be accurate with a heavy traffic load. Second, the system performance is evaluated in terms of the average delay for each packet rather than the average delay for each file transmission in [45].

The system and scheduling models are the same as those in Section 11.2.1. In our analysis, we make the following assumptions:

- When a queue is empty, a dummy packet is assumed to be in the queue. This system is usually called the *dominant system*.
- Dynamic packet scheduling is usually allowed in time-slotted networks, which makes delay analysis extremely difficult. For simplicity, we assume that the

system is not time-slotted. After finishing transmitting a packet, the base station can immediately serve another packet.

- We assume that all users have the same channel statistics and arrival traffic statistics.
- For the arrival traffic, all packets are assumed to have the same length L. In addition, the packet inter-arrival time for each user is assumed to be independently, identically, and exponentially distributed with rate λ_1. The total arrival rate is $\lambda = M\lambda_1$.
- For channel fading, we assume that each user experiences i.i.d. block-fading that is constant while a packet is being served, but is independent across different packet transmission durations.

11.3.1   Asymptotic distribution of service time

The service time S for transmitting a packet can be expressed as

$$
\begin{aligned}
S &= S(\Gamma) \\
&= \frac{L}{R} \\
&= \frac{L}{B} \frac{1}{\log_2(1 + \beta\Gamma)}.
\end{aligned}
$$

Thus, the distribution of the service time is

$$
F_S(s) = 1 - F_\Gamma(S^{-1}(s)),
$$

and its inverse function is

$$
\begin{aligned}
F_S^{-1}(x) &= S(F_\Gamma^{-1}(1 - x)) \\
&= \frac{L}{B} \frac{1}{\log_2(1 + \beta F_\Gamma^{-1}(1 - x))}.
\end{aligned}
$$

According to the MSC rule in the multi-user-symmetric environment, the base station should serve the user with the strongest channel condition. This is equivalent to serving the user who needs the shortest service time. Since the system performance in the scenario of a large number of users can be obtained by extreme value theory [50], we focus on the properties of the limit distribution of the random variable

$$
S_{min,M} = \min_{i \in \mathcal{M}} S_i,
$$

where S_i is the service time for user i.

Similar to the LTD theorem, we state the result about asymptotic distribution of service time as the following theorem:

THEOREM 11.5   *Assume that all users' SNRs, $\{\Gamma_1, \Gamma_2, \ldots, \Gamma_M\}$, are i.i.d. random variables with a distribution $F_\Gamma(\gamma)$ such that $\omega(F_\Gamma) = \infty$, and $f_\Gamma(\gamma) = F'_\Gamma(\gamma)$*

as well as $F_\Gamma''(\gamma)$ exist and $f_\Gamma(\gamma) \neq 0$ for all $x_1 \leq x < \infty$, where x_1 is some real number. If

$$\lim_{\gamma \to \infty} \frac{d}{d\gamma} \left[\frac{1 - F_\Gamma(\gamma)}{f_\Gamma(\gamma)} \right] = 0, \tag{11.37}$$

then the distribution of service, $F_S(s) = 1 - F_\Gamma(S^{-1}(s))$, belongs to the domain of the attraction of the Gumbel distribution (minima). In addition,

$$c_M = \frac{L}{B} \frac{1}{\log_2\left(1 + \beta F_\Gamma^{-1}\left(1 - \frac{1}{M}\right)\right)}, \tag{11.38}$$

$$d_M = \frac{L}{B} \frac{1}{\log_2\left(1 + \beta F_\Gamma^{-1}\left(1 - \frac{1}{M}\right)\right)} - \frac{L}{B} \frac{1}{\log_2\left(1 + \beta F_\Gamma^{-1}\left(1 - \frac{1}{Me}\right)\right)}. \tag{11.39}$$

$$\lim_{M \to \infty} \frac{d_M}{c_M} = 0. \tag{11.40}$$

The proof is omitted since it is very similar to that of the LTD theorem. The fact that $c_M \to 0$ as $M \to \infty$, together with (11.40), implies that $S_{min,M}$ is approximately a constant when M is large enough; i.e.,

$$S_{min,M} \approx c_M - E_0 d_M.$$

In the case of Rayleigh fading, we have

$$c_M = \frac{L}{B} \frac{1}{\log_2(1 + \beta\gamma_0 \ln(M))},$$

$$d_M = \frac{L}{B} \frac{1}{\log_2(1 + \beta\gamma_0 \ln(M))} - \frac{L}{B} \frac{1}{\log_2\left(1 + \beta\gamma_0 \left(1 + \ln(M)\right)\right)}.$$

Therefore, we obtain the service rate for one packet.

11.3.2 Average waiting time

Because the stochastic characteristics of packet arrivals and wireless channels of different users are symmetrical, we only need to consider the delay performance of a specific user. Under the MSC rule, each queue (user) is equally served with probability $\frac{1}{M}$. As a result, the time needed to transmit one packet, X, which is an integer multiple of $S_{min,M}$, has a geometric distribution,

$$\mathbb{P}(X = nS_{min,M}) = \left(\frac{1}{M}\right)\left(1 - \frac{1}{M}\right)^{n-1},$$

where n is an integer.

Due to the Poisson arrivals, each queue can be modeled as an M/G/1 queue with server vacations [44]. In equilibrium, the mean waiting time in a queue, W_q, can be decomposed into the expected residual service time T_{res} plus the average service time of packets in the queue $\mathbb{E}\{X\}N_q$, where N_q is the average queue length. Applying Little's law, $N_q = \lambda_1 W_q$, we have

$$W_q = T_{res} + \mathbb{E}\{X\}\lambda_1 W_q$$

$$= T_{res} + \rho W_q,$$

where $\rho = \lambda S_{min,M} = M\lambda_1 S_{min,M}$. Therefore, the mean waiting time in the queue is given by

$$W_q = \frac{T_{res}}{1 - \rho}. \qquad (11.41)$$

Using the results of M/G/1 queues with vacations in [44], the expected residual service time T_{res} can be obtained as

$$T_{res} = \frac{\lambda_1 \mathbb{E}\{X^2\}}{2} + (1 - \rho)\frac{\mathbb{E}\{V^2\}}{2\mathbb{E}\{V\}},$$

where the length of a server vacation, V, is equal to $S_{min,M}$. Thus, it follows that

$$T_{res} = \frac{\lambda_1(2M - 1)MS_{min,M}^2}{2} + (1 - \rho)\frac{S_{min,M}}{2}$$

$$= \frac{(2M - 1)\rho S_{min,M}}{2} + (1 - \rho)\frac{S_{min,M}}{2}.$$

The mean waiting time in the single-carrier system, W_{single}, includes the mean waiting time in the queue and the service time of transmitting one packet. Thus,

$$W_{\text{single}} = \frac{T_{res}}{1 - \rho} + \mathbb{E}\{X\}$$

$$= \frac{(2M - 1)\rho S_{min,M}}{2(1 - \rho)} + \left(M + \frac{1}{2}\right)S_{min,M}. \qquad (11.42)$$

11.4   Asymptotic performance analysis of multi-carrier networks

In a multi-carrier network with the same bandwidth B, the scheme assigns each sub-channel to the user with the best channel condition, which can be expressed as

$$m(k) = \arg\max_{i \in \mathcal{M}}\{\Gamma_i[k]\},$$

where $m(k)$ represents the user scheduled at subcarrier k, and $\Gamma_i[k]$ is the SNR of user i at subcarrier k. Let $F_\Gamma(\gamma)$ be the distribution of channel fading at each subcarrier. Other assumptions here are the same as in the single-carrier network in Sections 11.2 and 11.3.

11.4.1   Asymptotic throughput analysis

Note that there is no assumption on the correlation among subcarriers. The data rate at subcarrier k is given by

$$R_{\max}[k] = \max_{i \in \mathcal{M}} \frac{B}{K}\log_2(1 + \beta\Gamma_i[k]). \qquad (11.43)$$

Then, the total throughput is given by

$$R_{\text{total}} = \sum_{k=1}^{K} R_{\max}[k].$$

Thus,

$$\begin{aligned} \mathbb{E}\{R_{\text{total}}\} &= K\mathbb{E}\{R_{\max}[k]\} \\ &= \mathbb{E}\{\max_{i \in \mathcal{M}} B\log_2(1 + \beta\Gamma)\}, \end{aligned}$$

where Γ is distributed with $F_\Gamma(\gamma)$. It follows from Theorem 11.4 that with large M,

$$\mathbb{E}\{R_{\text{total}}\} \approx a_M + E_0 b_M, \tag{11.44}$$

where a_M and b_M are determined by (11.18) and (11.19). Therefore, the multi-carrier network has the same asymptotic throughput as the single-carrier network with the same bandwidth.

11.4.2 Asymptotic delay analysis

Similarly in the single-carrier system, each user has a probability $\frac{1}{M}$ of occupying a subcarrier. We consider an "extreme" scenario where the channel fluctuations are independent across the subcarriers, and the number of subcarriers $K \to \infty$. At subcarrier k, the resulting data rate is a random variable given by (11.43). According to the strong law of large numbers, as $K \to \infty$ and $M \ll K$, the total throughput is obtained as

$$\begin{aligned} R_{\text{total}} &= \frac{B}{K}\sum_{k=1}^{K} \max_{i \in \mathcal{M}} \log_2(1 + \beta\Gamma_i[k]) \\ &\to B\mathbb{E}\{\max_{i \in \mathcal{M}} \log_2(1 + \beta\Gamma)\}, \text{ as } K \to \infty, \end{aligned}$$

where Γ is distributed with $F_\Gamma(\gamma)$. Therefore,

$$R_{\text{total}} \approx a_M + E_0 b_M,$$

and the service time is

$$S'_{min,M} = \frac{L}{R_{\text{total}}}.$$

Since each user occupies a bandwidth of B/M, the MSC scheduling results in a traditional FDM system with the fixed service rate R/M for each user. In other words, the service time for one packet is $MS'_{min,M}$. Based on the results in [44], the mean waiting time in the queue can be expressed as

$$\begin{aligned} W_q &= \frac{\lambda_1 M^2 S'^2_{min,M}}{2(1-\rho)} + \frac{S'_{min,M}}{2} \\ &= \frac{M\rho S'_{min,M}}{2(1-\rho)} + \frac{S'_{min,M}}{2}. \end{aligned}$$

Therefore, the average waiting time in the multi-carrier network, W_{multi}, is given by

$$W_{\text{multi}} = \frac{M\rho S'_{min,M}}{2(1-\rho)} + \left(M + \frac{1}{2}\right) S'_{min,M}. \tag{11.45}$$

The structure of the average waiting time expression for multi-carrier networks in (11.45) is similar to that for single-carrier networks in (11.42). We will compare them in the next subsection.

11.4.3   Delay performance comparison

As the number of users $M \to \infty$,

$$S_{min,M} \to c_M$$
$$= \frac{L}{B} \frac{1}{\log_2\left(1 + \beta F_\Gamma^{-1}\left(1 - \frac{1}{M}\right)\right)},$$
$$S'_{min,M} \to \frac{1}{a_M}$$
$$= \frac{L}{B} \frac{1}{\log_2\left(1 + \beta F_\Gamma^{-1}\left(1 - \frac{1}{M}\right)\right)}.$$

In other words, both single- and multiple-carrier networks have the same asymptotic throughput

$$B\log_2\left(1 + \beta F_\Gamma^{-1}\left(1 - \frac{1}{M}\right)\right).$$

However, each system has a different delay performance for bursty traffic. When the traffic load is light,

$$\lim_{\rho \to 0}\left(\lim_{\substack{M \to \infty \\ \text{fixing } \rho}} \frac{W_{\text{single}}}{W_{\text{multi}}}\right) = 1.$$

When the traffic load is heavy,

$$\lim_{\rho \to 1}\left(\lim_{\substack{M \to \infty \\ \text{fixing } \rho}} \frac{W_{\text{single}}}{W_{\text{multi}}}\right) = 2. \tag{11.46}$$

This is because multi-carrier networks can provide smoother service rates by exploiting frequency diversity.

In the above analysis, we do not take the time correlation in channel fading into account. This issue will be discussed in a descriptive manner as follows. We consider the extreme case where the channel correlation time goes to infinity. In a single-carrier system using the MSC rule, the average waiting time will become infinite with a heavy traffic load. However, note that in the delay performance analysis of multicarrier networks, we exploit channel independence among subcarriers instead of channel independence across different packet transmission durations. Thus, for a multi-carrier network in a highly frequency-selective environment, the channel time correlation does not affect the average waiting time, and the average waiting time is always equal to (11.45). Therefore, when channel fading is slow and highly frequency-selective, the multi-carrier network *greatly* outperforms the single-carrier network in terms of delay performance. More accurately, if the number of subcarriers in the multi-carrier network is large, the average waiting time in the multi-carrier network is half that in the single-carrier network when the traffic load is heavy, or is considerably less than half with slow-fading.

Figure 11.2 shows that the average waiting time of different systems in the Rayleigh fading environment with the same bandwidth when there are 100 users. Since the MSC scheduling can improve the throughput through multi-user diversity, both single-carrier and multi-carrier networks with this scheduling provide a substantial delay performance improvement, compared to the traditional time division multiple access (TDMA). In the simulation, we consider a simple time correlation model of fading as follows. The channel is block-faded; the channel remains constant within a block, but is independent across different blocks. The length of a block is the coherence time of the channel, which is an integer multiple of $S_{min,M}$. A longer coherence time indicates a slow-fading rate. When the coherence time equals 1, the fading model is the same as that assumed in Section 11.2. It is concluded from Figure 11.2 that the analytical result (11.42) fits the simulation curve very well, and that slow-fading seriously impairs the delay performance of single-carrier networks.

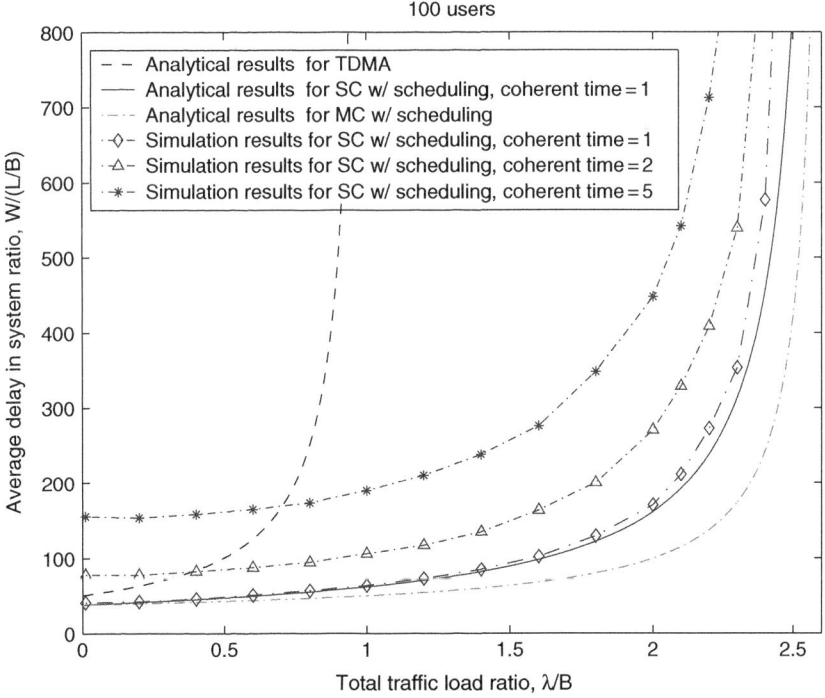

Figure 11.2 Average waiting time versus traffic load. $\beta \gamma_0 = 1$, and $M = 100$

11.5   Summary

Using extreme value theory, we have introduced asymptotic average throughput and delay analyses for MSC scheduling with a general fading distribution in both single-carrier and multi-carrier networks, which not only have concise expressions, but also provide accurate results. This asymptotic analysis shows that the use of the simple scheduling techniques and the feedback of CSI can significantly improve bandwidth efficiency. We have also extended the analysis into a scenario in which different users experience different path losses. The results shows that normalized-SNR-based scheduling can obtain the same multi-user diversity gain as that in the homogeneous case while maintaining access-time proportional fairness. Although multi-carrier networks with channel-aware scheduling have the same throughput as single-carrier networks, multi-carrier networks can provide better delay performance than single-carrier networks. This work is beneficial to QoS provisioning for channel-aware scheduling.

Part III

Distributed cross-layer optimization

12 Overview

Wireless is a shared medium and communication performance is affected not only by individual communication links but also by the interaction among the links that reuse the same frequency in the entire network. Interference is one of the major factors limiting system spectral efficiency, especially as wireless networks move toward more aggressive frequency-reuse scenarios in future wireless networks. For users that interfere with each other heavily, advanced medium access control (MAC) schemes can be used to allocate orthogonal network resources. The design of distributed medium access is essential in avoiding signaling overhead, ensuring network scalability, and determining overall network performance. Furthermore, the quality of a wireless channel varies both in time and for the user. To fully exploit network diversity, channel-aware medium access schemes should be used to adapt data transmission and resource assignment based on the states of wireless channels. The difference of channel-aware medium access schemes from traditional MAC protocols is that channel-aware MACs schedule users with favorable channel conditions to transmit with optimized link adaptation based on channel state information (CSI). By exploiting the channel variations across users, channel-aware MAC substantially improves network performance through exploiting multi-user diversity, whose gain increases with the number of users.

12.1   Design objective

As discussed in Section 5.1, the goal of distributed MAC protocols is to remove as many idle and collision states as possible. The ideal MAC protocol is able to remove all idle and collision states. Therefore, we define the complete resolution of network contention as follows:

DEFINITION 12.1   *The contention in a network is completely resolved:*

(i)   *if all links that have won the contention can transmit without collision;*
(ii)   *if any additional link that has not won the contention transmits, it will collide with at least one link that has won the contention.*

In the above definition, the first item means the removal of all collision states and the second of all idle states. Thus, complete resolution results in states in which network capacity is fully exploited.

With awareness of channel state information, an additional goal is to exploit multi-user diversity in a distributed way and enable users with better channel states to win the contention with higher probabilities, thus to maximize network performance. This will be illustrated further in the following section.

12.2   Distributed multi-user diversity

Multi-user diversity refers to the difference in certain characteristics of different users in wireless networks. The characteristics of interest can be channel fading, co-channel interference, traffic requirement, queueing status, adjacent interferers, etc. The exploitation of these differences can lead to significant network performance improvement. As shown in Part II, multi-user diversity can be exploited using opportunistic scheduling in centralized networks. With opportunistic scheduling, the user with the best metric of interest among all candidates can be scheduled. The more users present in the network, the more likely it is that there exists a user with a very good channel at any given time. Hence, the total throughput tends to increase with the number of users by exploiting multi-user diversity.

A simple example is max-signal-to-noise ratio (SNR) scheduling in a cellular network. With max-SNR scheduling, the base station always schedules the user with the highest SNR, resulting in the highest network throughput. The disadvantage of max-SNR scheduling is the lack of fairness among all users in the network. This is because users closer to the base station will have much higher opportunities of being scheduled while those at the cell edge may never be scheduled or served. Therefore, more advanced scheduling algorithms are needed to balance network performance and fairness. In Part II, we introduced state-of-the-art centralized techniques to exploit the multi-user diversity in wireless networks.

Multi-user diversity not only exists in centralized wireless networks but also in distributed ones. A simple example is illustrated in Figure 12.1, where users A_1 and B_1 are

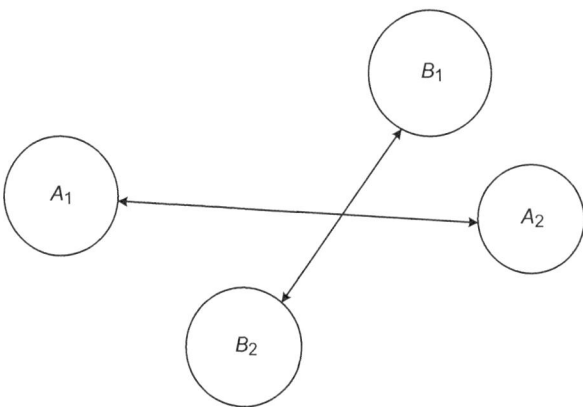

Figure 12.1 An example of distributed multi-user diversity

sending data to A_2 and B_2 respectively at the same time. Their channels are independent from each other. Assume both channels have 50% of time in GOOD states and another 50% in BAD ones. With GOOD channel states, 200 Mbit/s is achieved and with BAD ones, only 50 Mbit/s is achieved. First let's take a look at the performance of centralized scheduling. Assume there is a central controller that schedules the transmission of each user. If the round robin scheduler is used, the average network throughput is

$$T = 0.5 * (0.5 * 50 + 0.5 * 200) + 0.5 * (0.5 * 50 + 0.5 * 200) = 125 \text{Mbit/s}. \quad (12.1)$$

To exploit multi-user diversity, the central controller schedules the user with the GOOD state. If both links have GOOD states or both have BAD ones, round robin is used and they are scheduled randomly with an equal probability of $\frac{1}{2}$. Then the average network throughput is

$$T = 0.5 * 0.5 * 200 + 0.5 * 0.5 * 200 + 0.5 * 0.5 * (0.5 * 200 + 0.5 * 200)$$
$$+ 0.5 * 0.5 * (0.5 * 50 + 0.5 * 50) = 162.5 \text{Mbit/s}. \quad (12.2)$$

In this example, exploiting the multi-user diversity in centralized resource management results in an immediate 30% gain in throughput.

Now let us take a look at distributed resource management. Without considering multi-user diversity, we can use slotted Aloha as the random access method and each user sends a packet with probability $\frac{1}{2}$ in each time slot. Packet transmission succeeds when only one user is sending. In addition, when a user decides to send a packet, the channel has the same probability of being GOOD or BAD. The average network throughput is

$$T = 0.5 * 0.5 * (0.5 * 200 + 0.5 * 50) + 0.5 * 0.5 * (0.5 * 200 + 0.5 * 50)$$
$$= 62.5 \text{Mbit/s}. \quad (12.3)$$

To exploit multi-user diversity, we observe that the probability of each user sending a packet is $\frac{1}{2}$. The channel of each user has $\frac{1}{2}$ probability of being GOOD. If we can associate the random access with the channel state in a way such that each user sends packets only when the channel is GOOD, the contention probability will still be $\frac{1}{2}$. Since the channels of the two users are independent from each other, the packet transmissions are also independent. The average network throughput will be

$$T = 0.5 * 0.5 * 200 + 0.5 * 0.5 * 200$$
$$= 100 \text{Mbit/s}. \quad (12.4)$$

In this example, exploiting the multi-user diversity, in distributed resource management, results in an immediate 60% gain in throughput.

12.3   Approaches

We have just seen the power of exploiting multi-user diversity in distributed random access protocols. The following lists questions we need to further investigate and answer in order to fully take advantage of multi-user diversity in distributed wireless networks:

- In the above example, to exploit multi-user diversity, we have assumed we can associate the packet transmission with the channel state in a way such that each user sends packets only when the channel is GOOD. How shall we make the association in practical wireless networks? The example also assumes the probability that the channel in the GOOD state is the same as the contention probability. What should we do if the two probabilities are different?

- The example assumes the channel has binary states. In practice, the channel state is continuous. How do we exploit multi-user diversity in this case?

- The example assumes only two users. How do we design the protocol when there are multiple users? What should we do if different users have different channel statistics?

- The example assumes that only one user can transmit at a time. This works for networks such as the uplink access of a single-cell cellular network or wireless local area network (WLAN). What should we do for other types of networks of more complicated topologies, e.g. wireless ad hoc, sensor networks?

- Comparing (12.2) and (12.4), the ideal central scheduler performs 62.5% better than the distributed scheduler. The performance loss of the distributed scheduler is due to the collisions inherent in random access protocols. The question is: can we design a distributed random access protocol in a way such that its performance is comparable to that of centralized schedulers?

- The example assumes fixed transmission power and only one transmission at a time. In practice, there might be simultaneous transmissions located at different places that interfere with each other. The power should be varied depending on the channel and network interfering scenarios. This further complicates the meaning of GOOD channel states. How can we exploit the power control in a distributed way such that the whole network has the best performance?

We will answer the above questions in the following chapters. In general, wireless channels are continuous and different users will experience different channel and interference environments. With the traditional open system interconnection (OSI) model, the network protocols are divided into separate layers. The implementation of one layer is independent of that of any other layer. This greatly simplifies the design and analysis. For example, if a change is made in one layer, it will not affect other layers. In Part II, we introduced many distributed MAC protocols. As we can see, none of them relies on the operation of the physical layer. In other words, the physical layer is transparent to MAC. This simplifies the MAC and physical (PHY) designs. The drawback of the separate design is the loss of information since different layers know almost nothing about each other. This information can instead be used to further improve the network, as has been illustrated in the above example. The OSI model was originally designed for wireline communications. Once a wireline network is deployed, all physical layer channels are determined and the upper layer can be configured accordingly. Then no changes are necessary as the channels will not change. So the static wireline channel indicates little or no information in the physical layer and decoupled protocol design will provide good performance. However in wireless networks, both wireless channels and interference environments may vary significantly at different time, location, or frequency. This indicates a huge amount of information in the physical layer and ignoring the information in the higher-layer design would deteriorate network performance substantially. In this part, our objective is to show how to exploit this information in a distributed way in the MAC design such that the network capacity can be fully exploited.

We will illustrate how to design distributed channel-aware medium access protocols that completely resolve the contention of networks of any type. In addition, we will demonstrate how to design distributed channel-aware scheduling protocols that perform the same as channel-aware centralized algorithms, which is the ultimate goal of all distributed designs. To achieve the goal, we start by discussing channel-aware slotted Aloha in both uplink single-cell cellular networks and then networks of any topologies, such as ad hoc networks, so that the readers will gain some preliminary understanding about how to integrate channel state information in MAC protocol design for improved network performance. Then we introduce the optimal channel-aware medium access protocol that completely resolves the contention of networks of any types, and performs the same as channel-aware centralized algorithms. These protocols can be used in various networks to improve their spectrum efficiency. We will present a detailed example, illustrating how to implement the channel-aware access protocol in cellular networks to improve the downlink performance of cell-edge users, who usually experience heavy co-channel interference from adjacent cells. The distributed MAC schemes are designed to allocate orthogonal resources to users that heavily interfere with each other and who cannot send data simultaneously. There are also users in the network that are relatively far away from each other and the channel supports concurrent communications. For these users, power control is essential in determining the overall network performance. At the end of this part, we will also introduce distributed power control schemes for both real-time and elastic types of data traffic.

13 Opportunistic random access: single-cell cellular networks

Random access algorithms provide the means to share network resources among users under distributed control. Traditional contention-based random access methods include pure, slotted, and reservation Aloha schemes, *carrier sense multiple access* (CSMA) and CSMA with collision avoidance schemes, *multiple access with collision avoidance for wireless* (MACAW) schemes, and so on [54, 199]. These existing medium access control (MAC) approaches do not consider channel state information (CSI) in the contention protocol. The drawback is illustrated in Figure 13.1. When MAC decides to transmit a frame, the channel may be in a deep fade. MAC may not transmit even though the channel is in a good state, which wastes channel resources. With opportunistic random access design, MAC frames will be transmitted when the wireless channel is good. Furthermore, when there are multiple users in the network, different users usually experience different channel states. Especially, users have peak channel conditions at different time periods. This effect can be called multi-user diversity and the aim is to exploit this multi-user diversity such that users with better or peak channel conditions win the contention for channel access. The more users in the network, the more likely there is to be a user with a good channel at any time. Therefore, the total throughput will increase with the number of users and consideration of channel states in the MAC design will significantly improve network performance. This chapter focuses on exploiting multi-user diversity in a single-cell network. For downlink communications, the base station can easily schedule the transmission of all users based on their CSI, as it is the central controller and can decide who to serve each time it sends a packet. The uplink is much more complicated and would be the focus. We will study random access for uplink communications of a single-cell scenario. We will discuss joint MAC and physical (PHY) designs, with which the MAC contention will effectively take advantage of the multi-user diversity in the PHY layer, to achieve significant performance improvement compared to traditional separate designs.

13.1 Channel-aware Aloha

With opportunistic random access, each user exploits its own CSI to decide the contention behavior and the contention rule is designed such that users with better channel states have higher probabilities of success in contention. A simple yet very effective

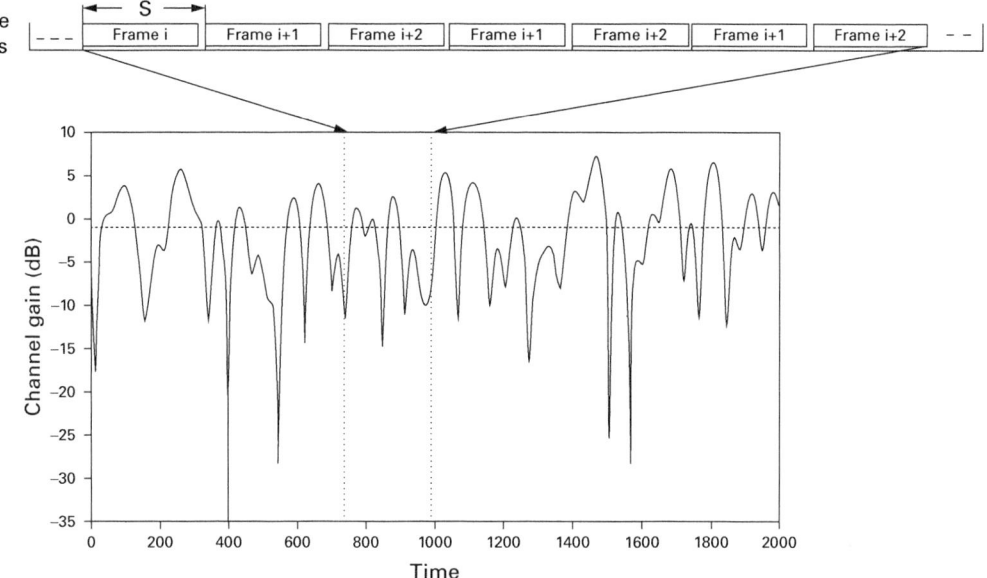

Figure 13.1 The impact of fading on MAC transmission

opportunistic random access is called channel-aware Aloha and is proposed in [209]. We will introduce this approach briefly in the following paragraphs.

The goal of channel-aware Aloha design is to exploit multi-user diversity gains in a distributed way. As an initial try, consider an uplink network where N users are sending packets to a single receiver, such as an access point in a wireless local area network (WLAN) or a base station in a cellular network (Figure 13.2). Assume block-fading channels, i.e. the channel stays constant in each time slot and may vary in different slots. At time t, the received signal, $y_n[t]$, is given by

$$y_n[t] = \sqrt{H_n[t]}x_n[t] + N_n[t], \tag{13.1}$$

where $x_n[t]$ is the transmitted signal, $H_n[t]$ is the channel power gain, and $N_n[t]$ is additive white Gaussian noise. Assume that the channel gains of each user in each timeslot are i.i.d. random variables with probability density $f(h)$. For a Rayleigh fading channel,

$$f(h) = e^{-\frac{h}{h_0}}\frac{1}{h_0}, \tag{13.2}$$

where h_0 is the average channel gain. Correspondingly, the distribution function is

$$F(h) = 1 - e^{-\frac{h}{h_0}}. \tag{13.3}$$

Assume each user knows its own channel gain and its distribution function.

The drawback with a centralized approach is that the scheduler must know the channels of all users. Therefore, all users have to estimate their channels and send the channel

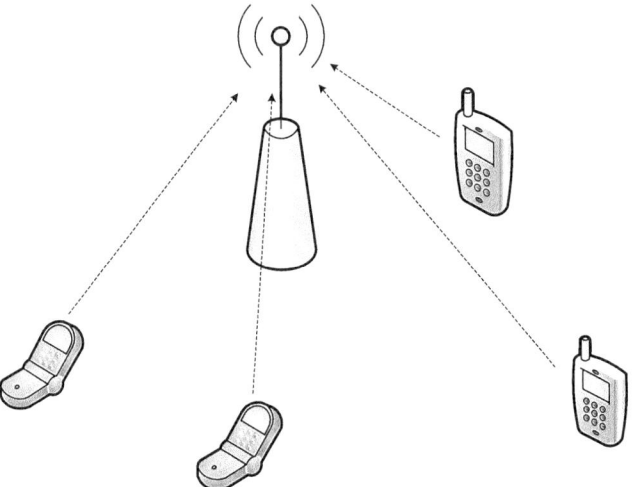

Figure 13.2 An example of uplink access

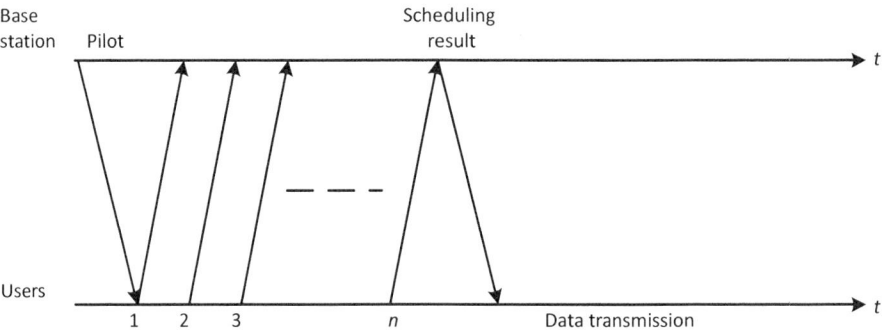

Figure 13.3 Drawback of centralized scheduling

knowledge to the scheduler. The overhead and delay in obtaining channel knowledge may limit network performance. For example, assume each user, having made their estimate, sends the channel estimate to the central scheduler. Each user uses an orthogonal signal, e.g. a time slot τ in time division multiple access (TDMA), to send the estimate to the central scheduler, as shown in Figure 13.3. The time overhead used to collect all channel estimates grows linearly with the number of users, n, in the network and the channel utilization efficiency can be expressed by

$$\eta = \frac{T}{n\tau + \tau_o + T},$$

(13.4)

where τ_o is the time used for both the channel estimation and the broadcast of scheduling decision and T for data transmission. When there is a large number of users in the network, the centralized scheduling will have very low efficiency in channel utilization and have zero efficiency when n approaches infinity. Furthermore, when the number of users is large, the time used for collecting channel estimates will be so large that it may exceed the channel coherence time, which degrades network performance. So the centralized approach has poor scalability, and when there are many users delay in obtaining the channel knowledge will be the bottleneck and limit network performance. In the following, we will assume there is no centralized controller in the network and each user has knowledge of its own channel. Here we assume that the uplink and downlink channels are reciprocal, as in a time division duplex (TDD) system. The base station will broadcast a reference signal to all users and each user can measure its own channel based on this reference signal. In this way, only a fixed one-half a round trip time is needed to get channel information, even with an infinite number of users.

13.1.1   Protocol design and parameter optimization

Channel-aware Aloha is a simple variation of the typical slotted Aloha protocol. The transmission of each user will be based on their channel gains, which are assumed to be fixed in each time slot and may vary randomly between slots. Same as slotted Aloha, if there is a collision, i.e. two users transmit at the same time, no data will go through. In the design of [209], after the transmission in each slot, the receiver will broadcast a three-state feedback,

$$(0, 1, e), \tag{13.5}$$

indicating whether the slot was idle, it was a successful transmission, or a collision had happened.

In the typical slotted Aloha system, each user sends a packet in each slot independently with probability p. With channel-aware Aloha, each user sends a packet only when its channel is above a threshold H_0. Since the channel is random inherent for each user, as shown in Figure 13.1, and the channels of different users are independent from each other, the transmissions of different users are also randomized. With threshold H_0, the transmission probability p of each user is

$$p = 1 - F(H_0). \tag{13.6}$$

In other words, to achieve a contention probability, p, the desired threshold should be

$$H_0 = F^{-1}(1 - p). \tag{13.7}$$

The transmission probability determines the network throughput. The total network throughput is given by

$$S(p, N) = \sum_{i=1}^{N} p(1 - p)^{N-1} R(p), \tag{13.8}$$

where $R(p)$ is the data rate achieved if a successful transmission happens. The above expression consists of two parts. The first part, $p(1 - p)^{N-1}$, is the probability of a successful transmission, where p stands for the condition that the channel gain of a user is above H_0 and $(1 - p)^{N-1}$ is the probability that the channel gains of all other users are below H_0. The second part is the achieved data rate given a successful transmission, $R(p)$. $R(p)$ needs to be optimized with respect to the transmission probability p. This is because p determines how often the user will transmit packets and therefore the resource consumption in the PHY layer, which is shown below. Usually wireless transmissions are subject to two constraints: the peak and average power constraint. Define the transmission power to be P. For the peak power constraint,

$$P \le P_m. \tag{13.9}$$

For the average power constraint,

$$\mathrm{E}P \le P_a. \tag{13.10}$$

If the transmission power in time slot n is P_n, the average power constraint would be

$$\lim N \to \infty \sum_{n=1}^{N} P_n \le P_a. \tag{13.11}$$

If the user does not transmit in a slot n, $P_n = 0$. The impact of the transmission probability p on the PHY power allocation and therefore the data rate R can be illustrated by the following example.

If a user can transmit at the rates of the Shannon capacity and the user allocates power such that the received power is always P_r, the data rate would be

$$R(p) = W \log \left(1 + \frac{P_r}{N_o W} \right), \tag{13.12}$$

where W is the signal bandwidth and N_o is the noise power spectral density. To achieve the fixed reception power P_r, the transmission power should be

$$P_t = \frac{P_r}{h}. \tag{13.13}$$

This can be called channel inversion. With this power allocation policy, the power consumption of each user is determined by the reception power P_r, which also determines the achieved data rate. Since the power consumption needs to meet both the peak and average power constraints, we have

$$\frac{P_r}{h} \le P_m \tag{13.14}$$

and

$$\mathrm{E} \left(\frac{P_r}{h} \right) = \int_{F^{-1}(1-p)}^{\infty} f(h) \frac{P_r}{h} \mathrm{d}h \le P_a, \tag{13.15}$$

i.e.

$$P_r \le \frac{P_a}{\int_{F^{-1}(1-p)}^{\infty} f(h) \frac{1}{h} \mathrm{d}h}. \tag{13.16}$$

While P_m and P_a are device parameters that cannot be changed, P_r and p are the two design parameters that need to be optimized. As shown in (13.12), the data rate is a function of P_r, and to achieve the highest data rate P_r should be the maximal possible value. We have

$$P_r = \min \left(P_m h, \frac{P_a}{\int_{F^{-1}(1-p)}^{\infty} f(h)\frac{1}{h}\mathrm{d}h} \right). \tag{13.17}$$

Therefore, P_r is uniquely determined by (13.14) and (13.15), in which only the transmission probability p is the design parameter. So, clearly, the rate R is a function of transmission probability p and is uniquely determined by p. This gives the rate and transmission probability relationship for channel inversion. For other types of power allocation policies, other corresponding relationships can be discovered with similar rationales.

Now come back to the network throughput in (13.8). Note that the two parts in (13.8) can be interpreted as the behaviors of two layers, *MAC* and *PHY*, respectively. p is determined by the *MAC* contention probability and $R(p)$ is determined by the PHY layer resource management approaches, e.g. modulation, coding, power allocation, etc. They are closely related to each other. Based on the above analysis, the network throughput is uniquely determined by the transmission probability p. Finding a global optimal p across MAC and PHY layers is not a trivial task because of the non-linearity in p. Let us turn to a suboptimal approach, which indeed performs very close to the global optimal one.

We choose p such that the first part in (13.8) is maximized, i.e.

$$p = \arg \max Np(1-p)^{N-1}. \tag{13.18}$$

It can be easily found that the optimal contention probability is

$$p = \frac{1}{N}. \tag{13.19}$$

Correspondingly the optimal threshold is

$$H_0 = F^{-1}\left(1 - \frac{1}{N}\right) \tag{13.20}$$

and with channel inversion, the power allocation is

$$P_r = \min \left(P_m h, \frac{P_a}{\int_{F^{-1}(1-\frac{1}{N})}^{\infty} f(h)\frac{1}{h}\mathrm{d}h} \right). \tag{13.21}$$

13.1.2   Performance analysis

$p = \frac{1}{N}$ is a suboptimal solution to maximizing the network throughput. The global optimal solution needs to consider $R(p)$, which is affected by many other factors.

Consider only the peak power constraint P_m. The total network throughput is given by

$$S(p,N) = Np(1-p)^{N-1}\mathbf{E}\left[W \log \left(1 + \frac{P_m h}{N_o W} \right) \right], \tag{13.22}$$

where the expectation should be the average over all possible channel realizations that have gains above H_0. The global optimal transmission probability p^* should be the value maximizing (13.22). The proposition below, proved in [209], shows that $p = \frac{1}{N}$ is a good approximation of the optimal contention probability and it approaches the global optimum as N goes to infinity.

PROPOSITION 13.1 *For any finite N, $p^* = \frac{\alpha(N)}{N}$, where $0 < \alpha(N) < 1$, and $\alpha(N) \to 1$ as N approaches infinity.*

Therefore, $p = \frac{1}{N}$ is a good approximation of the global optimal transmission probability and it approaches the global optimum as N approaches infinity.

Now let us compare the performance of channel-aware Aloha with that of centralized scheduling. With a centralized scheduler, the optimal scheduling approach is to schedule the user with the highest channel gain. Assume again only the peak power constraint. The average network throughput achieved by this optimal centralized scheduler is then

$$S_c(N) = \mathbf{E}\left(R(P_m \max_n H_n)\right).$$ (13.23)

The following proposition, proved in [209], shows the ratio of the throughput of channel-aware Aloha to that of the optimal centralized scheduler converges to $\frac{1}{e}$ as N goes to infinity.

PROPOSITION 13.2 $\lim_{N \to \infty} \frac{S(p,N)}{S_c(N)} = \lim_{N \to \infty} \left(1 - \frac{1}{N}\right)^{N-1} = \frac{1}{e}.$

Indeed, the performance loss of channel-aware Aloha is due to the contention inherent in the Aloha protocol. This can be seen by looking at $R(p)$, which equals

$$R(p) = \frac{S(p,N)}{Np(1-p)^{N-1}}.$$ (13.24)

Therefore,

$$\lim_{N \to \infty} \frac{R(p)}{S_c(N)} = \lim_{N \to \infty} \left(1 - \frac{1}{N}\right)^{N-1} \bigg/ \left(\left(1 - \frac{1}{N}\right)^{N-1}\right) = 1.$$ (13.25)

$R(p)$ is the average transmission rate if there is no collision, which approaches the average rate in the optimal centralized scheme. Therefore, the performance loss of using distributed channel knowledge in slotted Aloha is due to the contention inherent in the Aloha protocol.

13.2 Opportunistic splitting algorithms

In the previous section, we have seen that the throughput with channel-aware Aloha increases at the same rate as the optimal centralized scheduler. The only penalty of using channel-aware Aloha is because of the contention inherent in the Aloha protocol. To further improve network performance, the contention needs to be reduced. An effective way of reducing the contention is proposed in [210] and in the following, we discuss this scheme briefly.

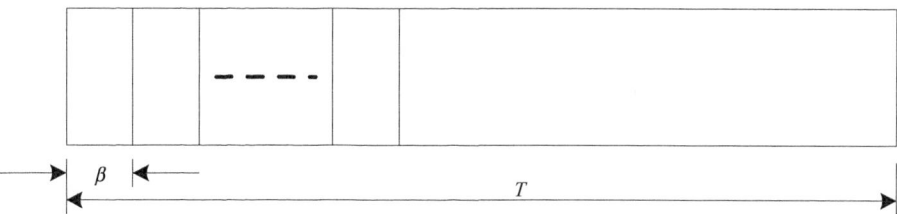

Figure 13.4 Frame structure

The frame structure is illustrated in Figure 13.4. In the beginning of each frame, the time is divided into some mini-slots with length β. These mini-slots are used to resolve the contention of all users in the network. β should be designed as small as possible. The minimum value of β can be the maximum round trip time between all users and the base station so that each user can send a mini-reservation packet and detect if a collision occurs. Let T denote the frame length within which the channel stays static. T is therefore smaller than the coherence time of the channel. Assume there are K mini-slots and n users competing for channel access. n is known to all users in the network. The access algorithm is designed such that after using several mini-slots for the communications between all users and the base station, the user with the best channel state is found for channel access in each frame.

The splitting algorithm can be described briefly as follows. Two thresholds, H_l and H_h, are introduced for the purpose of filtering out the best user. These thresholds are the same for all users in each mini-slot and may vary in different mini-slots. Initially set $H_l = 0$ and $H_h = \infty$. Therefore, the channel gains of all users fall between H_l and H_h. In each mini-slot, H_l or H_h may be modified. At the beginning of mini-slot k, only users whose channel gains satisfy $H_l < k < H_h$ will send a short signal to the base station for access contention. Then the base station will broadcast a $(0, 1, e)$ feedback, indicating that if the contention state in this mini-slot is idle (0), success (1), or collision (e). If there is only one user with the channel gain $H_l < k < H_h$, only the user with the best channel gain transmitted in the mini-slot and 1 is broadcasted. In this case, this user will continue to send data in the remaining time of the frame slot and all other users remain silent. If 0 or e is broadcasted, all users will adjust the two thresholds H_l and H_h in the same way and repeat the algorithm until either a success feedback is received or all mini-slots are used. The basic adjustment algorithm in [210] is described in Algorithm 13.1.

In the threshold update algorithm, $H_l = F^{-1}(1 - \frac{1}{n})$ and $H_h = \infty$, indicating a contention probability of $\frac{1}{n}$ for all users in the first mini-slot. This setting minimizes the probability of a collision in the first mini-slot. H_{ll} records the largest used H_l in the past that is smaller than the channel gain of at least one user. At any slot, if a collision occurs and the feedback is e, the range $H_l < h < H_h$ is split into two parts, by recording the original H_l to be $H_{ll} = H_l$ and redefining H_l using the function $H_l = split(H_l, H_h)$, and the new two parts are $H_l < h < H_h$ and $H_{ll} < h < H_l$. Users in the upper part will send access requests in the next mini-slot. If the feedback is 0, it indicates all users

Algorithm 13.1: *Threshold update*

1. $H_l = F^{-1}(1 - \frac{1}{n}), H_h = \infty, H_{ll} = 0$, and $k = 1$;
2.   **while** *feedback* $\neq 1$ and $k \leq K$
3.   **if** feedback=e
        (∗ a collision happens and the threshold needs to be increased so that less users contend. ∗)
4. $H_{ll} = H_l$ and H_l =split(H_l, H_h);
5.   **else** feedback=0
(∗ no user contended and the threshold needs to be decreased so that some user may contend. ∗)

6. $H_h = H_l$;
7.   **if** $H_{ll} \neq 0$
8. $H_l = split(H_{ll}, H_h)$;
9.   **else**
10. $H_l = lower(H_l)$,
11. $k = k + 1$.

have channel gains below H_l and therefore reset $H_h = H_l$. In addition, there are two possibilities. The first is there has been at least one collision before. In this case, the highest channel gain lies between H_{ll} and H_l. Again we can use the split() function to divide $H_{ll} < h < H_l$ into two parts and let the new contention range be the upper one. The second possibility is that there has never been any collision before, i.e. $H_{ll} = 0$. This indicates that all users have their channel gains below H_l and therefore H_l is decreased using the lower() function.

It can be seen that the split() and lower() functions determine the contention probability in the next mini-slot and therefore the network performance. In [210], the split function is designed such that each user will contend with probability 0.5 in the next/mini-slot because if a collision occurs, the most likely scenario is that two users were involved in this collision, as will be proved in the next chapter. This is a good estimate as no user knows how many users were involved in the previous collision. To achieve the 0.5 contention probability, the function is given by

$$split(H_l, H_h) = F^{-1} \left(\frac{F(H_l) + F(H_h)}{2} \right). \tag{13.26}$$

When the lower() function is used, it is known that all n users will be involved in the contention in the next mini-slot. The new H_l can be chosen such that the probability of success in the next slot, given an idle feedback received in the current slot, is maximized and is given by

$$lower(H_l) = F^{-1}(F(H_l)(1 - 1/n)). \tag{13.27}$$

Compared to the optimal centralized scheduler, max-signal-to-noise ratio (SNR) scheduler, the performance loss of this opportunistic splitting algorithm is due to the time, i.e. the number of mini-slots multiplied by the length of each mini-slot, required to find the user with the best channel. It is shown in [210] that by using this simple splitting algorithm, the average number of mini-slots required is upper bounded by 2.5 regardless of the number of users in the network. The length of each mini-slot should be determined by the maximum system round-trip time and the frame length, i.e. the channel coherence time. In typical wireless systems, the round-trip time is usually far smaller than the channel coherence time and therefore the overhead from using mini-slots is very limited. Therefore, the opportunistic splitting algorithm achieves almost the same performance as the max-SNR centralized scheduler. Note that this optimum performance is achieved assuming high synchronization capability of the whole network. In practice, imperfect synchronization will lead to certain performance losses.

14 Opportunistic random access: any network topology

In this chapter, we consider schemes for distributed cross-layer optimization of multi-channel random access by exploiting local channel state and traffic information. We consider the most generic network setting, where users are not necessarily within the transmission ranges of all others; therefore, when a user is transmitting, it may only interfere with some users. One example is an ad hoc wireless network. Later we will also show that it is possible to apply this technology in cellular networks to improve the throughput of cell-edge users. We also consider generic traffic distribution in the network and each user may choose to send packets to or receive packets from different users simultaneously. An example is illustrated in Figure 14.1, where arrows indicate traffic flows and circles transmission ranges of different users. To begin with, in this chapter, we will introduce the optimal channel-aware Aloha for this type of wireless network. A discussion of the optimal distributed channel-aware medium access control (MAC) will be given in the next chapter.

14.1 Network model

Consider multi-channel wireless networks. The whole band is divided into K subchannels. All channels between pairs of users are assumed to be reciprocal, i.e. when no interference exists, user A can receive a signal from user B if and only if user B can receive a signal from user A with the same channel gain. However, the interference environments at user A and user B may be different since they are at different locations. Each user has knowledge of its own channel state information (CSI) and makes independent transmission control decisions, including whether to transmit given the CSI, what data rate to use, and where to transmit, etc. Each user applies the same transmission control policy. In order to avoid an onerous signaling burden, no communication pair has instantaneous cooperation, such as exchange of CSI, transmit power, or subchannel selections.

All users are not necessarily within the transmission ranges of the others, which means that some users may not be able to receive packets from others due to weak received signal power. For simplicity, we assume those that can communicate with each other experience isotropic channels, i.e. channel power gains of different links are independent and identically distributed with probability density function, $f(h)$, and

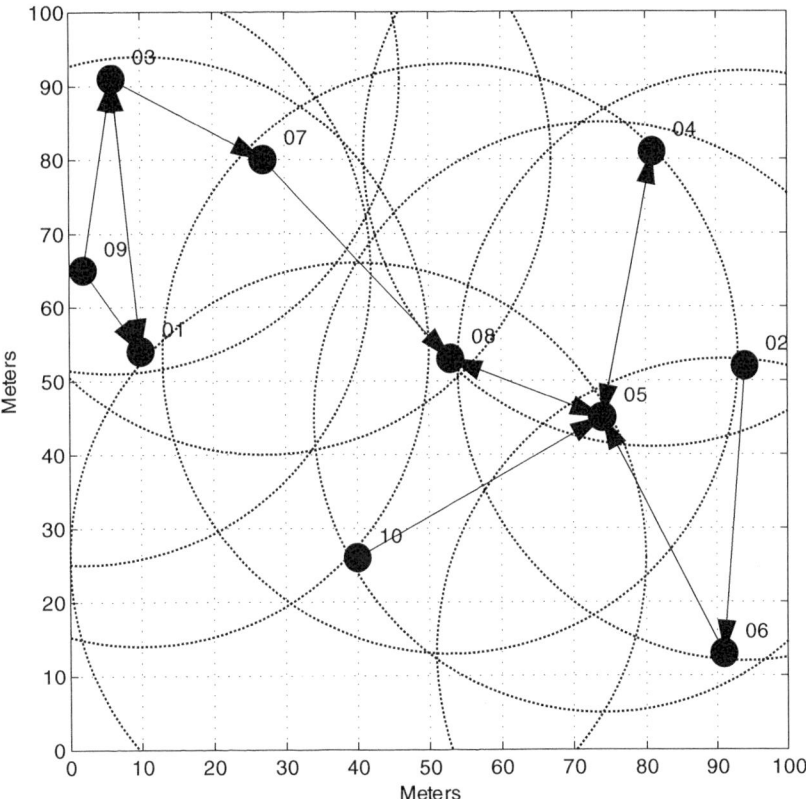

Figure 14.1 A network example

distribution function, $F(h)$. No capture is assumed for signal reception, i.e. the receiver cannot receive any signal successfully if any of its interfering neighbors, which are within the transmission range of the receiver, is transmitting simultaneously. A user cannot transmit and receive simultaneously on the same subchannel; however, it may transmit on a set of subchannels and receive on a different set of subchannels at the same time. Each user may choose to send packets to or receive packets from different users on different channels, and we assume that the links that carry traffic are backlogged, i.e. they always have packets to transmit.

During transmission, each user is subject to both average and instantaneous power constraints [76]. The average power constraint is due to heat accumulation and overall power consumption, while the instantaneous power constraint comes from the limited linear range of amplifiers. Two power allocation policies will be considered. In the first one, called channel inversion, each user transmits with just sufficient power to keep the received power constant so that the signal can be reliably detected. In the second, called adaptive modulation and power allocation, each user can vary both the modulation and transmit power during each transmission time slot to maximize throughput.

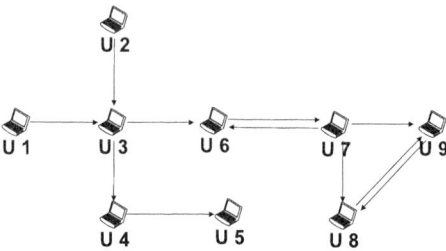

Figure 14.2 Network architecture example

14.2   Optimal design rules

In this section, we describe the wireless network model, and introduce a channel-aware multi-channel random access scheme. The characteristics of the scheme are analyzed, after which a criterion for cross-layer design is provided.

Define the wireless network as a directed graph $G(\mathcal{V}, \mathcal{E}, \mathcal{L})$, where $\mathcal{V}$, $\mathcal{E}$, and $\mathcal{L}$ are the set of active users, the set of all links over all K subchannels, and the set of links available for communication. We define $\mathcal{N}_i$ as the interfering neighbor set of user i. Each user may choose to send packets to or receive packets from several users, and $\mathcal{T}_i$ denotes the set of users receiving packets from i and $\mathcal{S}_j$ the set of users sending packets to j.

Figure 14.2 shows an example topology of a wireless network. The users are on a grid with unit spacing, and the transmission range is $\sqrt{2}$. The set of links available for communication is $\mathcal{L} = \{(1,3), (1,2), (1,4), (2,3), (2,6), (3,4), (3,5), (3,6), (4,6), (4,5), (5,6), (5,7), (5,8), (6,7), (6,8), (7,8), (7,9), (8,9)\}$. The arrows show the traffic flows in the network. For example, since $(4,6) \in \mathcal{L}$, any transmission by users 4 or 6 will be received by the other though they may not have packets to send to each other. So users 4 and 6 constitute an interfering pair and they interferer with the packet reception of each other. Observing user 3, it is easy to see that $\mathcal{T}_3 = \{4,6\}$, $\mathcal{S}_3 = \{1,2\}$, while $\mathcal{N}_3 = \{1,2,4,5,6\}$.

Slotted Aloha is a typical random access scheme. In slotted Aloha, the MAC layer makes transmission decisions based on the buffer occupancy and quality of service (QoS) requirement, and does not utilize the knowledge of the physical (PHY) layer at all. Hence, when the MAC decides to transmit a frame, the channel may be in a deep fade, but the physical layer still carries out the transmission, and causes a waste of power. The MAC layer may decide not to transmit even though the channel power gain is high, because it does not have this information from the PHY layer; this leads to wasted opportunity. With channel knowledge, the sender will transmit only when the channel power gain is above a certain threshold.[1] Therefore, we introduce the following *decentralized optimization for multi-channel random access* (DOMRA).

---
[1] Channel gains may be interferred with either due to CSI feedback or via channel reciprocity.

DOMRA: *User i (i ∈ V) decides to send packets to user j on subchannel k when the following conditions are satisfied:*

1   *User i has packets to send to j, j ∈ $\mathcal{T}_i$.*
2   *On subchannel k, link (i, j) has the best channel power gain, $h_{(i,j)_k} = \max_{l \in \mathcal{T}_i}\{h_{(i,l)_k}\}$.*
3   *The channel power gain is above a threshold, $h_{(i,j)_k} \geq \overline{H}_{(i,j)_k}$, where $\overline{H}_{(i,j)_k}$ is predetermined for link $(i, j)_k$. The transmission is then optimized according to $\overline{H}_{(i,j)_k}$, CSI and capability constraints.*

In DOMRA, each user transmits on the link with the best channel power gain provided that the gain is above a predetermined threshold. Proper choice of thresholds $\{\overline{H}_{(i,j)_k} | (i, j)_k \in \mathcal{E}\}$ and data transmission rates of all traffic flows, i.e. power allocation, will be determined in the following paragraphs so that overall network performance is optimized from certain perspectives. Further discussions of the optimality of using a threshold to control random access can be found in [214, 215].

As pointed out before, while many existing channel-aware schemes such as [209, 210, 11, 77] assume that each user has only one traffic flow to send and is within the transmission range of all other users, the DOMRA will provide solutions to networks in which users are not necessarily within the transmission ranges of all other users, and each user could send packets to or receive packets from different users simultaneously on different subchannels.

14.2.1   MAC layer analysis

According to the above transmission policy and the homogeneity assumption, the probability of a transmission on link $(i, j)_k \in \mathcal{E}$ is given by

$$p_{(i,j)_k} = \frac{1}{|\mathcal{T}_i|}\left(1 - F^{|\mathcal{T}_i|}(\overline{H}_{(i,j)_k})\right), \tag{14.1}$$

where $|\cdot|$ denotes the number of elements in the respective set. The proof of (14.1) is given below.

Proof   Since in each transmission time slot, user i sends packets to user j on subchannel k only when this subchannel has the best channel gain among all users in $\mathcal{T}_i$, and the subchannel power gain is above $\overline{H}_{(i,j)_k}$, we get the following:

$$p_{(i,j)_k} = \Pr\left\{h_{(i,j)_k} = \max_{a \in \mathcal{T}_i}(h_{(i,a)_k}), h_{(i,j)_k} \geq \overline{H}_{(i,j)_k}\right\}$$

$$= \Pr\left\{h_{(i,j)_k} = \max_{a \in \mathcal{T}_i}(h_{(i,a)_k})\right\} \cdot \Pr\left\{h_{(i,j)_k} \geq \overline{H}_{(i,j)_k} \Big| h_{(i,j)_k} = \max_{a \in \mathcal{T}_i}(h_{(i,a)_k})\right\}$$

$$= \frac{1}{|\mathcal{T}_i|}\Pr\left\{\max_{a \in \mathcal{T}_i}(h_{(i,a)_k}) \geq \overline{H}_{(i,j)_k}\right\} = \frac{1}{|\mathcal{T}_i|}\left(1 - \prod_{a \in \mathcal{T}_i}\Pr\left(h_{(i,a)_k} < \overline{H}_{(i,j)_k}\right)\right)$$

$$= \frac{1}{|\mathcal{T}_i|}\left(1 - F^{|\mathcal{T}_i|}(\overline{H}_{(i,j)_k})\right).$$

□

The probability that user i transmits on subchannel k is

$$p_{i_k} = \sum_{j \in \mathcal{T}_i} P_{(i,j)_k} = \sum_{j \in \mathcal{T}_i} \frac{1}{|\mathcal{T}_i|} \left(1 - F^{|\mathcal{T}_i|}(\overline{H}_{(i,j)_k}) \right). \tag{14.2}$$

Hence, the throughput on link $(i,j)_k$ is

$$T_{(i,j)_k} = R_{(i,j)_k} P_{(i,j)_k} (1 - p_{j_k}) \prod_{a \in \mathcal{N}_j, a \neq i} (1 - p_{a_k}), \tag{14.3}$$

where $R_{(i,j)_k}$ is the average data rate given that the user has decided to transmit on link $(i,j)_k$, and depends on the modulation and power allocation policy. $(1 - p_{j_k}) \prod_{a \in \mathcal{N}_j, a \neq i}(1 - p_{a_k})$ is the probability that neither user j nor its neighboring users except user i will transmit on subchannel k, which means successful transmission on link $(i,j)_k$.

For example, in Figure 14.2, the transmission from user 3 to user 6 on subchannel k succeeds only when neither user 6 nor its neighbors excluding user 3, i.e. users in $\mathcal{N}_6 \backslash \{3\} = \{2, 4, 5, 7, 8\}$, transmit. Hence, the throughput from user 3 to user 6 on subchannel k is $T_{(3,6)_k} = P_{(3,6)_k}(1 - p_{6_k})(1 - p_{2_k})(1 - p_{4_k})(1 - p_{5_k})(1 - p_{7_k})(1 - p_{8_k})R_{(i,j)_k}$.

14.2.2　Physical layer analysis

The average transmit power on link $(i,j)_k$ is the average of transmit power over all time slots, whether or not transmission happens on this link. According to the ergodicity of the channel, it is the average of transmit power over all channel states. Hence, we have

$$\mathbf{E}\{P_{(i,j)_k}\} = \int_0^\infty \Pr\{H_{(i,j)_k} = h, \text{and user } i \text{ transmits on } (i,j)_k\} P_{(i,j)_k}(h) \mathrm{d}h$$

$$= \int_0^\infty \Pr\{H_{(i,j)_k} = h\} \Pr\{\text{user } i \text{ transmits on } (i,j)_k | H_{(i,j)_k} = h\} P_{(i,j)_k}(h) \mathrm{d}h$$

$$= \int_{\overline{H}_{(i,j)_k}}^\infty f(h) F^{|\mathcal{T}_i| - 1}(h) P_{(i,j)_k}(h) \mathrm{d}h = \frac{1}{|\mathcal{T}_i|} \int_{\overline{H}_{(i,j)_k}}^\infty P_{(i,j)_k}(h) \mathrm{d}F^{|\mathcal{T}_i|}(h), \tag{14.4}$$

where $\mathbf{E}\{\}$ denotes expectation; $P_{(i,j)_k}(h)$ is the transmit power on link $(i,j)_k$ when the channel has power gain h and it depends on modulation and power allocation policy. For example, in order to achieve a constant *signal-to-noise ratio* (SNR) at the receiver, $P_{(i,j)_k}(h)$ is allocated such that $P_{(i,j)_k}(h) = \frac{P_r}{h}$, where P_r is the received power level satisfying the SNR requirement. According to the average power constraint, we have

$$\sum_{j \in \mathcal{T}_i, k=1,\dots,K} \mathbf{E}\{P_{(i,j)_k}\} \leq P_a, \qquad \forall i, j \in \mathcal{V}. \tag{14.5}$$

In existing channel access protocols, there are usually several subchannels to be selected for utilization. For example, the IEEE 802.11b physical layer [94] has 14 subchannels, 5 MHz apart in frequency, all of which have the same transmission capability. However, typically there is only one single radio frequency (RF) chain, and the peak constraint on the instantaneous transmit power has to be satisfied for the total combined transmission. We have the instantaneous power constraint

$$\sum_k \left(\max_{h,j} P_{(i,j)_k}(h) \right) \leq P_m, \qquad \forall i,j \in \mathcal{V}. \tag{14.6}$$

Given power allocation $P_{(i,j)_k}(h)$, the achieved average data rate given that a user has decided to transmit on link $(i,j)_k$ is

$$
\begin{aligned}
R_{(i,j)_k} &= \mathbf{E}\left\{ R(\eta(h)) | \text{user } i \text{ transmits on } (i,j)_k \right\} \\
&= \int_{\overline{H}_{(i,j)_k}}^{\infty} R(\eta(h)) \Pr\{H_{(i,j)_k} = h | \text{user } i \text{ transmits on } (i,j)_k\} dh \\
&= \int_{\overline{H}_{(i,j)_k}}^{\infty} R(\eta(h)) \frac{A}{B} dh,
\end{aligned}
\tag{14.7}
$$

where

$$A = \Pr\{H_{(i,j)_k} = h, \text{user } i \text{ transmits on } (i,j)_k\} = f(h) F^{|\mathcal{T}_i|-1}(h), \tag{14.8}$$

and

$$
\begin{aligned}
B &= \Pr\{\text{user } i \text{ transmits on } (i,j)_k\} \\
&= \int_{\overline{H}_{(i,j)_k}}^{\infty} \Pr\{H_{(i,j)_k} = g, \text{user } i \text{ transmits on } (i,j)_k\} dg \\
&= 1 - F^{|\mathcal{T}_i|}(\overline{H}_{(i,j)_k}).
\end{aligned}
\tag{14.9}
$$

Hence,

$$R_{(i,j)_k} = \frac{\int_{\overline{H}_{(i,j)_k}}^{\infty} R(\eta(h)) dF^{|\mathcal{T}_i|}(h)}{1 - F^{|\mathcal{T}_i|}(\overline{H}_{(i,j)_k})}, \tag{14.10}$$

where $\eta(h) = \frac{hP_{(i,j)_k}(h)}{N_o W/K}$ is the received SNR, N_o is noise spectral density, W is the total system bandwidth, and $R(\eta)$ is the instantaneous data rate when the channel has SNR η.

If channel capacity is achieved in *additive white Gaussian noise* (AWGN) channels,[2] $R(\eta) = W \log_2(1 + \eta)$. Assuming continuous rate M-ary quadrature amplitude modulation (M-QAM) and given the *bit error rate* (BER) requirement, $R(\eta)$ can be expressed as $R(\eta) = W \log_2(1 + \frac{3\eta}{-2\ln(5BER)})$ according to [83]. It is easy to see that in both cases, $R(\eta)$ is strictly concave in η. In general, we assume that $R(\eta)$ is continuously differentiable with first-order derivative $R'(\eta)$ positive and strictly decreasing in η.

14.2.3   Criterion for cross-layer design

When optimizing multi-user networks, we have to take both overall network throughput and fairness into consideration. A very commonly discussed fairness criterion is max–min fairness [64]. When max–min fairness is achieved, the throughput of a certain link

---

[2] In slow-fading channels, channel varies slightly within each packet. With sufficiently long packet length, ideal coding can be applied to achieve channel capacity

cannot be increased without simultaneously decreasing the throughput of another link which already has smaller throughput. Usually, max–min fairness just implies equal sharing of channel resources on each link, which compromises the overall throughput of the wireless network a lot since different links usually have different transmission conditions. Hence, we consider proportional fairness, the objective of which is to maximize the product of throughput of all links, or the geometric average [165]. As pointed out in [72], a vector of throughputs $T = (T_1, T_2, \ldots, T_n)$ is proportional fair if it satisfies required constraints, and for any other feasible vector $\overline{T}$, the aggregate of proportional changes is non-positive, i.e. $\sum_{i=1}^{n} \frac{\overline{T}_i - T_i}{T_i} \leq 0$. Some analysis has been given in [165] from a game-theoretic standpoint and it is shown that a strategy achieving proportional fairness satisfies certain axioms of fairness and is a Nash arbitration strategy [148]. With proportional fairness, the network will be operated at Pareto equilibrium, which corresponds to the situation where no user can improve its throughput without affecting at least one user adversely.

Define transmission control of the whole network as $\mathcal{C} = \{\overline{\mathcal{H}}, \mathcal{P}\}$, where $\overline{\mathcal{H}}$ is the set of predetermined channel power gain thresholds and $\mathcal{P}$ is the set of power allocation policies. With the constraints in (14.5) and (14.6), the optimal configuration of the whole network, $\mathcal{C}^* = \{\overline{\mathcal{H}}^*, \mathcal{P}^*\}$, that achieves proportional fairness among all subchannels carrying traffic flows will be

$$\mathcal{C}^* = \arg \max_{\{\overline{\mathcal{H}}, \mathcal{P}\}} \sum_{(i,j)_k \in \mathcal{E}, j \in \mathcal{T}_i} \ln(T_{(i,j)_k}), \tag{14.11a}$$

subject to

$$\sum_{j \in \mathcal{T}_i, k=1,\ldots,K} \frac{1}{|\mathcal{T}_i|} \int_{\overline{H}_{(i,j)_k}}^{\infty} P_{(i,j)_k}(h) \mathrm{d}F^{|\mathcal{T}_i|}(h) \leq P_a, \tag{14.11b}$$

and

$$\sum_k \left(\max_{h,j} P_{(i,j)_k}(h) \right) \leq P_m, \tag{14.11c}$$

where throughput $T_{(i,j)_k}$ is given by (14.3). Define utility $U_{(i,j)_k} = \ln(T_{(i,j)_k})$. Problem (14.11) aims to maximize overall network utility subject to individual power limits.

14.3   Low-complexity MAC

In the previous section, we have discussed a criterion for cross-layer design. The optimization of (14.11) depends on the threshold configuration, $\overline{\mathcal{H}}$, power allocation, $\mathcal{P}$, and modulation policy. The global optimization of the problem is difficult and computationally expensive, and requires complete network knowledge for each user. Therefore, in this section, we find a suboptimal solution, which only needs decentralized neighborhood information.

From (14.11), we have

$$\mathcal{C}^* = \arg\max_{\{\mathcal{H},\mathcal{P}\}} \sum_{(i,j)_k \in \mathcal{E}, j \in \mathcal{T}_i} \left(\ln\left(P_{(i,j)_k}(1 - p_{j_k}) \prod_{a \in \mathcal{N}_j, a \neq i} (1 - p_{a_k}) \right) + \ln\left(R_{(i,j)_k} \right) \right),$$

(14.12)

which reveals two ways to improve overall system performance. One way is to reduce the probability of collisions in the whole network, whose effect is captured by the term $P_{(i,j)_k}(1 - p_{j_k}) \prod_{a \in \mathcal{N}_j, a \neq i}(1 - p_{a_k})$. The other is to allocate power properly so that the achieved data rate of each individual user can be maximized. Hence, we decompose it into two related problems, and find a suboptimal transmission control policy. The solution is to find the optimal MAC layer transmission control $\overline{\mathcal{H}}^*$ to resolve collisions in the whole network while guaranteeing that proportional fairness can be formulated by

$$\overline{\mathcal{H}}^* = \arg\max_{\mathcal{H}} \sum_{(i,j)_k \in \mathcal{E}, j \in \mathcal{T}_i} \left(\ln\left(P_{(i,j)_k}(1 - p_{j_k}) \prod_{a \in \mathcal{N}_j, a \neq i} (1 - p_{a_k}) \right) \right).$$

(14.13)

Given the MAC transmission decision, in order to maximize the mean physical layer throughput within power capability, the optimal power allocation $\mathcal{P}_i^*$ of user i is formulated by

$$\mathcal{P}_i^* = \arg\max_{\mathcal{P}_i} \sum_{j \in \mathcal{T}_i, k} R_{(i,j)_k},$$

(14.14a)

subject to (14.11b)

$$\sum_{j \in \mathcal{T}_i, k=1,\ldots,K} \frac{1}{|\mathcal{T}_i|} \int_{\overline{H}_{(i,j)_k}^*}^{\infty} P_{(i,j)_k}(h) \mathrm{d}F^{|\mathcal{T}_i|}(h) \leq P_a,$$

(14.14b)

and (14.11c)

$$\sum_k \left(\max_{h,j} P_{(i,j)_k}(h) \right) \leq P_m,$$

(14.14c)

where $\{\overline{H}_{(i,j)_k}^*\}$ is the solution of (14.13) and $R_{(i,j)_k}$ is given by (14.10). Although problem (14.11) has been decomposed into (14.13) and (14.14) to resolve network collisions and improve individual transmission capability respectively, these two problems are closely coupled through $\overline{\mathcal{H}}^*$.

When optimizing the network with proportional fairness in (14.13), all users are assumed to transmit at the same data rate once the channel power gain is above a certain threshold. Problem (14.13) turns out to be similar to the problem of finding a distributed access control strategy to achieve proportional fairness in traditional Aloha networks [113] and [149]. By applying techniques used in [113] and [149], the optimal transmission probability is readily achieved

$$p_{(i,j)_k}^* = \frac{1}{|\mathcal{S}_i| + \sum_{m \in \mathcal{N}_i} |\mathcal{S}_m|}.$$

(14.15)

Combining (14.1) and (14.15), Theorem 14.1 follows immediately, and the proof is omitted.

THEOREM 14.1   *The optimal predetermined channel power gain threshold for any link $(i,j)_k \in \mathcal{E}$ where $j \in \mathcal{T}_i$, $\overline{H}^*_{(i,j)_k}$, as defined in (14.13), is given by*

$$\overline{H}^*_{(i,j)_k} = F^{-1}\left[\left(1 - \frac{|\mathcal{T}_i|}{|\mathcal{S}_i| + \sum_{m \in \mathcal{N}_i}|\mathcal{S}_m|}\right)^{\frac{1}{|\mathcal{T}_i|}}\right]. \tag{14.16}$$

From threshold (14.16), the optimal threshold of user i is independent of the receiver j but depends on the neighborhood information of user i itself, including the number of users receiving packets from user i, $|\mathcal{T}_i|$, the number of users sending packets to user i, $|\mathcal{S}_i|$, and the total number of users sending packets to the interfering neighbors of user i, $\sum_{m \in \mathcal{N}_i}|\mathcal{S}_m|$. The first two are local information, while $|\mathcal{S}_m|'s, m \in \mathcal{N}_i$, is information about interfering neighbors. The number of flows each interfering neighbor receives, i.e. $|\mathcal{S}_m|$ for all $m \in \mathcal{N}_i$, can be obtained through broadcasting the number of interfering neighbors whenever this number changes. Since this knowledge needs to be broadcast to notify the interfering neighbors, we call it *two-hop knowledge*. The broadcasting of this two-hop knowledge incurs only a trivial signaling overhead since only when either a traffic session or the network topology varies will this broadcasting be triggered. Besides, some form of two-hop knowledge is typical in many protocols, such as routing information discovery in mobile ad hoc networks [28, 101]. Hence, it can be easily obtained.

Consider user 7 in Figure 14.2. It is easy to see that $|\mathcal{T}_7| = 3$, $|\mathcal{S}_7| = 1$, and the two-hop knowledge of interfering neighbors $|\mathcal{S}_5| = 1$, $|\mathcal{S}_6| = 2$, $|\mathcal{S}_8| = 2$, and $|\mathcal{S}_9| = 2$. Hence, for all $j \in \mathcal{T}_7$ and $k = 1, \ldots, K$, $\overline{H}^*_{(7,j)_k} = F^{-1}\left[(1 - \frac{3}{1+7})^{1/3}\right] = F^{-1}(0.855)$. If the channel is experiencing Rayleigh fading with average power gain h_a, $\overline{H}^*_{(7,j)_k} = 1.931h_a$. Hence, since there are many traffic flows in the neighborhood of user 7, it transmits only when the channel has very good condition.

As we can see above, the optimal threshold can be obtained through two-hop knowledge. In the following, we consider two special applications.

(i)   Transmission control with one-hop knowledge

To avoid the broadcast of signaling, assume no user has two-hop knowledge, and it needs to be estimated to get approximation of the optimal thresholds. Since the transmission of each interfering neighbor $j \in \mathcal{N}_i$ can be detected by user i, $|\mathcal{T}_j|$ is available, user i can approximate $|\mathcal{S}_i| + \sum_{m \in \mathcal{N}_i}|\mathcal{S}_m|$, the total number of received traffic flows within the interfering range of user i, to be $|\mathcal{T}_i| + \sum_{m \in \mathcal{N}_i}|\mathcal{T}_m|$, the total number of transmitted traffic flows user i can detect. Hence, instead of (14.16), the transmission threshold with one-hop knowledge, i.e. local knowledge, is

$$\overline{H}^*_{(i,j)_k} = F^{-1}\left[\left(1 - \frac{|\mathcal{T}_i|}{|\mathcal{T}_i| + \sum_{j \in \mathcal{N}_i}|\mathcal{T}_j|}\right)^{\frac{1}{|\mathcal{T}_i|}}\right]. \tag{14.17}$$

Since the approximation in (14.17) is not always accurate, there might be some performance degradation. Approximation error happens when there exists undetectable traffic flows that are sent either into or out of the interfering range of user i.

(ii) Transmission control for one-hop networks

Assume that all users are within the transmission range of each other, i.e. this is a one-hop network. A simple example is the uplink transmissions of different users to the access point in the *wireless local-area network* (WLAN), and at most one traffic flow within the network can succeed in transmission in one transmission slot on one subchannel. Define $n = |\mathcal{S}_i| + \sum_{m \in \mathcal{N}_i} |\mathcal{S}_m|$ for any user i, then n is the same for all users and represents the total number of traffic flows in the network. During any time slot on each subchannel, at most one traffic flow within the network can send data successfully. The transmission threshold is given by

$$\overline{H}^*_{(i,j)_k} = F^{-1}\left[\left(1 - \frac{|\mathcal{T}_i|}{n}\right)^{\frac{1}{|\mathcal{T}_i|}}\right]. \tag{14.18}$$

If each user has only one traffic flow to send, i.e. $|\mathcal{T}_i| = 1$, the transmission threshold is

$$\overline{H}^*_{(i,j)_k} = F^{-1}\left[\left(1 - \frac{1}{n}\right)\right], \tag{14.19}$$

which is the same as the transmission control in [209]. [209] has demonstrated that the total throughput for such a system achieves a fraction, $(1 - \frac{1}{n})^{n-1}$, of its counterpart's throughput with an optimum centralized scheduler. The throughput reduction is due to the inherent contention in random access.

14.4 Optimal PHY operation

14.4.1 Physical layer optimization with channel inversion

Consider a simple transmitter adaptation technique, channel inversion [25], which maintains a constant received power level so that the signals can be reliably received during each traffic session. Once the MAC decides to transmit with channel power gain h, the transmit power is directly given by $P_t = P_r/h$, where P_r is the received power level. Different traffic flows may have different received power levels, P_r, according to the power allocation strategy. The reliable transmission data rate is given by $R(P_r)$. According to the assumption in Section 14.2.2, $R(P_r)$ is strictly concave in P_r since the average noise power is constant on each subchannel.

From (14.4), the average transmit power on link $(i,j)_k$ is

$$\mathbf{E}\{P_{(i,j)_k}\} = \frac{1}{|\mathcal{T}_i|} \int_{\overline{H}_{(i,j)_k}}^{\infty} \frac{P_{r(i,j)_k}}{h} \mathrm{d}F^{|\mathcal{T}_i|}(h). \tag{14.20}$$

Hence, the instantaneous received power is

$$P_{r(i,j)_k} = |\mathcal{T}_i|\mathbf{E}\{P_{(i,j)_k}\}\left(\int_{\overline{H}_{(i,j)_k}}^{\infty} \frac{\mathrm{d}F^{|\mathcal{T}_i|}(h)}{h}\right)^{-1}. \tag{14.21}$$

Define by $\mathcal{P}_{ri} = \{P_{r(i,j)_k}|(i,j)_k \in \mathcal{E}, j \in \mathcal{T}_i\}$ the set of the received power configuration of user i. According to (14.10), the average data rate is $R_{(i,j)_k} = \mathbf{E}\{R(\eta(h))\} = R(P_{r(i,j)_k})$. The problem in (14.14) is equivalent to

$$\mathcal{P}_{ri}^* = \arg\max_{\mathcal{P}_{ri}} \sum_{j \in \mathcal{T}_i, k} R(P_{r(i,j)_k}), \tag{14.22a}$$

subject to

$$\sum_{j \in \mathcal{T}_i, k} \frac{1}{|\mathcal{T}_i|} \int_{\overline{H}_{(i,j)_k}^*}^{\infty} \frac{P_{r(i,j)_k}}{h} \mathrm{d}F^{|\mathcal{T}_i|}(h) \leq P_a, \tag{14.22b}$$

and

$$\sum_{k} \left(\max_{j} \frac{P_{r(i,j)_k}}{\overline{H}_{(i,j)_k}^*}\right) \leq P_m. \tag{14.22c}$$

The above power allocation problem is solved by Theorem 14.2, which is proved below.

THEOREM 14.2 *Assuming the strict concavity of the data rate function $R(P_r)$, (14.22) has unique globally optimal reception power levels $P_{r(i,j)_k}^*$ on any link $(i,j)_k \in \mathcal{E}$ where $j \in \mathcal{T}_i$*

$$P_{r(i,j)_k}^* = \min\left(\frac{P_a}{K}\left(\int_{\overline{H}_{(i,j)_k}^*}^{\infty} \frac{1}{h}\mathrm{d}F^{|\mathcal{T}_i|}(h)\right)^{-1}, \frac{P_m \overline{H}_{(i,j)_k}^*}{K}\right), \tag{14.23}$$

in which $\overline{H}_{(i,j)_k}^*$ is determined by Theorem 14.1.

Proof According to (14.16), we can see that $\overline{H}_{(i,j)_k}^*$ is independent of j and k. Hence, the first constraint of (14.22) is

$$\sum_{j \in \mathcal{T}_i, k=1,\dots,K} P_{r(i,j)_k} \leq P_a |\mathcal{T}_i| \left(\int_{\overline{H}_{(i,j)_k}^*}^{\infty} \frac{1}{h}\mathrm{d}F^{|\mathcal{T}_i|}(h)\right)^{-1}.$$

Since data rate function $R()$ is assumed to be a strictly concave function,

$$\sum_{j \in \mathcal{T}_i, k} R(P_{r(i,j)_k}) \leq |\mathcal{T}_i|KR\left(\frac{\sum_{j \in \mathcal{T}_i, k} P_{r(i,j)_k}}{K|\mathcal{T}_i|}\right).$$

The equation holds if and only if $P_{r(i,j)_k}$ is the same value for all $j \in \mathcal{T}_i$ and k. Hence, for optimal solution,

$$\max_{j} \frac{P_{r(i,j)_k}^*}{\overline{H}_{(i,j)_k}^*}$$

is the same for all $k = 1, ..., K$, and the second constraint of (14.22) is equivalent to

$$P^*_{r(i,j)_k} \leq \frac{P_m \overline{H}^*_{(i,j)_k}}{K}.$$

Then, it is easy to see that when the first constraint in (14.22) takes effect, the optimal solution is the first term in (16.4), while when the second constraint takes effect, the optimal solution is the second term in (16.4). Hence, (16.4) satisfies both constraints, and the objective value will be maximized when one constraint takes effect while satisfying the other constraint. $\square$

Whenever MAC decides to transmit, the physical layer always executes the transmission. However, when $\overline{H}^*_{(i,j)_k}$ is very small, (16.4) turns out to be very small and the physical layer has extremely low throughput due to the penalty of allowing transmission on deeply faded channels. Hence, $\overline{H}^*_{(i,j)_k}$ should be further modified by the physical layer to avoid transmitting on deeply faded channels. Observing (14.15), $p^*_{(i,j)_k}$ can be $1, \frac{1}{2}, \frac{1}{3}$, etc. Assuming a Rayleigh channel with average power gain h_a and one traffic flow is carried, the corresponding thresholds are $0, 0.69h_a, 1.10h_a$, etc. Hence, transmission on deeply faded channels is possible only when $p^*_{(i,j)_k} = 1$. Thus, define $\overline{H}_o$ as

$$\overline{H}_o = \arg\max_{\overline{H}} R\left(\frac{P_a}{K \int_{\overline{H}}^{\infty} \frac{1}{h} f(h) \mathrm{d}h}\right)(1 - F(\overline{H})), \tag{14.24}$$

which leads to the maximum physical layer throughput when the physical layer is required to transmit under any channel conditions. If $\overline{H}^*_{(i,j)_k}$ determined by Theorem 14.1 is less than $\overline{H}_o$, then substitute it with $\overline{H}_o$. This lowers $p_{(i,j)_k}$ slightly since the channel is not deeply faded most of the time. The revision effectively improves link performance but impacts trivially on network performance, and we do not need to further improve the thresholds of other users to adapt to this change for the sake of optimality in (14.13), which otherwise incurs an additional signaling overhead.

With channel inversion, the instantaneous transmit power allocation $\mathcal{P}^*$ is

$$P^*_{(i,j)_k}(h) = \begin{cases} \frac{P^*_{r(i,j)_k}}{h} & h \geq \overline{H}^*_{(i,j)_k} \\ 0 & \text{otherwise} \end{cases}. \tag{14.25}$$

14.4.2 Physical layer optimization with adaptive modulation and power allocation

Consider ideal physical layer transmissions. Each user can vary both the transmit power and rate to achieve the best transmission performance. According to (14.10) and (14.14), the power allocation strategy can be formulated by

$$\mathcal{P}^*_i = \arg\max_{\mathcal{P}_i} \sum_{j \in \mathcal{T}_{i,k}} \frac{\int_{\overline{H}^*_{(i,j)_k}}^{\infty} R(\eta(h)) \mathrm{d}F^{\mathcal{T}_i}(h)}{(1 - F^{\mathcal{T}_i}(\overline{H}^*_{(i,j)_k}))}, \tag{14.26a}$$

subject to (14.11b)

$$\sum_{j \in \mathcal{T}_{i,k}} \frac{1}{|\mathcal{T}_i|} \int_{\overline{H}^*_{(i,j)_k}}^{\infty} P_{(i,j)_k}(h) \mathrm{d}F^{|\mathcal{T}_i|}(h) \leq P_a, \tag{14.26b}$$

and (14.11c)

$$\sum_k \left(\max_{h,j} P_{(i,j)_k}(h) \right) \leq P_m. \tag{14.26c}$$

The optimal solution of (14.26) is given in Theorem 14.3.

THEOREM 14.3   *Assume the data rate function $R(\eta)$ to be continuously differentiable and the first-order derivative $R'(\eta)$ is positive and strictly decreasing. For any link $(i,j)_k \in \mathcal{E}$, where $j \in \mathcal{T}_i$, (14.26) has a unique globally optimal power allocation given by: if $P_m < \frac{P_a}{1 - F^{|\mathcal{T}_i|}(\overline{H}^*_{(i,j)_k})}$, $P^*_{(i,j)_k}(h) = \frac{P_m}{K}$ for $h \geq \overline{H}^*_{(i,j)_k}$; otherwise,*

$$P^*_{(i,j)_k}(h) = \begin{cases} \frac{P_m}{K} & v^* < R'\left(\frac{hP_m}{n_o W}\right)\frac{hK}{n_o W}, \\ 0 & v^* \geq R'(0)\frac{hK}{n_o W}, \\ R'^{-1}\left(\frac{v^* n_o W}{hK}\right)\frac{n_o W}{Kh} & \text{otherwise,} \end{cases} \tag{14.27}$$

for $h \geq \overline{H}^*_{(i,j)_k}$. $R'^{-1}()$ is the inverse function of $R'()$. $v^* \geq 0$ is uniquely determined by

$$\int_{\overline{H}^*_{(i,j)_k}}^{\infty} P^*_{(i,j)_k}(h) \mathrm{d}F^{|\mathcal{T}_i|}(h) = \frac{P_a}{K}, \tag{14.28}$$

where $\overline{H}^*_{(i,j)_k}$ is given by Theorem 14.1.

Proof   It can be easily proven that the problem is strictly convex of $P_{(i,j)_k}(h), j \in \mathcal{T}_i, k = 1, \ldots, K$. Using the Lagrangian method and obtaining the Karush–Kuhn–Tucker (KKT) condition [171] for optimal power allocation, the local maximum in Theorem 14.3 can be readily obtained. The solution is globally optimal since for convex optimizations, KKT conditions are both necessary and sufficient for a local minimum to be a global minimum. If, in addition, the objective function is strictly convex, the globally optimal solution is unique. The detailed proof is omitted.   □

Observing (14.27), when $v^* \geq R'(0)\frac{hK}{n_o W}$, the channel is deeply faded and although the MAC layer decides to transmit, the physical layer further optimizes the transmission and decides not to transmit.

For example, assume the data rate function to be $R(\eta) = \frac{W}{K}\ln(1 + \eta)$. The power allocation when $P_m \geq \frac{P_a}{1 - F^{|\mathcal{T}_i|}(\overline{H}^*_{(i,j)_k})}$ is given by

$$P^*_{(i,j)_k}(h) = \begin{cases} \frac{P_m}{K} & \frac{1}{v^*} - \frac{n_o W}{Kh} > \frac{P_m}{K} \\ 0 & \frac{1}{h} \geq \frac{K}{v^* n_o W} \\ \frac{1}{v^*} - \frac{n_o W}{Kh} & \text{otherwise} \end{cases} \tag{14.29}$$

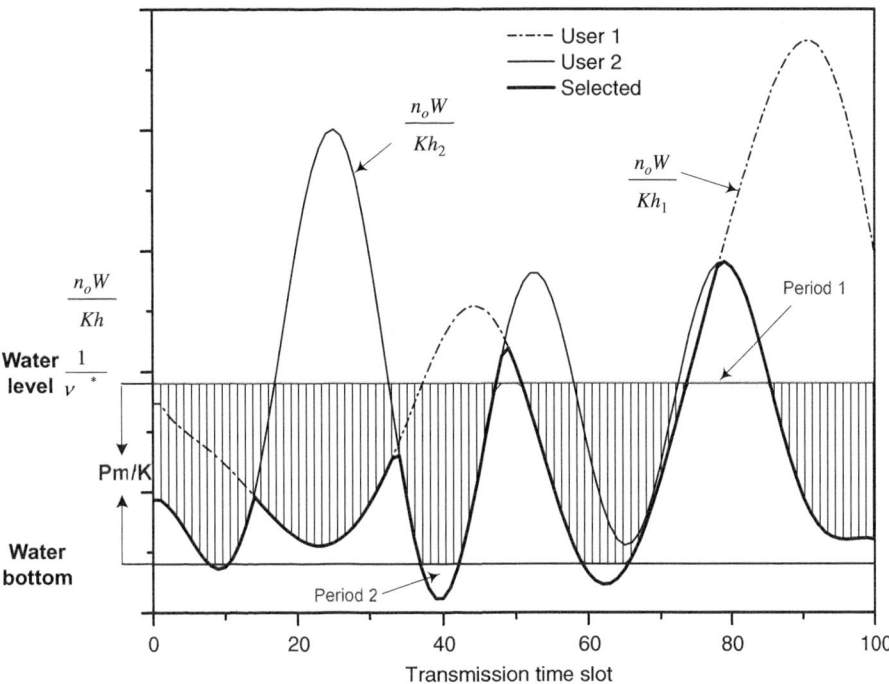

Figure 14.3 Capability limited water-filling over time

for $h \geq \overline{H}^*_{(i,j)_k}$, which is similar to the well-known water-filling power allocation scheme [92, 98, 81]. Since the power allocation scheme has a maximum instantaneous power constraint, we call it capability-limited water-filling.

According to (14.27), power will be optimally distributed over both time and all subchannels. Figure 14.3 illustrates the capability-limited power allocation of a user that is transmitting data to users 1 and 2 on a subchannel by using (14.29), and the striped parts in the figure represent the amount of power allocated. The power allocation during 100 transmission time slots is shown. We assume that $\frac{v^* n_o W}{K} > \overline{H}^*_{(i,j)_k}$ here. According to the transmission policy, the user always selects the destination with better channel power gains. As indicated by "Period 1" in Figure 14.3, there are no transmissions when subchannels of both users 1 and 2 are deeply faded. In "Period 2" in Figure 14.3, although the channel conditions are so good that higher data rates can be achieved, the actual data rate is limited by the instantaneous transmission capability P_m.

When $\frac{v^* n_o W}{K} < \overline{H}^*_{(i,j)_k}$, since the MAC decides to transmit only when $h > \overline{H}^*_{(i,j)_k}$, the physical layer will always transmit when the MAC wants to transmit according to (14.29). Assuming large P_m, the power allocation is always $\frac{1}{v^*} - \frac{n_o W}{Kh}$. Then according to (14.28), the water level is

$$\frac{1}{v^*} = \frac{\frac{n_o W}{K} \int_{\overline{H}^*_{(i,j)_k}}^{\infty} \frac{1}{h} dF^{|\mathcal{T}_i|}(h) + \frac{P_a}{K}}{1 - F^{|\mathcal{T}_i|}(\overline{H}^*_{(i,j)_k})}. \tag{14.30}$$

We can always use (14.30) to approximate the water level since with large probability, most transmissions will fall within the normal working ranges of the transmitter.

14.5   System performance

In this section, we first demonstrate DOMRA performance in a network with random topologies. Then we further show how closely DOMRA performs to the globally optimum solution.

14.5.1   Network performance improvement

Consider a network with random topologies and compare the average performance of all simulation trials. In each trial, users are randomly dropped and uniformly distributed in a square area with side length of 100 meters. Each user has a transmission range of 40 meters and selects neighboring users randomly for data transmission. A network topology in one trial has been given in Figure 14.1. Different schemes will be implemented to provide detailed performance comparisons.

Single-channel network

Assume that the network operates with one channel. For simplicity, assume a Rayleigh fading channel and $R(P) = W \ln(1 + \frac{hP}{WN_0})$. We will compare the performance with the channel-aware Aloha in [209], and the optimal traditional Aloha in [113], which does not consider cross-layer optimizations. For traditional Aloha transmissions, in order to make the comparison meaningful, the same average power constraint and instantaneous power constraint are enforced. Since there is no cooperation between MAC and the physical layer, the physical layer assumes that it keeps on transmitting except when the channel is deeply faded. In order to satisfy power constraints, the transmission threshold is chosen so that the average data rate is maximized, i.e. $\overline{H} = \arg\max_H (1 - F(H))R(P_r)$ subject to the instantaneous power constraint (14.6), and P_r is given by (14.21). The threshold is found through linear search.

Figure 14.4 shows the aggregate utility comparison of the whole network when the channel has different average channel gains. The "TwoHop" curve represents the result of DOMRA when each user has two-hop information of the neighboring users, while the "OneHop" curve represents the result when each user has only one-hop information. As we can see, with only one-hop knowledge, the system has slight performance degradation as compared with the transmissions when two-hop knowledge is available. Curve "QIN" shows the performance of [209], which assumes that each user has the knowledge of how many users there are in the whole network. Curve "Traditional" shows the result using the traditional optimal Aloha. As shown in Figure 14.4, with the advantage of cross-layer design, the scheme in this chapter outperforms traditional optimal Aloha greatly. In addition, by exploiting the neighborhood information of each user, the method also outperforms the existing channel-aware Aloha in [209]. This is

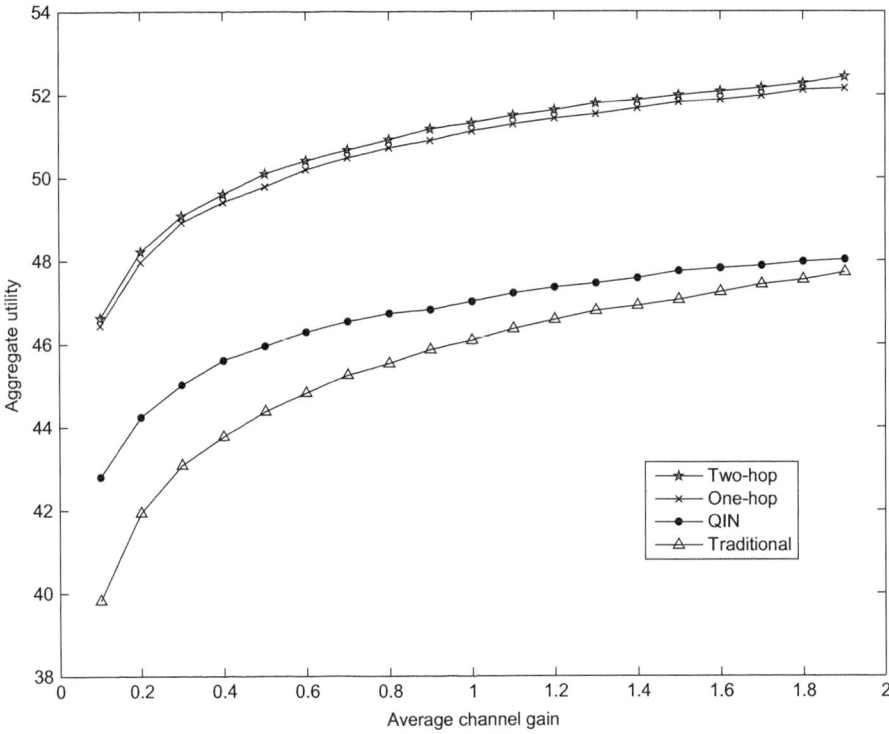

Figure 14.4 Network aggregate utility comparison. $P_m = 50$ dBm, $P_a = 43$ dBm, $W = 100$ Hz, and $N_o = 0.001$ W/Hz

due to the consideration of the inhomogeneous traffic spatial distribution in the scheme and the channels are better utilized.

Multichannel network

Consider the same wireless network configurations as those in the single-channel network scenario except that there are five subchannels. Besides implementing schemes in the single-channel network scenario for multi-channel environment, we also run the CAMCRA proposed in [77]. During each each transmission slot, CAMCRA chooses c subchannels with the c most significant gains, where $c = \max\left(1, \lfloor \frac{\text{subchannel number}}{\text{user number}} \rfloor\right)$. Then the method in [209] is applied on each subchannel given that each user knows how many users are using the subchannel. Since the number of users in each subchannel is a random variable, it is proposed in [77] to use $\max\left(1, \frac{\text{user number}}{\text{subchannel number}}\right)$ as an estimate. As shown in Figure 14.5, the CAMCRA in [77] has slight performance improvement as compared with the channel-aware Aloha in [209] because of exploitation of multichannel diversity. However, these two schemes do not perform well when the network has arbitrary spatial traffic distribution. DOMRA with either two-hop or one-hop information significantly outperforms these existing schemes due to exploitation of multi-

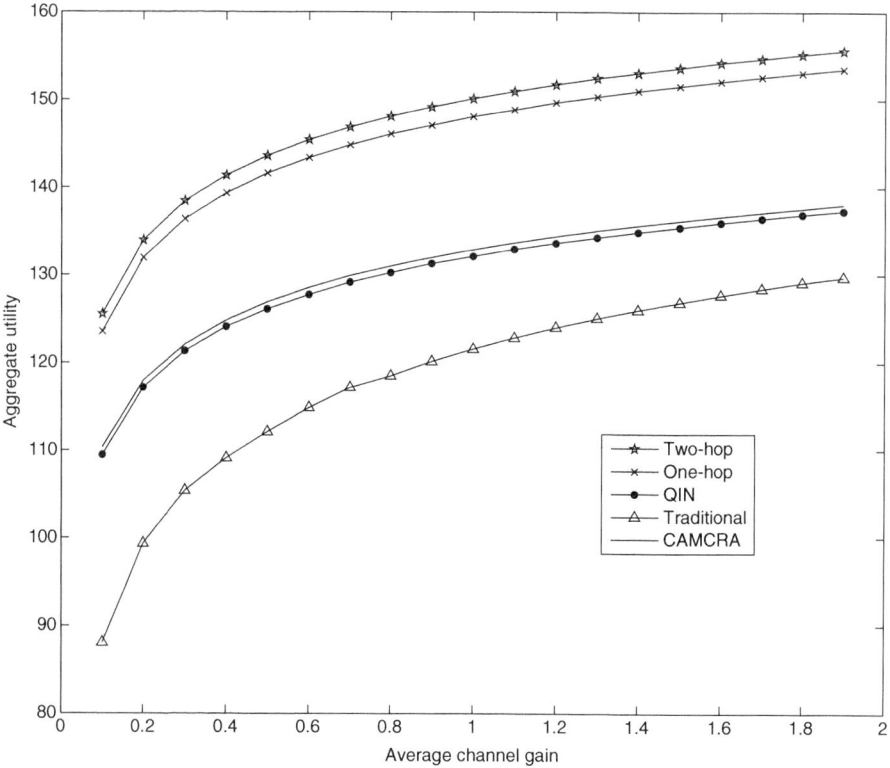

Figure 14.5 Five-channel network aggregate utility comparison. $P_m = 50\,\mathrm{dBm}, P_a = 43\,\mathrm{dBm}$, $W = 100\,\mathrm{Hz}, N_o = 0.001\,\mathrm{W/Hz}$

Figure 14.6 A simple network topology

user diversity and proper adaptive transmission settings and power allocation according to inhomogeneous traffic spatial distribution in the network.

14.5.2   Suboptimality gap

Problem (14.11) is decomposed into subproblems (14.13) and (14.14) to obtain feasible suboptimal control policy. In order to show the suboptimality gap, we exhaustively search for the global optimum in (14.11), and run a simple network topology to reduce search complexity. As shown in Figure 14.6, arrows indicate traffic flows. User 3 is sending traffic to n receivers, who are all out of the transmission ranges of users 1

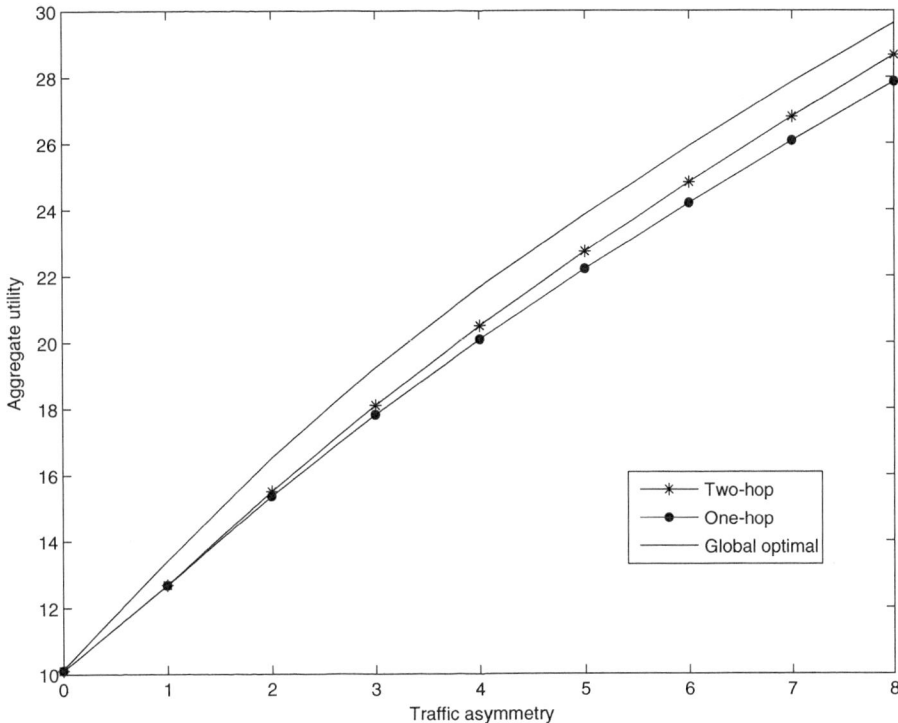

Figure 14.7 Aggregate utility gap to the global optimum. $P_m = 50\,\text{dBm}, P_a = 43\,\text{dBm},$
$W = 100\,\text{Hz}, N_o = 0.001\,\text{W/Hz}$

and 2. User 1 can communicate with 2, but not 3, while user 2 can communicate with both. When n is zero, the traffic distribution is symmetric in the network. The larger the number n, the more asymmetric the traffic distribution is. We call n *traffic asymmetry*, and vary it from 0 to 8. Figure 14.7 compares network aggregate utility and shows the suboptimality gap. While the global optimum can only be obtained through floods of broadcast of complete network knowledge, our decomposition technique yields a feasible suboptimal decentralized solution, which requires limited (two-hop knowledge case), or no (one-hop knowledge case) signaling overhead. Besides, the scheme in this chapter performs closely to the global optimum, and even reaches the global optimum when the traffic is symmetric.

15 Optimal channel-aware distributed MAC

Decentralized optimization for multi-channel random access (DOMRA) is based on slotted Aloha and once a collision happens, the entire data frame has to be dropped. Although DOMRA is the optimal channel-aware Aloha, it is not the optimal channel-aware medium access control (MAC). The performance can be further improved through scheduling users in a distributed way to avoid the collision of data transmission. The goal of all distributed MAC design is to transmit as much as possible while avoiding as many collisions as possible. With awareness of channel state, an additional goal is to exploit multi-user diversity and enable users with better channel states to win the contention with higher probabilities and thus to maximize network performance, as has been illustrated in Section 12.2.

Network performance is upper bounded by that of central schedulers. One fundamental question is: "Can distributed random access algorithms achieve the performance of centralized algorithms, and how to do it?" This chapter will introduce a feasible solution to these questions. We want the solution to be applicable to as many types of wireless networks as possible, e.g. cellular networks, ad hoc networks, sensor networks, and so on. We consider a very generic network model, where the network can have arbitrary topologies. In addition, the spatial traffic distributions can also be arbitrary in the sense that users can receive traffic from or send traffic to different users, and different communication links may interfere with each other. To make the access protocol more generic, we consider heterogeneous channels, where the random channel gains of different links may have different distributions.

The solution will be called channel-aware distributed medium access control (CAD-MAC). With CAD-MAC, each frame is divided into contention and transmission periods. The contention period is used to resolve the conflicts of all users while the transmission period is used to send data in collision-free scenarios. So CAD-MAC is a reservation-based MAC. For the contention period, a multi-stage channel-aware Aloha scheme is introduced to enable users with relatively better channel states to have higher probabilities of contention success. Then, we will further analyze the performance and show that with limited semi-static information exchange, CAD-MAC completely resolves the contention of networks of any type and size, and performs close to centralized schedulers. More importantly, ideal channel estimation is impossible in practice. However, we will show that CAD-MAC is robust to any uncertainty in non-ideal channel estimation.

15.1 System description

Consider a network where users are not necessarily within the transmission ranges of all others, i.e. some users may not be able to receive packets from others. Such an example is given in Figure 15.1. In the figure, the arrows indicate traffic flows and dashed circles, marked by italic numbers, denote the transmission ranges of the corresponding users. For simplicity, we assume that all users are synchronized and the transmission and interference ranges are identical, which are indicated by circles in Figure 15.1. The interference scenario in Figure 15.1 is summarized in Table 15.1. Note that the scheme described in this chapter can also be applied in networks where different users may have different transmission and interference ranges. Each transmitter has knowledge of its own channel state information (CSI) and makes an independent decision on its transmission. Since the transmitter knows the CSI, the transmitter will choose a modulation and coding scheme (MCS) according to the channel state to achieve good

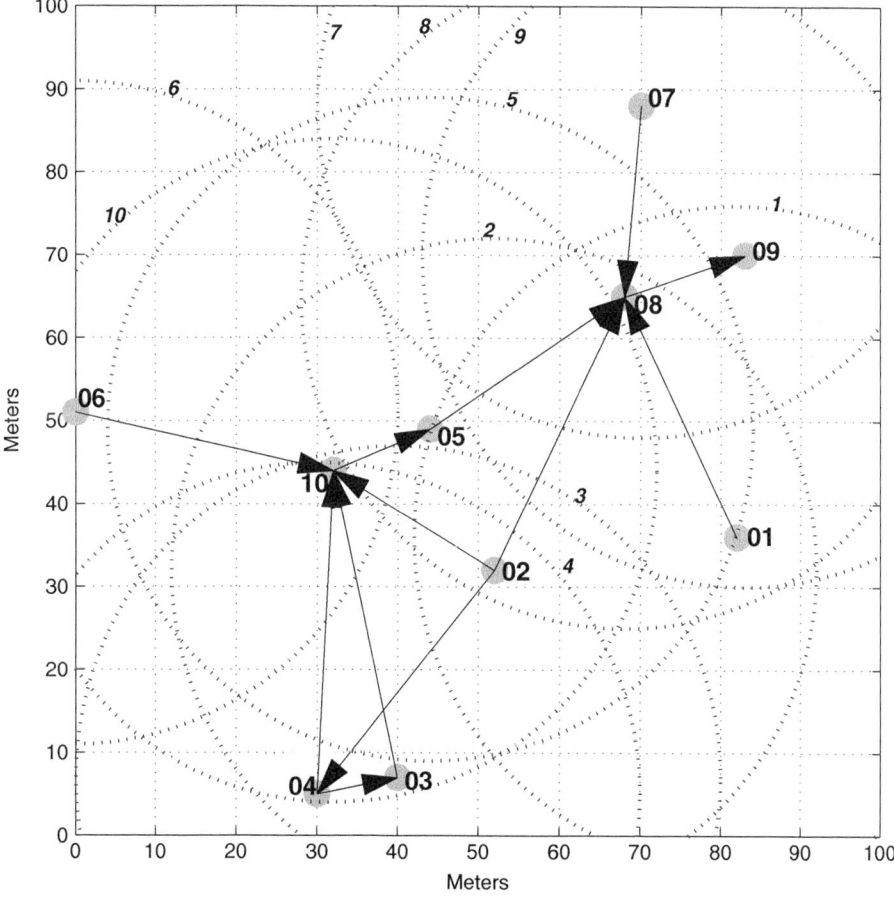

Figure 15.1 A network example

Table 15.1 Interference scenario

Users	Users within the interference range
1	2, 8, 9
2	1, 3, 4, 5, 8, 10
3	2, 4, 10
4	2, 3, 10
5	2, 8, 10
6	10
7	8, 9
8	1, 2, 5, 7, 9
9	1, 7, 8
10	2, 3, 4, 5, 6

link performance. For example, when the channel is in a deep fade, a low-rate MCS can be used to achieve robustness. However, since the transmitter cannot predict collisions with other users, its MCS selection is based on the assumption of no collision. As a result, a collision with other users destroys the robustness of selected MCS, and thus the transmission fails. Therefore, a receiver cannot decode a packet successfully if the channel is simultaneously used by another user within its interference range, i.e. a collision happens. For simplicity, assume a collision model for packet reception, i.e. if there is no collision, the reception is always successful; otherwise, the reception fails. Furthermore, assume that channels between pairs of users are reciprocal. In other words, when there is no interference, user 1 can receive a signal from user 2 if and only if user 2 can receive a signal from user 1 with the same channel gain. Each user may choose to send packets to or receive packets from different users and each traffic flow always has data to send.

The backoff-after-collision approach in most existing random access schemes can resolve contention. However, it ignores channel and multi-user diversity in wireless communications, and deferring transmission without considering channel variation may result in data communications in deep fades and waste of channel resources. To fully exploit network diversity, we should use channel CSI to control the random access such that a user with a favorable channel condition has a higher probability of contention success. Furthermore, data transmission should follow immediately after contention resolution as otherwise the channel may change to an unfavorable state. Besides, the contention should be resolved fast enough so that the channel does not vary much when the data are transmitted. We will take these into consideration in designing our new distributed random access scheme in the following sections. Since this novel scheme uses channel knowledge, it is called *channel-aware distributed medium access control* (CAD-MAC).

In CAD-MAC, the channel access time is divided into frame slots of length, T_f, as shown in Figure 15.2. Block fading is assumed [163], i.e. the frame length is smaller than the channel coherence time and the channel state remains constant within each frame slot, and may change from frame to frame. Let h_{ij} be the channel gain of Link (i, j), the one from user i to j, with probability density function $f_{ij}(h)$ and distribution

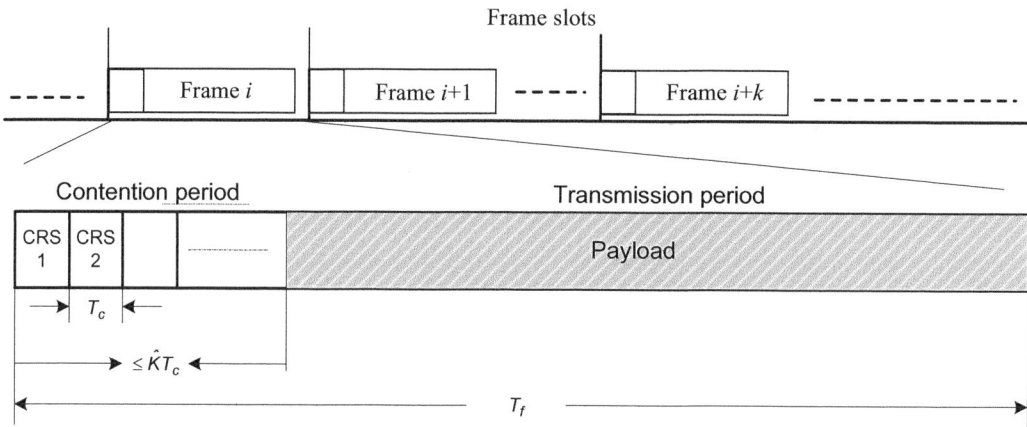

Figure 15.2 Channel-aware distributed medium access control

function $F_{ij}(h)$. Both $f_{ij}(h)$ and $F_{ij}(h)$ are assumed to be continuous to facilitate our discussion. For Rayleigh fading channels,

$$f_{ij}(h) = \frac{1}{\bar{h}_{ij}} e^{-\frac{h}{\bar{h}_{ij}}} \quad \text{and} \quad F_{ij}(h) = 1 - e^{-\frac{h}{\bar{h}_{ij}}}, \tag{15.1}$$

where $\bar{h}_{ij}$ is the average channel gain of Link (i,j). Channel gains of different links are assumed to be independent, but may have different distributions. The channel states of successive frames can be either correlated or independent and the operation of CAD-MAC does not rely on this correlation. Instead, the operation of CAD-MAC at frame slot t is determined only by the channel states in the current frame slot and the distribution $F_{ij}(h)$ of all links.

As shown in Figure 15.2, each frame slot consists of both contention and transmission periods. The contention period is used to resolve the collisions in a distributed way and users that have won the contention will transmit data with rate adaptation according to channel states in the transmission period while other interfering links keep silent. The contention period is further divided into *contention resolution slots* (CRSs) of length T_c, each for one contention resolution. Users may send access requests in each CRS and the contention is based on channel-aware Aloha. Unlike backoff-after-collision approaches, users will not back off several CRSs if the requests of different users collide. Instead, users may go on sending requests in each of the following CRSs depending on channel states and contention results in previous CRSs. The design of the contention protocol in each CRS is to assure that after sufficient CRSs, only one user in each local area, with the relatively best channel state within this area, will win the contention and transmit data. In this way, multi-user diversity is exploited effectively. For example, one contention resolution result in Figure 15.1 would be that links $(4,3)$ and $(7,8)$ win and send data while all other links keep silent. In this case, no other links should transmit as, otherwise, they would collide with either link $(4,3)$ or $(7,8)$.

The number of CRSs necessary for resolving all network contentions may vary across frames and depends on channel states and interfering scenarios. However, if a user has used $\widehat{K}$ CRSs and the contention has not been resolved, it will stay silent in this frame. Hence, the contention period has a variable time length that is no larger than $\widehat{K}T_c$. In Section 15.4, we will show that $\widehat{K}$ can be chosen such that $\widehat{K}T_c$ is far smaller than the frame length and the signaling overhead is trivial.

15.2   Channel-aware medium access control

In this section, we define the detailed contention rule. To summarize, at the beginning of each CRS, each active user compares its channel gain with its specific threshold, which may be updated in each CRS. If the channel gain is higher than the threshold, it will send a contention signal. Since wireless channel gains are random, the contention is randomized for all users. At the end of each CRS, there are three possible outcomes for each contending user: success, failure, or unknown. A success outcome implies that the contention signal experiences no collision and the user wins contention. A failure outcome implies that the user should keep silent in this frame slot to avoid interference to users that have succeeded. Users with unknown outcomes remain active and continue their contention in the following CRSs.

There are two types of contention in the network: type-I and type-II, which denote the contention among links with the same transmitter and with different transmitters, respectively. For example, the contention between links $(2,4)$, $(2,8)$, and $(2,10)$ in Figure 15.1 is type-I and that between links $(2,4)$ and $(4,3)$ is type-II. We do not consider the case that two users are sending traffic to each other since the reciprocal channel between them is always the same and they can negotiate to share the channel, e.g. in a static time division fashion, through control signaling, or use a slightly improved random access protocol. In Section 15.4, we will discuss the impact of this case briefly.

The type-I contention can be resolved by the transmitter as it has the CSI of all its links and can perform centralized scheduling. In this case, if all users experience independently and identically distributed (i.i.d.) channels, the max-signal-to-noise ratio (SNR) scheduler can be used [56], i.e. the user with the highest channel gain is scheduled to achieve the best performance. This is also fair since users will be scheduled with an equal probability. However, when the channels are not i.i.d., the max-SNR scheduler may result in fewer transmission opportunities for users with lower average channel gains. To resolve this issue, we can use the following scheduler, named the self-max-SNR scheduler. With the self-max-SNR scheduler, user i chooses link (i,j) that satisfies

$$j = \arg \max_l F_{il}(h_{il}), \tag{15.2}$$

where $F_{il}(h_{il})$ is the probability that the channel gain of link (i,l) is smaller than h_{il}. For example, suppose user 1 is sending traffic to users 2 and 3, and both links $(1,2)$ and $(1,3)$ are experiencing Rayleigh fading but with different average channel gains, 1 and 10, respectively. With self-max-SNR scheduler, user 1 will choose the link with higher

$F_{1l}(h_{1l}) = 1 - e^{-\frac{h_{1l}}{\bar{h}_{1l}}}$, i.e. the link with higher relative channel gain, $\frac{h_{1l}}{\bar{h}_{1l}}$. In this case, while the max-SNR scheduler will select link $(1, 3)$ much more often than $(1, 2)$, the self-max-SNR scheduler selects both links with equal probability and is fair. In general, the link with the highest $F_{il}(h_{il})$ is the one with the best instantaneous channel condition relative to its own average channel condition. Hence, criterion (15.2) effectively exploits the instantaneous multi-user diversity. Furthermore, $F_{il}(h_{il})$ is uniformly distributed between 0 and 1 for all links (i, l) from user i. Hence, these links have the same probability of being scheduled and the scheme is fair. Note that when all channels are i.i.d., the self-max-SNR scheduler is the same as the max-SNR scheduler. The self-max-SNR scheduler is different from the distribution fair scheduling rule in [211], which maximizes the expected sum of the distribution functions of all users.

Users selected by the self-max-SNR scheduler are involved in the type-II contention and further contend with each other for channel access in a distributed way. Hence, type-II contention is always resolved after type-I contention. In type-II contention, a link with a better channel state should have a higher probability of success. The contention period is used to resolve this type of contention. The basic idea is to resolve the contention in each CRS, and, after one CRS, links with higher gains are selected in a distributed way to continue contention in the following CRSs. Finally, only one link is selected within each local area and its interferers are informed to keep silent in the current frame slot. To facilitate the design of type-II contention, REQUEST, BUSY, SUCCESS, IDLE, and OCCUPIED signals are defined as follows:

- **REQUEST:** sent by a transmitter to request access;
- **BUSY:** sent by a receiver to deny access;
- **SUCCESS:** sent by a receiver to allow access;
- **IDLE:** sent by a receiver to petition for access;
- **OCCUPIED:** sent by a transmitter to prevent neighbors from data reception in this frame.

To resolve type-II contention, each CRS consists of the following three steps of signaling exchange, which are illustrated in Figure 15.3. The following procedure is used by all users involved in type-II contention, where we assume user i has chosen j after resolving type-I contention.

1. *Transmitters send REQUEST*: If user i has neither received a BUSY signal from j nor detected a SUCCESS signal destined to others, i.e. is active, and

$$h_{ij} > \widehat{H}_{ij}[k], \tag{15.3}$$

where $\widehat{H}_{ij}[k]$ is a predetermined threshold that is adjusted CRS-by-CRS, then it sends REQUEST to user j. The choice of the threshold $\widehat{H}_{ij}[k]$ will be described in Section 15.3.

2. *Receivers notify BUSY, SUCCESS, or IDLE*:

   - **BUSY:** User j responds BUSY if it receives REQUEST correctly and has received OCCUPIED in the previous CRSs.

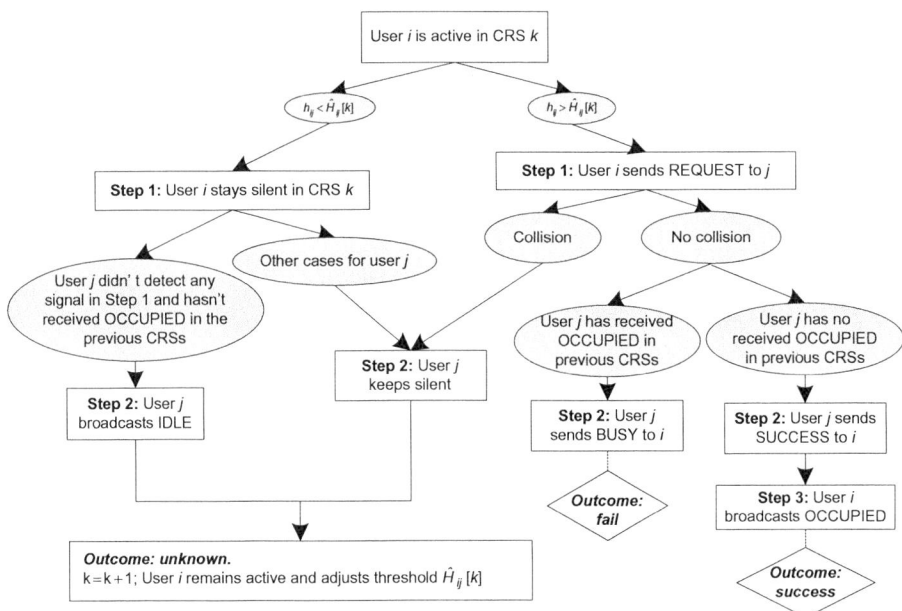

Figure 15.3 Signaling exchange of CAD-MAC between users i and j in one CRS

- **SUCCESS:** User j responds SUCCESS if the REQUEST is received correctly and no OCCUPIED signals have been received in the previous CRSs.
- **IDLE:** User j broadcasts IDLE to all users that want to send traffic to user j if no OCCUPIED signals have been received in the previous CRSs and no signals detected at Step 1.

Note that the BUSY or SUCCESS feedback is sent only when there is no collision, i.e. the contention succeeds.

3. *Transmitters broadcast OCCUPIED and start sending data*: If user i has received SUCCESS, it succeeds in the contention and broadcasts OCCUPIED in this step to notify those within its transmission range that they should not receive data in this frame slot.

Five typical contention processes are illustrated in Figure 15.4, where the solid arrows indicate signals between the observed pair of users and the empty arrows indicate signals sent from or detected by the interfering neighbors. As an example, observe the contention among only links $(6, 10)$, $(10, 5)$, and $(8, 9)$ in Figure 15.1. If all three links have good channel gains and send REQUEST in CRS 1, only user 9 receives REQUEST without collision and it sends back SUCCESS to user 8 at the second step while users 5 and 10 remain silent. At the third step, user 8 broadcasts OCCUPIED. Then CRS 2 starts. Users 6 and 10 may still send REQUEST, depending on the adjusted threshold. Suppose both send and only user 5 receives a collision-free REQUEST. At the second step, user 5 responds BUSY to user 10. Nothing happens at Step 3. In CRS 3, only user

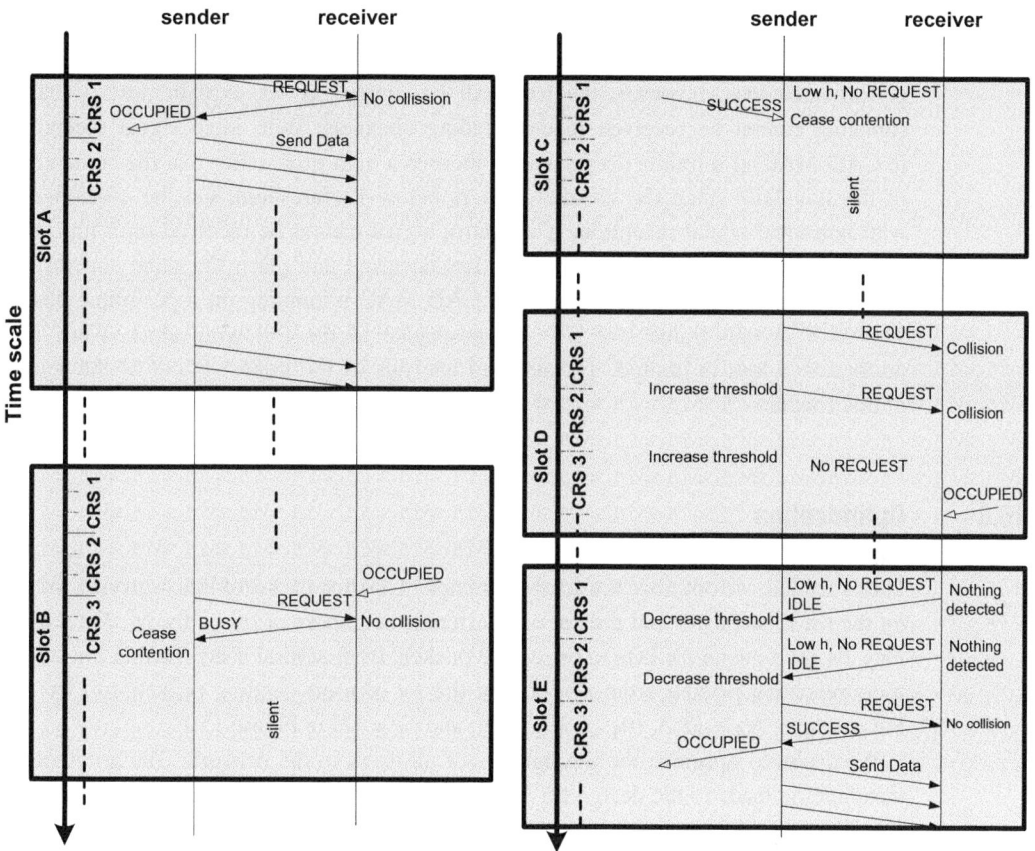

Figure 15.4 Flowcharts of typical access contention

6 may still send REQUEST and user 10 will respond SUCCESS. Finally, links $(6, 10)$ and $(8, 9)$ send data in this frame slot.

REMARK 1   *At Step* 2, *the BUSY or SUCCESS signals can always be received by user i correctly. This can be justified as follows. Suppose links (i_1, j_1) and (i_2, j_2) succeed in their contention and users j_1 and j_2 are sending BUSY or SUCCESS signals to i_1 and i_2, respectively. User i_1 does not interfere with j_2 and hence cannot receive any signal from j_2 since the channel is assumed to be reciprocal. Hence, user i_1 can receive the BUSY or SUCCESS signal without interference from j_2. Similarly user i_2 also receives the BUSY or SUCCESS signals correctly.*

REMARK 2   *At Step* 3, *the OCCUPIED signals may collide. However, as only the OCCUPIED signals are broadcasted in this step, and, if any signal is detected, it will be the OCCUPIED signal.*

In the following, we assume all control signals are received correctly if there is no collision. According to the design, errors in control signal reception do not affect the operation of CAD-MAC. This is because if an error occurs in receiving a control signal, the two sides of the link will act as if this signal was not sent and the network will continue to run accordingly. These errors may cause some performance loss; however, the loss will be limited as we explain next. Usually signaling cannot be received in deep fading channels. With ideal signal reception in CAD-MAC, if a link is deeply faded, there is a high probability that the contention of the link fails since the channel gain is below its threshold. On the other hand, with non-ideal signal reception, if a control signal cannot be received on a link, the transmitter may assume that a collision has occurred and reach the same conclusion that the contention failed. Furthermore, CAD-MAC is running on a per-frame basis. The error in one frame slot will not propagate to the following slots. Thus, the performance loss is limited. The detailed analysis of the impact is out of the scope of this chapter.

15.3   Optimization

In this section, we optimize the access parameters. Define the set of links carrying traffic by the set $\mathcal{L} = \{(i,j)\}$ and the set of interfering neighbors of user i by $\mathcal{N}_i$. Each user may choose to send packets to or receive packets from several users. Define the set of users receiving packets from i by $\mathcal{T}_i$ and the set of users sending packets to j by $\mathcal{S}_j$. For example, $\mathcal{N}_4 = \{2, 3, 10\}$, $\mathcal{T}_4 = \{3, 10\}$, and $\mathcal{S}_4 = \{2\}$ in Figure 15.1.

We desire to optimize the throughputs of all users in the network. The arithmetic-mean metric leads to the design for sum throughput maximization, but assures no fairness since some users may have zero throughput. The geometric-mean metric [136] takes both throughput and fairness among all users into consideration. Therefore, we will find the thresholds in (15.3) to maximize the geometric mean of the throughputs of all links, i.e.

$$\{\widehat{H}_{ij}^*[k]\} = \arg\max_{\{\widehat{H}_{ij}[k]\}} \prod_{(i,j)} T_{ij} = \arg\max_{\{\widehat{H}_{ij}[k]\}} \sum_{(i,j)} \log(T_{ij}), \tag{15.4}$$

where T_{ij} is the average throughput of link (i,j). By maximizing the geometric mean, proportional fairness can be achieved among all links in the network [165]. As shown in [165], a strategy achieving proportional fairness satisfies certain axioms of fairness and is a Nash arbitration strategy [148]. With proportional fairness, the network will be operated at a Pareto equilibrium, which corresponds to the situation where no user can improve its throughput without affecting at least one user adversely.

It is not feasible to globally optimize (15.4), because, after the contention in each CRS, new local knowledge is collected according to receiver feedback and the detection of signals broadcasted from neighboring users. This knowledge is generally different from one CRS to another and cannot be obtained in advance. To fully exploit this knowledge, the contention will be optimized sequentially, i.e. in a CRS-by-CRS way, and use newly collected knowledge to improve the contention behaviors afterward.

In the following, define the probability that user i sends a REQUEST to user j in CRS k by $p_{ij}[k]$. The overall probability that user i sends REQUESTs to other users in CRS k is

$$p_i[k] = \sum_{j \in \mathcal{T}_i} p_{ij}[k]. \tag{15.5}$$

15.3.1   CRS 1

We first optimize CRS 1. The throughput on link (i,j) out of CRS 1 is

$$T_{ij}[1] = R_{ij} p_{ij}[1](1 - p_j[1]) \prod_{m \in \mathcal{N}_j, m \neq i} (1 - p_m[1]), \tag{15.6}$$

where R_{ij} is the average data rate of payload transmission; $(1 - p_j[1]) \prod_{m \in \mathcal{N}_j, m \neq i} (1 - p_m[1])$ is the probability that neither user j nor the neighboring users except user i transmit, which means the successful contention of link (i,j) in CRS 1. In Figure 15.1, the transmission from user 2 to user 4 succeeds only when neither user 4 nor its neighbors excluding user 2, i.e. users in $\mathcal{N}_4 \backslash \{2\} = \{3, 10\}$, transmit. Hence, $T_{2,4}[1] = R_{2,4}p_{2,4}[1](1 - p_4[1])(1 - p_3[1])(1 - p_{10}[1])$.

The contention probability for CRS 1 is given by

$$\{p_{ij}^*[1]\} = \arg \max_{\{p_{ij}[1]\}} \sum_{(i,j) \in \mathcal{L}} \log(T_{ij}[1]). \tag{15.7}$$

Both $\log(p_{ij}[1])$ and $\log\left(1 - p_i[1]\right) = \log\left(1 - \sum_{j \in \mathcal{T}_i} p_{ij}[1]\right)$ are strictly concave functions of $p_{ij}[1]$. Hence, $\sum_{(i,j) \in \mathcal{L}} \log(T_{ij}[1])$ is strictly concave in $\{p_{ij}[1]\}$ and a unique global optimal $\{p_{ij}^*[1]\}$ can be determined by setting the first-order derivative of the objective function to be zero. The optimal contention probability can be readily obtained after some mathematical manipulations and is

$$p_{ij}^*[1] = \frac{1}{|\mathcal{S}_i| + \sum_{m \in \mathcal{N}_i} |\mathcal{S}_m|}, \tag{15.8}$$

which is the reciprocal of the total number of received traffic flows within the interference range of user i. Intuitively, as the number of interfered users increases, $p_{ij}^*[1]$, the contention probability of user i should decrease. Similar results can be found in [113] and [149] which design distributed access control strategies to achieve proportional fairness in traditional Aloha networks.

The threshold should be chosen to satisfy the contention probability in (15.8). According to Section 15.2, the contention probability of link (i,j) is

$$\begin{aligned} p_{ij}[1] &= \Pr\{(i,j) \text{ is chosen}; h_{ij} > \widehat{H}_{ij}[1]\} \\ &= \int_{\widehat{H}_{ij}[1]}^{\infty} f_{ij}(h) \Pr(j = \arg \max_{l \in \mathcal{T}_i} F_{il}(h_{il})) \mathrm{d}h \\ &= \int_{\widehat{H}_{ij}[1]}^{\infty} \Pr(F_{il}(h_{il}) < F_{ij}(h) : l \neq j) \mathrm{d}F_{ij}(h) \\ &= \frac{1}{|\mathcal{T}_i|} \left(1 - F_{ij}^{|\mathcal{T}_i|}(\widehat{H}_{ij}[1])\right), \end{aligned} \tag{15.9}$$

where $| \cdot |$ denotes the number of elements in the set.

From (15.9) and (15.8), the optimal threshold is

$$\widehat{H}_{ij}^*[1] = F_{ij}^{-1}(Y_{ij}[1]), \qquad (15.10)$$

where

$$Y_{ij}[1] = \left(1 - \frac{|\mathcal{T}_i|}{|\mathcal{S}_i| + \sum_{m \in \mathcal{N}_i} |\mathcal{S}_m|} \right)^{\frac{1}{|\mathcal{T}_i|}}. \qquad (15.11)$$

The optimal threshold (15.10) depends on the interference scenario of user i, i.e. the number of users receiving packets from user i, $|\mathcal{T}_i|$, the number of users sending packets to user i, $|\mathcal{S}_i|$, and the total number of users sending packets to the interfering neighbors of user i, $\sum_{m \in \mathcal{N}_i} |\mathcal{S}_m|$. The first two require only local knowledge while the third can be obtained through one-hop signaling exchange. This exchange incurs only a trivial signaling overhead since it needs to be triggered only when either a traffic session or the network topology changes sufficiently. Besides, this type of knowledge is typical in many protocols, such as routing discovery in mobile ad hoc networks [28, 101]. Hence, it can be readily obtained. Consider user 4 in Figure 15.1. $|\mathcal{T}_4| = 2$, $|\mathcal{S}_4| = 1$, $|\mathcal{S}_2| = 0$, $|\mathcal{S}_3| = 1$, and $|\mathcal{S}_{10}| = 4$. Hence, $\widehat{H}_{4,3}^*[1] = F_{4,3}^{-1}\left[(1 - \frac{2}{1+1+4})^{1/2} \right] = F_{4,5}^{-1}(0.667)$. If link $(4, 3)$ experiences Rayleigh fading with average gain h_a, $\widehat{H}_{4,5}^*[1] = 1.1 h_a$.

15.3.2   CRS k, $k > 1$

After one CRS, the thresholds of active users will be adjusted according to the contention result. If a collision happens for two or more users, then in the next CRS, these users will increase their thresholds properly to reduce the probability of collision. Otherwise, if no users send contention signals, then in the next CRS, these users will decrease their thresholds for higher contention probabilities. In the whole network, the thresholds of all users will be either decreased or increased in each CRS such that after sufficient CRSs, only one user within each local area has its channel gain above its threshold and this user has relatively the best channel among its adjacent users. Then this user wins the contention and transmits data.

From CRS 2, links whose transmitters have not been notified SUCCESS or BUSY continue the contention. The new threshold is chosen such that the contention probability is $p_{ij}[k]$. There are three possibilities for adjusting the threshold.

- *Adjustment* (AD) I (increase): If in the previous CRS, user i sent a REQUEST and no feedback is received, indicating a collision, all links involved in this collision should increase their thresholds to reduce the probability of collision. From previous knowledge, $h_{ij} > \widehat{H}_{ij}^*[k-1]$ and $h_{ij} < \widehat{H}_{ij}^M$, where $\widehat{H}_{ij}^M$ is the minimum threshold in all the previous CRSs such that $h_{ij} < \widehat{H}_{ij}^M$ and initially $\widehat{H}_{ij}^M = \infty$. The new threshold satisfies

$$\Pr\left(h_{ij} > \widehat{H}_{ij}^*[k] \, \middle| \, h_{ij} > \widehat{H}_{ij}^*[k-1], h_{ij} < \widehat{H}_{ij}^M \right) = p_{ij}[k]. \qquad (15.12)$$

Solving Equation (15.12) for $\widehat{H}_{ij}^*[k]$, we have

$$\widehat{H}_{ij}^*[k] = F_{ij}^{-1}(Y_{Iij}[k]), \tag{15.13}$$

where

$$Y_{Iij}[k] = (1 - p_{ij}[k])F_{ij}(\widehat{H}_{ij}^M) + p_{ij}[k] \cdot F_{ij}(\widehat{H}_{ij}^*[k-1]). \tag{15.14}$$

- AD II (decrease): If user i is still active and has received IDLE correctly from j in the previous CRS, indicating all contending users around user i, if any, including user i, have channel states below their thresholds, user i should decrease the threshold. Similar to the first case, the new threshold satisfies

$$\text{Pr}\left(h_{ij} > \widehat{H}_{ij}^*[k] \,\middle|\, h_{ij} < \widehat{H}_{ij}^*[k-1]; h_{ij} > \widehat{H}_{ij}^m\right) = p_{ij}[k], \tag{15.15}$$

where $\widehat{H}_{ij}^m$ is the maximum threshold in all the previous CRSs such that $h_{ij} > \widehat{H}_{ij}^m$ and initially $\widehat{H}_{ij}^m = 0$. Solving Equation (15.15), we have the decreased threshold

$$\widehat{H}_{ij}^*[k] = F_{ij}^{-1}(Y_{IIij}[k]), \tag{15.16}$$

where

$$Y_{IIij}[k] = p_{ij}[k] \cdot F_{ij}(\widehat{H}_{ij}^m) + (1 - p_{ij}[k])F_{ij}(\widehat{H}_{ij}^*[k-1]). \tag{15.17}$$

- AD III: In other cases, the threshold is kept the same, i.e.

$$\widehat{H}_{ij}^*[k] = \widehat{H}_{ij}^*[k-1]. \tag{15.18}$$

This usually happens when no REQUEST was sent and no IDLE was received in a previous CRS and user i temporarily quits the contention. In this case, user i would contend again only if it receives IDLE in the future CRSs.

Define all the competing links in CRS k by $\mathcal{L}[k]$. With the same approach as in CRS 1, where the optimal contention probability is given in (15.8), the optimal contention probability for $(i,j) \in \mathcal{L}[k]$ is

$$p_{ij}^*[k] = \frac{1}{|\mathcal{S}_i[k]| + \sum_{m \in \mathcal{N}_i[k]} |\mathcal{S}_m[k]|}, \tag{15.19}$$

where $\mathcal{S}_n[k]$ and $\mathcal{N}_n[k]$ are users that can contend in CRS k. A user may contend if and only if its threshold will be changed as in ADs I or II. However, who will adjust their thresholds is unknown to others and $p_{ij}^*[k]$ cannot be determined locally. Instead, we give a suboptimal approach as follows:

$$p_{ij}[k] = \begin{cases} \frac{1}{2}, & \text{AD I,} \\ p_{ij}[k-1], & \text{AD II.} \end{cases} \tag{15.20}$$

For AD II, an IDLE signal most likely indicates that no users sent REQUEST in the previous CRS and all active users remain active and still contend. Hence, the contention scenario is not changed and $p_{ij}[k]$ keeps the same. We assign one half for AD I because after the selection in CRS 1, in which the contending probability is (15.8), it is most likely that only one other link is contending with link (i,j) if a collision happens.

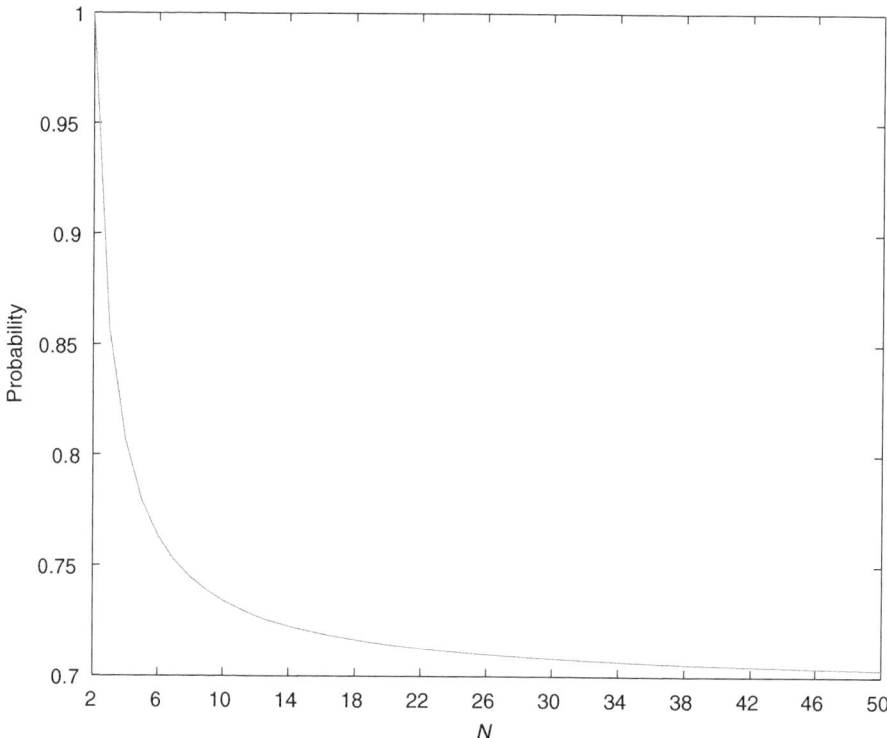

Figure 15.5 Probability that two users are involved in the collision once a collision happens

To further understand this, consider a network with N active links, each interfering with all others. Then initially, all links contend with probability $\frac{1}{N}$ according to (15.8). Figure 15.5 gives the probability that two users are involved in the collision once a collision happens; the number of users N varies from 2 to 50 and the probability is given below.

As there are N users in the network and each has contenting probability $\frac{1}{N}$, then the probability that only two users are involved in the collision once a collision happens is

$$
\begin{aligned}
&\text{Pr(two users in the collision} \,|\, \text{a collision happens})\\
&= \frac{\text{Pr\{two users involded in collision\}}}{\text{Pr\{a collision happens\}}}\\
&= \frac{\binom{N}{2}(\frac{1}{N})^2(1-\frac{1}{N})^{N-2}}{\sum_{k=2}^{N}\binom{N}{k}(\frac{1}{N})^K(1-\frac{1}{N})^{N-K}}\\
&= \frac{\frac{1}{2}\frac{N}{N-1}}{\frac{1}{(1-\frac{1}{N})^N}-1-\frac{N}{N-1}} \xrightarrow{N\to\infty)} \frac{1}{2(e-2)} \approx 0.70.
\end{aligned}
\tag{15.21}
$$

We can see this probability is large and converges to 0.7. Hence, it is most likely that only one other link is contending with link (i,j) if a collision happens, and $\frac{1}{2}$ is a good estimate.

From (15.8) and (15.20), the contention probability $p_{ij}[k]$ will always be $\frac{1}{|\mathcal{S}_i|+\sum_{m\in\mathcal{N}_i}|\mathcal{S}_m|}$ if nobody sends REQUEST in the first several CRSs until a collision happens and AD I is used. After a collision, if user i is still active, $p_{ij}[k]$ is always $\frac{1}{2}$ for simplicity.

With Rayleigh fading, the threshold in each CRS is

$$\widehat{H}_{ij}^*[k] = (-\overline{h}_{ij})\ln(1-X), \tag{15.22}$$

where X is given by (15.11), (15.14), or (15.17) depending on the contention state. The transmission condition in (15.3) equals

$$\frac{h_{ij}}{\overline{h}_{ij}} > -\ln(1 - Y_{ij}[1]). \tag{15.23}$$

While different links may have different channel distributions, the relative channel gain, $\frac{h_{ij}}{\overline{h}_{ij}}$, is identically and independently distributed (i.i.d.) for all links. Hence, the relative channel gain, rather than the absolute one, determines the contention of each link and users with higher relative channel gains have higher probabilities of contention success. Therefore, both throughput and fairness are optimized for all users.

15.4 Robustness analysis

In this section, we analyze the robustness of CAD-MAC. We say a link wins the contention if it transmits data in the transmission period. To simplify the analysis, we assume all signals are received correctly.

The following theorem states that CAD-MAC can completely resolve the network contention and is proved in Appendix A.9:

THEOREM 15.1   *With probability one, the contention of networks with any topology can be completely resolved by CAD-MAC if sufficient CRSs are allowed.*

Theorem 15.1 indicates that CAD-MAC solves both the hidden node and exposed node problems in wireless networking completely with probability one.

To illustrate Theorem 15.1, one example is given in Figure 15.6 where the two channels are i.i.d. When $h_{12} = h_{34}$, users 1 and 3 have the same update of thresholds and their REQUESTs always collide. However, the probability that $h_{12} = h_{34}$ is zero because the two channels are independently fading with continuous probability distribution function

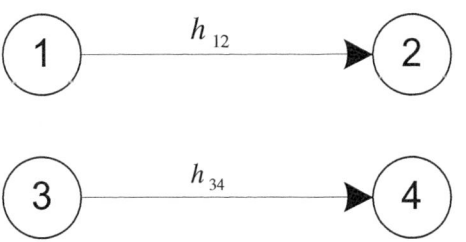

Figure 15.6 A network in which all interfere with others

$F_{ij}(h)$. Hence, with probability one the contention of this network can be completely resolved by CAD-MAC if sufficient CRSs are used. In Section 15.2, we assume that in the network no two users are sending traffic to each other at the same time. This is needed so that the contention can be resolved with probability one. To illustrate this, consider a network with only users 1 and 2, who are sending data to each other. If the channel is not reciprocal, the contention can be resolved with probability one. However, because of channel reciprocity, h_{12} and h_{21} are always the same. Users 1 and 2 have the same update of their thresholds and their REQUESTs always collide. The contention will not be resolved using CAD-MAC. This can be fixed by improving CAD-MAC slightly. For example, in the first step of each CRS, on top of the existing procedures, each user sends the REQUEST only if an independent random variable falls within a certain range with a probability $0 < p < 1$. Then the contention can be resolved with probability one.

Theorem 15.1 indicates that CAD-MAC achieves performance comparable to that of a centralized scheduler. Assume that in the transmission period, CAD-MAC uses the same rate adaptation as that of the centralized scheduler. Then compared to the centralized scheduler, CAD-MAC loses throughput because of the CRS overhead for resolving network contention. Define the throughputs of CAD-MAC and the centralized scheduler by $T_{CAD-MAC}$ and $T_{Centralized}$, respectively. Then we define the efficiency, γ, of CAD-MAC as follows:

$$\gamma = \frac{T_{CAD-MAC}}{T_{Centralized}} = 1 - \frac{\overline{K}T_c}{T_f}, \tag{15.24}$$

where $\overline{K}$ is the average number of CRSs necessary for completely resolving the network contention. In the following, we show that $\overline{K}$ is bounded regardless of the network type and size. To simplify the analysis, we assume in the following that a link contends again only if all neighbors of the receiver have resolved their contention and the receiver sends IDLE to the receiver since it can still receive data. Besides, we assume sufficient CRSs.

First, consider the case that each link interferes with all others and only one link wins the contention in each frame slot, such as in a network where all users send traffic to a common receiver or a small-scale ad hoc network where each user is within the transmission range of all others. For a network with N traffic flows, each interfering with all others, an upper bound of $\overline{K}_N$ is given by the following theorem and is proved in Appendix A.10:

THEOREM 15.2 *For a network with N links, each interfering with all others, the average number of CRSs necessary to completely resolve the network contention satisfies*

$$\overline{K}_N \leq \frac{\widehat{M}_N}{1 - (1 - \frac{1}{N})^N} + \frac{(1 - \frac{1}{N})^N}{(1 - (1 - \frac{1}{N})^N)^2}, \tag{15.25}$$

where

$$\widehat{M}_N = \sum_{n=1}^{N} \binom{N}{n} \left(\frac{1}{N}\right)^n \left(1 - \frac{1}{N}\right)^{N-n} (\log_2(n) + 1).$$

Furthermore,

$$\overline{K}_N < \overline{K}_\infty \le 2.43. \tag{15.26}$$

Based on Theorem 15.2, the following theorem gives a universal upper bound of $\overline{K}$ for any type of network and is proved in Appendix A.11.

THEOREM 15.3   *For a network of any type and size, the average number of CRSs necessary to completely resolve the contention satisfies*

$$\overline{K} < \frac{2.43 \cdot \overline{L}}{\beta}, \tag{15.27}$$

where the transmission coexistence factor, $\overline{L}$, is the average number of links that win the contention in one frame slot and the contention coexistence factor, β, is the average number of simultaneous resolutions in each CRS.

In Theorem 15.3, the contention coexistence factor, β, is related to the number of simultaneous resolutions that occur in each CRS. Here one resolution is the process where all links, among which only one link will win, adjust their thresholds using ADs I or II. Since multiple links may win in one frame slot, the resolutions that lead to the wins of these links may happen in the same CRS, and β characterizes this overlap. Obviously, both $\overline{L}$ and β depend on the distribution density and transmission range of all users.

For example, if each user interferes with all others and only one link wins, then $\overline{L} = 1$, $\beta = 1$, and $\overline{K} < 2.43$ as in Theorem 15.2. If a network consists of two groups of users and the communication within the different groups does not interfere with each other, then these two groups can resolve their contention within themselves to produce two winners. Consequently, $\overline{L} = 2$. Define the CRSs for a two-cell network using different frequency sets in the two cells to resolve the contention to be $\mathcal{K}_1 = \{1, 2, \ldots, k_1\}$ and $\mathcal{K}_2 = \{1, 2, \ldots, k_2\}$, respectively, where k_1 and k_2 are random and vary from one frame to another. Then the resolution overlaps from CRSs 1 to $\min\{k_1, k_2\}$ and there is only one resolution from CRSs $\min\{k_1, k_2\} + 1$ to $\max\{k_1, k_2\}$. Consequently, according to the proof,

$$\beta = \frac{\mathbf{E}(k_1 + k_2)}{\mathbf{E}(\max\{k_1, k_2\})}, \tag{15.28}$$

and

$$\overline{K} < \frac{4.86 \cdot \mathbf{E}(\max\{k_1, k_2\})}{\mathbf{E}(k_1 + k_2)}. \tag{15.29}$$

From Theorems 15.1 and 15.3, we have the following proposition:

PROPOSITION 15.4   *The efficiency of CAD-MAC satisfies*

$$\gamma > 1 - \frac{2.43 \cdot \overline{L}T_c}{\beta T_f}. \tag{15.30}$$

For a network where each user interferers with all others, the efficiency is

$$\gamma > 1 - \frac{2.43 \cdot T_c}{T_f}. \tag{15.31}$$

T_c and T_f are determined by the round-trip time of signal propagation and the channel coherence time, respectively. If $T_f \gg T_c$ as in slow-fading channels, CAD-MAC performs almost the same as the centralized scheduler that, in practice, cannot be implemented due to poor scalability and the huge overhead of CSI collection. For example, it is shown in [112] that the round trip time for 802.11 wireless *local area networks* (WLAN) is within 10 μs and for cellular networks, with 6 km radius, is within 50μs. The channel coherence time is hundreds of milliseconds in an indoor office or home environment and tens of milliseconds in cellular networks with 900 MHz carrier frequency and user speed 72 km/h [27]. Hence, in both WLAN and cellular networks, the efficiency of CAD-MAC is close to unity.

With non-ideal channel estimation, users have imperfect CSI $\{\widetilde{h}_{ij}\}$ and $\{\widetilde{F}_{ij}()\}$ instead of the true CSI $\{h_{ij}\}$ and $\{F_{ij}()\}$. The relationship between $\widetilde{h}_{ij}$ and h_{ij} or $\widetilde{F}_{ij}()$ and $F_{ij}()$ is not specified. Therefore, the uncertainty of the channel estimation can be arbitrary. Users run CAD-MAC based on the channel estimates $\{\widetilde{h}_{ij}\}$ and $\{\widetilde{F}_{ij}()\}$. As shown in the appendices, the proofs of Theorems 15.1, 15.2, and 15.3 are independent of the channel distribution of any user. Hence, they also hold for the operations of CAD-MAC based on $\{\widetilde{h}_{ij}\}$ and $\{\widetilde{F}_{ij}()\}$. This is because regardless of the type of channel estimate users have, their thresholds will be configured such that their contending probabilities are given in (15.8) or (15.20), which indeed determines how all users contend with each other. Furthermore, suppose the centralized scheduler compared in (15.24) is also based on the imperfect channel knowledge. Then the efficiency of CAD-MAC is still given by (15.24). Therefore, we have the following theorem about the robustness of CAD-MAC:

THEOREM 15.5   *The conclusions in Theorems 15.1, 15.2, and 15.3 and Proposition 15.4 hold when all users have imperfect channel knowledge and CAD-MAC is robust to any channel uncertainty.*

Theorem 15.5 indicates that CAD-MAC can be used even when the CSI is not available. In this case, the transmitter can assume a certain channel distribution, e.g. uniform distribution. In each frame slot, the transmitter first generates a random channel gain following the channel distribution and uses CAD-MAC to decide its random access policy. In this case, CAD-MAC works like traditional MAC approaches and the packet transmission will be independent of wireless channel states. However, whenever the transmitter is able to obtain a certain amount of CSI, CAD-MAC will be able to exploit this knowledge to improve network performance through enabling users with better channel states to access the channel with higher probabilities.

15.5   Simulation results

In this section, we demonstrate the performance of CAD-MAC in a network with random topologies. First, we illustrate how CAD-MAC operates given a network instance.

Then we show the probability mass function of the number of CRSs that are used to completely resolve the network contention. Finally, we compare CAD-MAC with DOMRA and the 802.11 request to send / clear to send (RTS/CTS) scheme, to demonstrate performance gain.

In each simulation trial, users are randomly and uniformly distributed in a square area with side length of 100 meters. Each user has a transmission range of 40 meters and selects neighboring users randomly for data transmission. The number of selected receivers is uniformly distributed between one and half of the number of neighboring users. A network topology in one trial is illustrated in Figure 15.1. The Rayleigh block-fading channel with average fading level, h_o, is assumed. Hence, the cumulative distribution function (CDF) is $F(h) = 1 - e^{-\frac{h}{h_o}}$. The data rate in each frame is given by $R(h) = W \log_2(1 + \frac{hP}{N_o})$, where $W = 100$ kHz is the system bandwidth, $P = 0.01$ W, is the transmit power, and $N_o = 0.0001$ W, is the noise power. The channel gains are independent with either the same or different averages. $h_o = 1$ for homogeneous channels; h_o is uniformly distributed between 0.5 and 1.5 for different links for heterogeneous channels. The length of each frame slot is 200 ms and each CRS is 2 ms long.

First, consider the network topology given in Figure 15.1 and let $h_o = 1$. The signaling flow of the contention process for a set of channel states in a frame slot is illustrated by Table 15.2, where blanks indicate no values or no actions. Each user chooses a receiver with the best channel gain, e.g. user 2 selects user 4. In the first CRS, links $(4, 3)$ wins access. Users $2, 3$, and 10 detect SUCCESS and decide to stop contention since some neighboring users will receive data in this frame slot. In the second CRS, only users $1, 5$, 7, and 8 contend but none sends REQUEST even though their thresholds are lowered. In the third CRS, only user 7 sends REQUEST and wins the contention. Hence, three CRSs completely resolve type-II contention and links $(4, 3)$ and $(7, 8)$ will send data in this frame slot. Note that this result also fully exploits the network capacity as transmission of any other users will produce interference and reduce the network throughput.

Figure 15.7 shows the probability mass function of the number of CRSs for completely resolving contention. To verify the impact of network load, we run simulations with 5, 10, 15, or 20 randomly distributed users. For each case, we run 1000 trials, each of which contains transmission of 5000 frame slots. We see that a heavier network load requires only slightly more CRSs. The average numbers of CRSs in these four cases are 2.35, 3.74, 4.92, and 6.00, while the corresponding standard deviations are 1.66, 2.39, 3.11, and 4.00, respectively. Theorem 15.3 is also verified. For example, when there are 20 users, $\overline{L} = 3.4$, $\beta = 1.34$, and $\overline{K} = 6.00 < 2.43 * 3.4/1.34 = 6.2$.

Figure 15.8 compares the throughput of CAD-MAC,DOMRA, and the 802.11 RTS/CTS scheme [115]. The channel is heterogeneous. DOMRA is a channel-aware slotted Aloha scheme and achieves near optimal network performance for Aloha-based channel-aware schemes. DOMRA exploits CSI but allows collisions of whole data packets. RTS/CTS has signaling negotiation ahead of data transmission to avoid data collisions but does not exploit CSI. CAD-MAC has the advantages of both and achieves performance comparable to that of centralized schedulers, as shown in Section 15.4. In Figure 15.8, significant performance improvement can be observed. Here, we vary the number of active users in the network and for each number of users, we run 5000

Table 15.2 Contention process for a set of channel states in Figure 15.1

User i	1	2	3	4	5	6	7	8	9	10
Receivers	8	4;8;10	10	3;10	8	10	8	9	9	5
Channel gains h	**0.66**	**1.36**;0.63;0.61	**0.91**	**2.98**;1.36	**0.49**	**1.33**	**0.94**	**0.23**		**0.11**
Selected receiver j	8	4	10	3	8	10	8	9		5
CRS 1 $H_{ij}[1]$	1.61	2.30	1.79	1.70	2.20	1.39	1.61	1.79		1.95
Step 1				REQ						
Step 2	IDL	TKN	TKN	SUC	IDL		IDL	IDL		TKN
Step 3				OCP						
CRS 2 $H_{ij}[2]$	1.02				1.56	1.39	1.02	1.19		
Step 1										
Step 2	IDL				IDL		IDL	IDL		
Step 3										
CRS 3 $H_{ij}[3]$	0.72				1.21	1.39	0.72	0.86		
Step 1							REQ			
Step 2	TKN				TKN		SUC	TKN		
Step 3							OCP			

TKN: detect SUCCESS of others and stop contention; REQ: send REQUEST; SUC: feed back SUCCESS; OCP: broadcast OCCUPIED; IDL: send IDLE to transmitter; BSY: feed back BUSY.

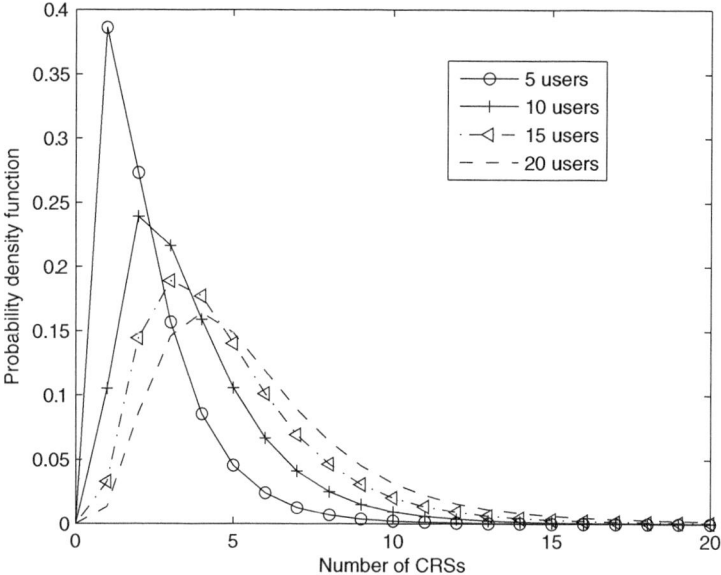

Figure 15.7 Probability mass function of the number of CRSs necessary for complete contention resolution

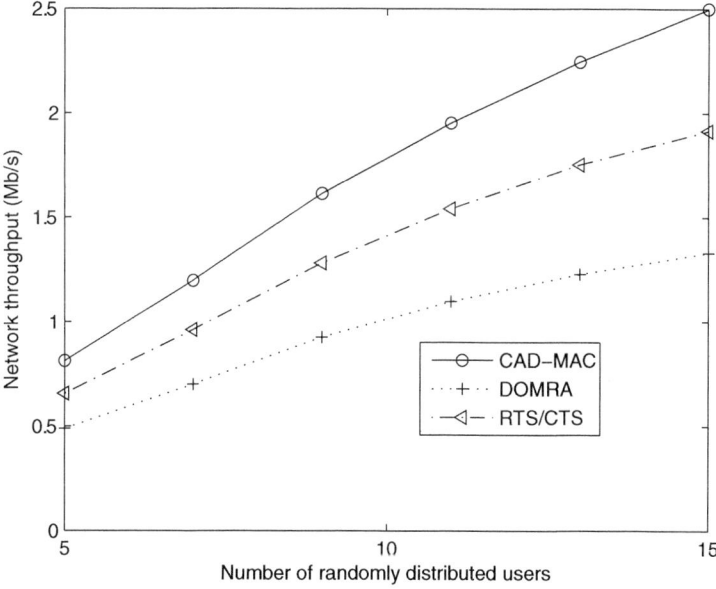

Figure 15.8 Network throughput comparison

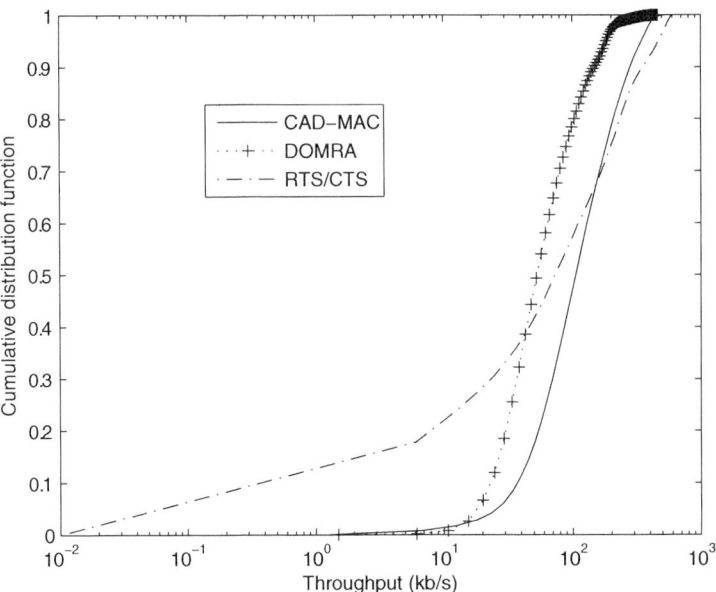

Figure 15.9 Cumulative distribution function of link throughput

trials, each with 1000 seconds of simulation time. When there are 15 active users, the throughput of CAD-MAC outperforms DOMRA by approximately 90% and RTS/CTS by approximately 30%. Figure 15.9 further compares the throughput CDFs of all links when there are 15 users in the network and the curves come from the statistics across all 5000 trials. In the CDF curves, lower throughput is the throughput of links with more competing neighbors as they have fewer chances of channel access, and vice versa. We can see that CAD-MAC improves the performance of almost all links compared with DOMRA. RTS/CTS is slightly better than CAD-MAC for about 35% of the links that have high throughput. However, CAD-MAC performs significantly better than RTS-CTS for the remaining 65% of the links. This is because the design of RTS/CTS does not have an inherent concept of fairness for networks of arbitrary topologies. When there is asymmetry in contention, RTS/CTS is unfair and biased toward giving more access opportunities to links with fewer competing neighbors, a phenomenon already observed in [15]. For links with more competing links, RTS/CTS allocates much fewer access opportunities and thus these links have poor throughput performance. The design of CAD-MAC achieves fairness in both type-I and type-II contention scenarios. It automatically balances the access opportunities among all links according to the spatial distribution of the traffic flows. If we look at the throughput of users in densely populated areas, i.e. those with the lowest throughput in Figure 15.9, CAD-MAC improves the throughput by about 1000 times compared to RTS/CTS. CAD-MAC, therefore, is a good solution for densely deployed wireless networks.

16 Opportunistic random access with intelligent interference avoidance

In the previous two chapters, we have introduced how to design optimal channel-aware medium access strategies that perform close to that of centralized schedulers. We have considered a very generic network model, where the network can be of an arbitrary topology. The solution designed can be applied in many types of wireless networks, e.g. cellular networks, ad hoc networks, sensor networks, and so on. To simplify the design, we have assumed a collision model, i.e., the transmission and interference ranges are fixed and a packet will be received correctly only if the receiver is within the transmission range and there are no other simultaneous transmissions within the interference range. In practice, there are no fixed transmission or interference ranges and the packet reception depends on the signal to interference plus noise ratio (SINR) of received signals. Therefore, some improvements are necessary to apply the channel-aware medium access control (MAC) schemes. As we can see from the designs, to run the protocols only the optimal thresholds in (14.16) for decentralized optimization for multi-channel random access (DOMRA) and (15.10) for channel-aware distributed medium access control (CAD-MAC) are related to the transmission and interference ranges. So the question would be how to determine the thresholds in practice, or, in other words, how to determine the severe interference relationship of all users in the network, as the thresholds are determined by the severe interferers. After determining the threshold, the data transmission and reception can follow normal communication procedures. To illustrate how to use the distributed channel-aware access protocols in practice, we will show the application of the DOMRA protocol in cellular networks to improve the performance of cell-edge users that experience heavy co-channel interference (CCI).

A commonly used method to avoid CCI is to assign different sets of channels to neighboring cells [14, 7] and a good summary of channel assignment can be found in [79]. One recent popular approach to reducing interference for cell-edge users is through fractional frequency reuse (FFR) [109, 63, 132, 20, 21]. With FFR, a lower frequency reuse is specified for users at cell edges, while full frequency reuse is applied for those at cell centers. This improves the throughput of cell-edge users since they experience lower levels of interference. To further improve frequency-reuse efficiency, CCI can be mitigated by advanced digital signal-processing techniques [97, 88, 100, 86]. However, these techniques have high complexity and therefore result in high costs for mobile terminals (MTs).

In this chapter, we will show that with a very simple threshold design all MTs will be able to recognize severe interferers in a dynamic and intelligent way. Using channel-aware Aloha as the downlink transmission protocol, severe interference can be randomized for improved network performance. This approach will be named *cochannel interference avoidance MAC* (CIA-MAC). It outperforms conventional schemes because of its high frequency-reuse and intelligent interference avoidance. The CIA-MAC scheme maintains backward compatibility and requires only minor changes to existing base stations (BSs), while no improvement is necessary for mobile terminals (MTs). Low overhead is added as the scheme requires only limited signaling coordination among BSs at a semi-static level. Only occasional cooperation is required when the network topology is changed and the instantaneous coordination at the packet level is not required. Although the use of randomization for collision avoidance is used extensively for uplink random access channels and wireless local area network (WLAN) systems, we will show it is also possible to use the principle for automatically controlling the level of downlink interference per link in cellular networks.

16.1   Intelligent interferer recognition

We only consider downlink transmission since complicated multi-user detection and CCI cancelation algorithms can be implemented at the BS for uplink CCI mitigation. The MTs at cell edges not only face the weakest signals but also suffer the largest amount of interference from neighboring cells. The CIA-MAC scheme targets performance improvement of these users. In general, users are categorized into two classes: those experiencing no or slight CCI and those suffering from severe CCI. The first class will be scheduled by the traditional centralized MAC and the second will be first accepted by the traditional call admission control policies and then scheduled by CIA-MAC.

Each MT measures the average *interference-to-carrier ratio* (ICR) of neighboring BS k, which is defined to be

$$ICR_k = \frac{E(h_k P_k)}{E(hP)}, \qquad (16.1)$$

where h and P are the channel power gain and the transmit power of the desired link while h_k and P_k correspond to those of the interfering link from BS k. $E()$ is the average of the past data and tracks slow-fading, i.e. it is a local mean and averages the effect of fast fading [79]. This definition of average will also apply in the following paragraphs.

Severe interferer: *If the ICR from neighboring BS k satisfies $ICR_k \geq \Gamma_m$, the transmission of BS k always causes the failure of packet reception, where Γ_m, called trigger, is a predetermined severe interference threshold. BS k is called a severe co-channel interferer.*

In Section 16.3, we will discuss the trigger selection and present examples.

If all BSs causing severe interference keep on transmitting, the packet receptions of the interfered MTs always fail. If BSs can collaborate, the interfering BSs may transmit in turn. However, this incurs huge signaling overhead. If there is no collaboration among BSs, we let BSs transmit randomly when identified as severe interferers.

Their transmission should be managed such that the overall network performance as well as the fairness among all users are jointly optimized. Hence, besides traditional MAC, a complementary MAC is used for optimizing the randomized transmission of the severe interferers. The new complementary MAC aims at improving cellular throughput through *co-channel interference avoidance*, and is, therefore, called CIA-MAC.

In the following, we only consider BSs and MTs controlled by CIA-MAC, and CIA-MAC will be optimized by cross-layer design to fully exploit the system capacity and multi-user diversity while maintaining fairness.

The following definitions will be used in the subsequent discussion:

- $\mathcal{B} = \{1, 2, \ldots, M\}$: set of BSs.
- $\mathcal{M} = \{1, 2, \ldots, N\} = \bigcup_{i \in \mathcal{B}} \mathcal{M}_i$: set of MTs. $\mathcal{M}_i$ is the set of MTs in the cell of BS i. Obviously, $\mathcal{M}_i \bigcup \mathcal{M}_j = \emptyset, \forall i \neq j$.
- $\mathcal{E} = \{(i, j) | i \in \mathcal{B}, j \in \mathcal{M}_i\}$: set of transmission links; (i, j) denotes the link from BS i to MT j.
- $\mathcal{N}_m = \{(i, j) | \forall (i, j) \in \mathcal{E}$, transmission at link (i, j) causes severe interference to MT $m\}$: set of links whose transmission will bring severe interference to MT m;
- $\mathcal{T}_{(i,j)} = \{m | \forall m \in \mathcal{M}$, MT m is severely interfered with by the transmission at link $(i, j)\}$: set of MTs severely interfered with by transmission at link (i, j).

Figure 16.1 demonstrates an example. The solid lines represent data transmission links and the dashed ones represent links from severe interferers. We have $\mathcal{B} = \{1, 2, \ldots, 7\}$, $\mathcal{M} = \{1, 3, 5, 7, 8, 9, 10\}$, and $\mathcal{E} = \{(1, 1), (2, 5), (3, 3), (6, 8), (7, 7)\}$. The transmission from BS 1 to MT 1 on channel 1 is severely interfered with by transmission from BS 2 to MT 5. Meanwhile, BS 1 also causes severe CCI to MTs 3, 7, and 8. Hence, $\mathcal{N}_1 = \{(2, 5), (3, 3)\}$ and $\mathcal{T}_{(1,1)} = \{3, 7, 8\}$.

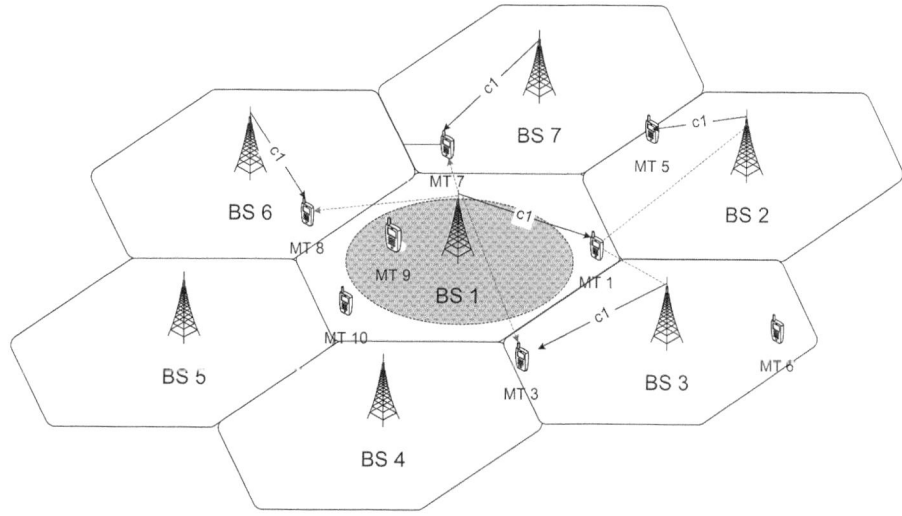

Figure 16.1 Co-channel interference in cellular networks with a reuse factor of one

16.2   Co-channel interference avoidance MAC

In this section, we describe the principle of CIA-MAC. Figure 16.2 shows the MAC transmission and the frame structure. Transmission time is divided into slots with length S. The MAC layer of each link independently sends a MAC frame at the beginning of each slot with probability p. Complete *channel state information* (CSI) is known and used to determine the MAC contention and the physical (PHY) link adaptation. To obtain CSI at a transmitter [191], the CSI can be estimated through pilots at the receiver and sent to the transmitter, or CSI is already available at the transmitter whenever the channel is reciprocal, such as in a time-division duplex system. Incomplete CSI results in some performance loss and the study of its impact is out of the scope of this chapter. Furthermore, assume ideal *cyclic redundancy check* (CRC). Any error inside a frame will result in the drop of the frame. Errors are uniformly and independently distributed. Each frame has L_f symbols, of which L_d symbols carry data. Once the MAC layer makes a decision to transmit a frame, the frame will be continuously transmitted by the PHY layer until the frame is sent out.

In a traditional network, MAC makes the transmission decision based on buffer status and *quality of service* (QoS) requirements and does not use PHY knowledge at all. When MAC decides to transmit, the physical channel may be in a deep fade, which wastes bandwidth and power resources. Alternatively, MAC may decide on no transmission while the channel is experiencing high gain. With cross-layer design, MAC decides whether to transmit or not according to channel information. We assume a block-fading channel [163], i.e. the channel state remains constant during each MAC frame. If the channel power gain at a time slot, h, is above a predetermined threshold $\overline{H}$, MAC sends a frame. As wireless channels are inherently random, the MAC transmission is also randomized and the threshold $\overline{H}$ determines the transmission probability. The thresholds $\overline{H}$ and the PHY transmission should be jointly optimized for all BSs subject to their power constraints. Each BS maximizes its throughput with both average power constraint P_a and instantaneous power constraint P_m while assuring fairness to the users in other cells.

DOMRA will be used to optimize the operations of this network. Define the probability cumulative distribution function of channel power gain as $F(h)$ and the cardinality

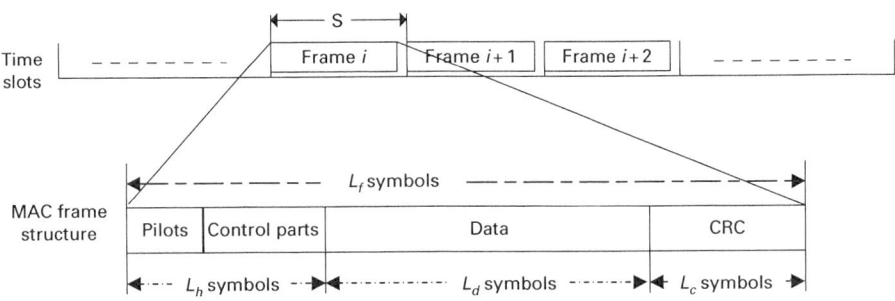

Figure 16.2 MAC transmission and frame structure

of $\mathcal{T}_{(i,j)}$ as $|\mathcal{T}_{(i,j)}|$. From Theorem 14.1, the optimal channel gain threshold for any link $(i,j) \in \mathcal{E}$, $\overline{H}^*_{(i,j)}$, that also assures proportional fairness among all CIA-MAC links, is

$$\overline{H}^*_{(i,j)} = F^{-1}\left(\frac{|\mathcal{T}_{(i,j)}|}{1 + |\mathcal{T}_{(i,j)}|}\right), \tag{16.2}$$

and the corresponding transmission probability is

$$p^*_{(i,j)} = \frac{1}{|\mathcal{T}_{(i,j)}| + 1}, \tag{16.3}$$

where $F^{-1}(\cdot)$ denotes the inverse function of $F(\cdot)$. From (16.2), the optimal threshold of link (i,j) depends only on $|\mathcal{T}_{(i,j)}|$, the number of MTs severely interfered with by transmission at link (i,j). This knowledge can be shared by the BSs of neighboring cells. $|\mathcal{T}_{(i,j)}|$ changes only when the severely interfered MTs have large status variations that result in the obvious changes of the ICR in (16.1) to go across the trigger. For example, an existing traffic session ends, a new one starts, or the movements of MTs either cause severe interference to a new MT or an existing MT no longer has severe interference. These variations will trigger the update of $|\mathcal{T}_{(i,j)}|$.

In the PHY layer, consider channel inversion [25] and each BS allocates transmit power to maintain a constant received power level so that signals can be reliably detected. Once MAC decides to transmit, the transmit power is given by $P_{(i,j)}(h) = P_r/h$, where P_r is the power level for reliable receiver detection. Given $\overline{\mathcal{H}}^*$ in (16.2), the received power level is optimized by Theorem (14.2)

$$P^*_{r(i,j)} = \min\left(\frac{P_a}{\int_{\overline{H}^*_{(i,j)}}^{\infty} \frac{1}{h} f(h) \mathrm{d}h}, P_m \overline{H}^*_{(i,j)}\right). \tag{16.4}$$

Figure 16.3 illustrates the flowchart of CIA-MAC. Each MT identifies the list of neighboring BSs causing severe interference by comparing their ICRs with the trigger Γ_m and reports the list to its home BS. The home BS communicates the list to other

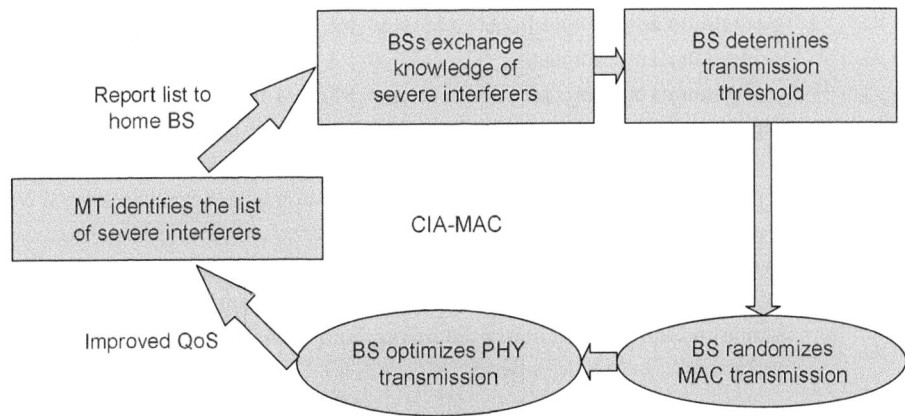

Figure 16.3 CIA-MAC flowchart

BSs and each BS knows the links on which it needs to randomize transmission. Then each BS determines a channel threshold $\overline{H}_{(i,j)}$ per CIA-MAC link (i,j) based on the number of links affected by the transmission on that particular link. A BS transmits on a CIA-MAC link only when the channel gain on the link exceeds its channel threshold and thus randomizes the transmission. The transmission power and modulation are optimized on each link separately. As shown in Figure 16.3, the operations of CIA-MAC are classified into two parts. The operations in the rectangles are semi-static and take place only when the severely interfered MTs have large status variation. With this trivial cost, the operations in the ovals automatically improve the QoS of all MTs experiencing severe interference.

16.3   Parameter optimization

In this section, we discuss the selection of the trigger.

16.3.1   Trigger selection

Each MT measures the ICR of each neighboring BS and CIA-MAC is used at that BS when the corresponding ICR is above the trigger, Γ_m. Hence, the trigger determines severe interferers and relates to the performance of CIA-MAC. We will choose the trigger to maximize the throughput of the whole network rather than that of any individual link.

From (14.3), the throughput of link (i,j) can be expressed as

$$T_{(i,j)} = p_{(i,j)} \prod_{l \in \mathcal{N}_j} (1 - p_l) \frac{RL_d}{S} \cdot (1 - p_F), \tag{16.5}$$

where p_l is the transmission probability on link l, $\prod_{l \in \mathcal{N}_j} (1 - p_l)$ is the probability that none of the severe interferers of MT j transmits, R is the average transmitted bits per symbol when the MAC of BS i decides to transmit at link (i,j) and depends on the modulation and power allocation policy, and p_F is the *frame-error rate* (FER) when no severe interferers transmit.

Optimum triggers are different for MTs with different interference scenarios. Figure 16.1 illustrates an example. MT 3 wants to judge BS 1. If BS 1 is judged to be a severe interferer, it will transmit with lower probability according to Equation (16.3). However, the impact of this variation to the packet receptions of MTs 1, 7, and 8 is unknown to MT 3. Hence, it is difficult for each MT to evaluate the variation of the overall network throughput that results from the judgement of severe interferers. Even assuming that the knowledge can be shared, different MTs have different interfering scenarios, and judging severe interferers and exchanging signaling would be daunting tasks across the whole network.

We will get one trigger for all to simplify the calculations. Consider a network in which each MT suffers interference by K neighboring BSs on average. The BS of each MT also brings severe CCI to K MTs in the neighboring cells on average. Empirical values can be assigned to K, e.g. $K = 3$ is a good choice based on our simulation observations for a reuse of one network. From Equation (16.3), the transmission probability of each BS is $\frac{1}{1+K}$ and the throughput is

$$T = \frac{1}{1+K}\left(1 - \frac{1}{1+K}\right)^K \frac{RL_d}{S}(1 - p_F) = \frac{K^K}{(1+K)^{K+1}}(1 - p_b)^{RL_d}\frac{RL_d}{S}, \qquad (16.6)$$

where the frame error rate is approximated by $p_F = 1 - (1 - p_b)^{L_d R}$ according to the assumption of uniform and independent error distribution in Section 16.2. The bit error rate (BER) is approximated by $p_b = P_e(\eta)$, where η is the average SNR and the BER function $P_e()$ depends on the modulation and coding. For example, the BER for coherently detected M-ary quadrature amplitude modulation (M-QAM) with Gray mapping over an additive white Gaussian noise channel can be well approximated by [24]

$$P_e(\eta) \approx 0.2 \exp\left(-\frac{1.5 G_c \eta}{M - 1}\right), \qquad (16.7)$$

where G_c is the coding gain and M is the modulation order.

With the traditional MAC, all BSs keep on transmitting. Each link has throughput

$$\widehat{T} = \frac{L_d \widehat{R}}{S} \cdot (1 - \widehat{p}_F) = \frac{L_d \widehat{R}}{S} \cdot (1 - \widehat{p}_b)^{L_d \widehat{R}}, \qquad (16.8)$$

where $\widehat{p}_F$ and $\widehat{p}_b$ are the average frame and bit error rates, and $\widehat{R}$ is the average number of bits transmitted per symbol in this mode. $\widehat{p}_b$, $\widehat{p}_F$, and $\widehat{R}$ are different from those in (16.6) since BSs have different transmission durations and signal receptions have different interference scenarios, which result in different power and modulation allocation approaches. CIA-MAC is triggered when it achieves better throughput, i.e. $T > \widehat{T}$, or

$$\frac{K^K}{(1+K)^{K+1}}(1 - p_b)^{L_d R}\frac{L_d R}{S} > (1 - \widehat{p}_b)^{L_d \widehat{R}}\frac{L_d \widehat{R}}{S}. \qquad (16.9)$$

Then we have

$$\widehat{p}_b > 1 - \left(\frac{K^K}{(1+K)^{K+1}}(1 - p_b)^{L_d R}\frac{R}{\widehat{R}}\right)^{\frac{1}{L_d \widehat{R}}}. \qquad (16.10)$$

Since $\widehat{p}_b = P_e(\widehat{\eta})$, where $\widehat{\eta}$ is the average SINR,

$$\widehat{\eta} < P_e^{-1}\left[1 - \left(\frac{K^K}{(1+K)^{K+1}}(1 - p_b)^{L_d R}\frac{R}{\widehat{R}}\right)^{\frac{1}{L_d \widehat{R}}}\right]. \qquad (16.11)$$

SINR and ICR follow the relationship

$$ICR = \frac{1}{\widehat{\eta}} - \frac{1}{\eta}. \qquad (16.12)$$

The trigger Γ_m follows immediately

$$\Gamma_m = \frac{1}{P_e^{-1}\left[1 - \left(\frac{K^K}{(1+K)^{K+1}}(1-P_e(\eta))^{L_dR\frac{R}{\widehat{R}}}\right)^{\frac{1}{L_d\widehat{R}}}\right]} - \frac{1}{\eta}. \tag{16.13}$$

Γ_m depends on the *SNR* and the BER function, both of which are known to each MT. Hence, Γ_m can be easily calculated to judge of severe interferers.

For fixed modulation, $R = \widehat{R}$. The trigger is

$$\Gamma_m = \frac{1}{P_e^{-1}\left[1 - \left(\frac{K^K}{(1+K)^{K+1}}\right)^{\frac{1}{L_d\widehat{R}}}(1-P_e(\eta))\right]} - \frac{1}{\eta}. \tag{16.14}$$

For normal data transmission, *SNR* is high and $P_e(\eta) \ll 1$, thus (16.14) is further simplified to be

$$\Gamma_m = \frac{1}{P_e^{-1}\left[1 - \left(\frac{K^K}{(1+K)^{K+1}}\right)^{\frac{1}{L_d\widehat{R}}}\right]} - \frac{1}{\eta}. \tag{16.15}$$

16.3.2   An alternate trigger mechanism using location knowledge

In the flowchart of Figure 16.3, each MT determines the list of severe interferers and reports the list to the home BS, which requires additional improvement MTs. In the following, we show how to enable BSs to determine the severe interferers to avoid the necessity of MT improvement.

We assume that BSs have the position knowledge of MTs in both the home cells and the neighboring cells. Note that large quantities of positioning techniques have been proposed in cellular networks [13, 46, 12, 10], hence it is practical to obtain the position knowledge of each MT and this knowledge can be shared among neighboring BSs. Furthermore, assume that each BS knows the average received signal power at a desired MT, which can be obtained through feedback or observation of link power control. This knowledge will also be shared among neighboring BSs.

We have shown that the optimal threshold (16.2) for each BS depends on the number of MTs suffering from severe interference by its transmission. This number can be obtained through cooperation among BSs. A BS located at coordinate (x_b, y_b) needs to determine whether it brings severe interference to the neighboring-cell MT at coordinate (x_m, y_m). The distance between them is

$$d_I = \sqrt{(x_b - x_m)^2 + (y_b - y_m)^2}, \tag{16.16}$$

which results in path loss $L(d_I)$. The average received signal power at the MT is P_s, while the interfering BS has the average transmit power P_I. According to (16.1), the average is

$$ICR = \frac{P_I h_I}{P_s}, \tag{16.17}$$

where h_I is the average channel power gain of this interfering link. However, h_I is unknown to the interfering BS and needs to be estimated. Radio propagation is characterized by three nearly independent phenomena: path loss variation with distance, slow log-normal shadowing, and fast multi-path fading [79]. Similar to the ICR in (16.1), h_I tracks slow-fading, i.e. it is a local mean and averages the effect of fast multi-path fading. Hence, we consider only the path loss and shadowing. Shadow represents the error between the actual and estimated path loss [79]. While the estimated path loss is determined by the radio path distance d, the shadowing/estimation error has been observed to be nearly independent of d and we assume the independence. Hence, we model the estimated average interference channel gain by two parts: the estimated path loss determined by path loss model $L(d)$ and the estimation error determined by the shadowing model. Shadows are generally modeled as being log-normally distributed and $10\log_{10}(h_I)$ has normal distribution with mean $-10\log_{10}(L(d))$ and standard deviation σ, where σ is independent of the radio path length d and typically ranges from 5 to 12 dB [79]. To ensure a detection probability β of severe interferers, the BS determines that the MT is suffering from severe interference when the probability of severe interference is above β, i.e.

$$Prob[\text{severe interference}] = Prob\,[ICR \geq \Gamma_m] \geq \beta, \tag{16.18}$$

which is equivalent to

$$Prob\left[10\log_{10}(h_I) \geq 10\log_{10}\left(\frac{\Gamma_m P_s}{P_I}\right)\right] \geq \beta. \tag{16.19}$$

Solving (16.19) yields the detector of severe interference as follows

$$\frac{P_I}{L(d_I)P_s} \geq \Gamma_m 10^{-\frac{\sigma Q^{-1}(\beta)}{10}} \triangleq \Gamma_b, \tag{16.20}$$

where $Q(x)$ is the right-tail probability of the standard normal distribution, i.e.

$$Q(x) = \int_x^{\infty} \frac{1}{\sqrt{2\pi}} e^{-\frac{t^2}{2}}\, dt. \tag{16.21}$$

With (16.20), each BS can detect how many MTs are suffering severe interference and determine transmission probability (16.3) as well as the threshold (16.2). In this case, no MT improvement is necessary for the functioning of CIA-MAC. Note that the selection of β determines how pessimistically or optimistically a severe interferer is judged.

16.4   Network performance

In this section, we show the relationship between the trigger and the SNR, verify the effectiveness of the trigger, and demonstrate the performance of CIA-MAC in a cellular network through comparison with the traditional MAC and a static FFR approach. In the simulated cellular system, the radius of each cell is 2 km and no sectoring is used. The thermal noise power is -104 dBm over the whole bandwidth. The carrier frequency is 900 MHz. BSs are 100 meters high with 8.2 dB antenna gain, while MTs have a height of 1.5 meters with 2.2 dB antenna gain. Path loss is given by the urban area Hata–Okumura model. Log-normal shadowing and Rayleigh fading are applied. Each MAC frame consists of 1000 symbols, in which 900 carry the payload.

16.4.1   Relationship of trigger and SNR

Considering uncoded 4-QAM modulation, the relationship between the trigger Γ_m and SNR when L_d has different values is illustrated in Figure 16.4. The amount of bits transmitted per MAC frame varies and is usually very large to fully exploit link capacity. For example, in 802.16e [63], each frame has a maximum length of 2048 bytes of payload followed by one CRC verification, i.e. 16384 bits of payload per frame. In *high-speed downlink packet access* (HSDPA) transmission of *universal mobile telecommunications system* (UMTS) [19], the size of a transport block followed by one CRC verification ranges widely from 15890 bits to 204000 bits. In general, we assume large $L_d R$ and illustrate the cases when L_d is 500, 1000, and 2000, respectively. The curves without markers are calculated through (16.14) while those with markers through (16.15). Figure 16.4 clearly shows that (16.15) is a good approximation of the threshold for high SNR. In the high-SNR region, the receiver can bear higher interference when signal power increases, yielding an increasing trend of the curve. In the low-SNR region,

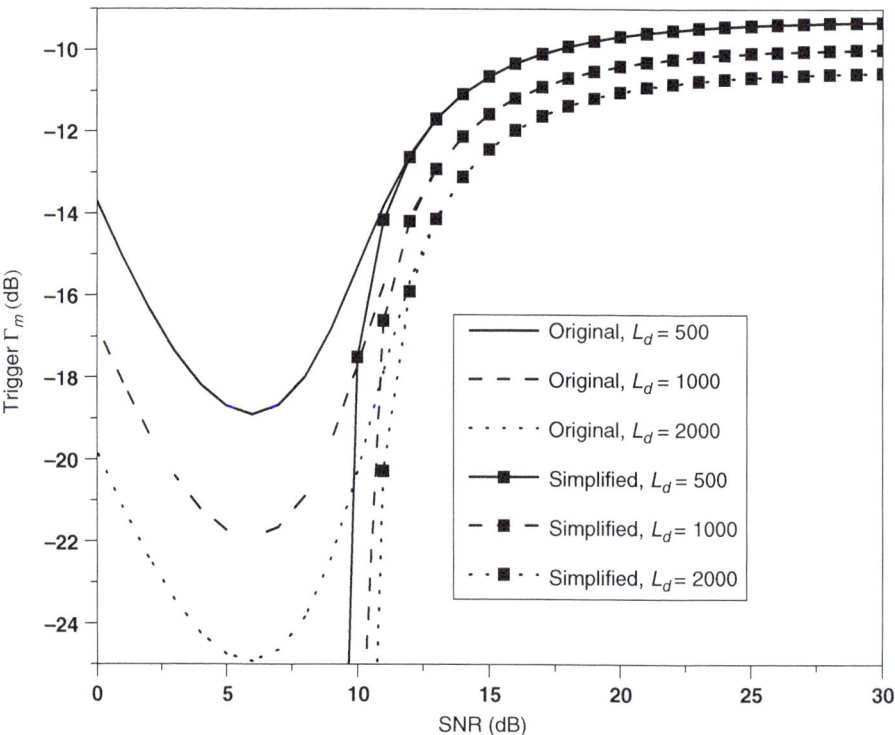

Figure 16.4 Trigger of severe co-channel interferer

when SNR goes lower, i.e., noise power goes higher, interference needs to have stronger power to impact more on frame reception than noise. This indicates the increasing trend of the curve as SNR goes lower in the low-SNR region. This also indicates that in the noise-dominated region, it is better to ignore interference, as suppressing it will not provide much advantage.

16.4.2   Performance improvement

In this section, we demonstrate the performance of CIA-MAC in a 19-cell cellular network. Each BS serves one user on a channel. Users are randomly dropped and uniformly distributed in each cell for each simulation trial.

CIA-MAC is implemented either with or without cross-layer design. The CIA-MAC with cross-layer design follows what we have discussed in this chapter and transmission happens only when channel power gain is above the threshold in Equation (16.2). For CIA-MAC without cross-layer design, each BS transmits randomly with probability given by Equation (16.3) and independently of channel states. We implement both trigger mechanisms, and in the second one the detection probability is set to be 0.9. We compare CIA-MAC with the traditional MAC and a static FFR, and the overall system bandwidth is the same for all of them. The network has a reuse factor of 1 for

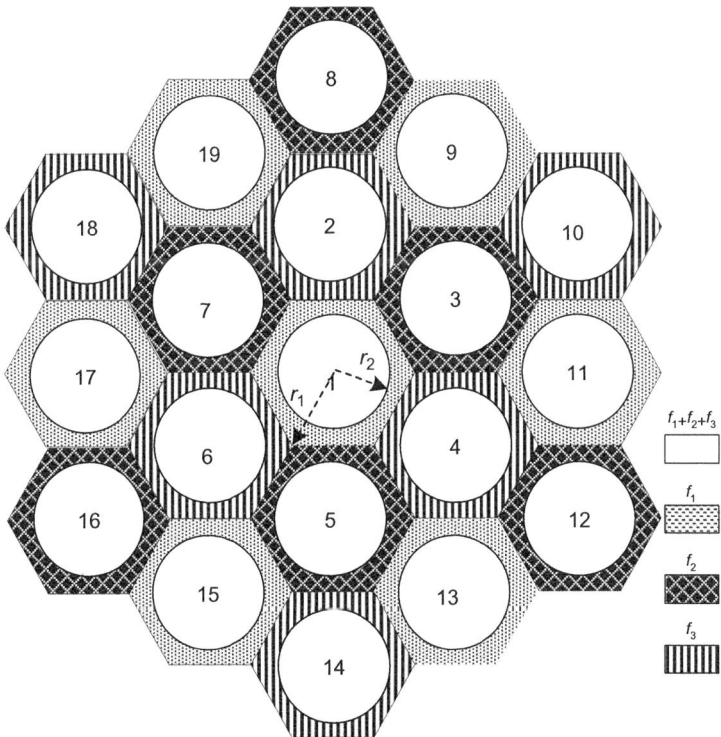

Figure 16.5 Cellular networks with fractional frequency reuse

both CIA-MAC and the traditional MAC. The traditional MAC keeps all BSs transmitting and users experience interference from all neighboring BSs. The static FFR reduces cell-edge interference through low-frequency reuse at cell edges [109, 63, 132, 20, 21]. Figure 16.5 illustrates the network frequency deployment of FFR in our simulation. The radius of each cell is $r_1 = 2$ km and the cell-center users, located within $r_2 = 2r_1/3$ from the BS, will transmit over the whole frequency band. For cell-edge communications, the whole frequency band is equally divided into three subbands, f_1, f_2, and f_3, and users at cell edges are assigned one of them according to the frequency deployment in Figure 16.5.

In Figures 16.6 and 16.7, we compare the SINR and throughput. Either uncoded 4-QAM or coded 4-QAM with different coding gains is used. The coding gain can be obtained through exploitation of receiver diversity or channel coding.

Figure 16.6 compares the *cumulative distribution functions* (CDFs) of average SINR. The SINR is the equivalent value after decoding when coded 4-QAM is used. For reference, the relationship between FER and SINR is also plotted with a bold curve. Observing the FER curve, when SINR is lower than 10 dB, most decoded frames have at least one bit in error and do not pass CRC, resulting in transmission outage. We compare the schemes with either uncoded 4-QAM or coded 4-QAM that has 8 dB coding gain. For CIA-MAC, only SINRs when no severe interferers transmit are averaged to produce the CDF curves. Curves with legend CIA-MAC$_b$ correspond to the performance of CIA-MAC when severe interferers are determined by the BSs according to (16.20). We can see that CIA-MAC$_b$ performs closely to CIA-MAC in both cases

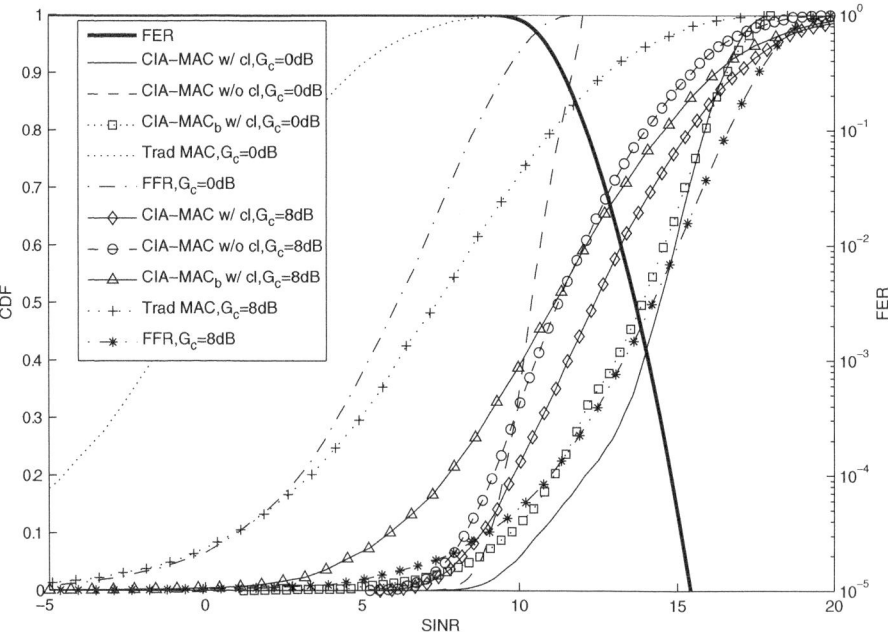

Figure 16.6 Cumulative distribution function of SINR

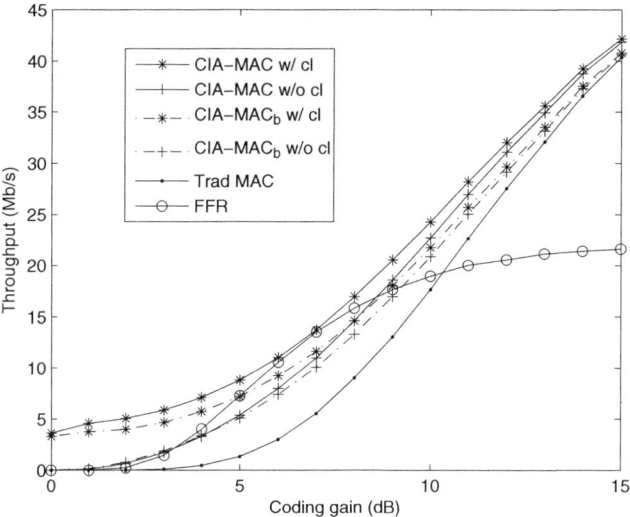

Figure 16.7 Throughput comparison (SNR=9 dB)

and while significantly reducing the improvement cost, BS judgement is effective in detecting severe interferers using position knowledge. Without coding, the traditional MAC suffers strong interference and the average SINRs of all simulation trials falls far below 10 dB. In this case, the network is completely in outage. With FFR, the average SINRs of all trials are significantly improved. However, interference from neighboring cells still affects the SINRs and most SINRs are less than 10 dB since the target SNR is only 12 dB. Amazingly, CIA-MAC has better SINR distribution than FFR, even the CIA-MAC without cross-layer design. This is because we only average SINRs when no severe interferers transmit, i.e. when transmission succeeds in CIA-MAC. Furthermore, CIA-MAC with cross-layer design achieves very high average SINR. Now observe the performance of schemes with coded 4-QAM that has 8 dB coding gain. We can see that the SINR performance of both FFR and the traditional MAC are improved by around 8 dB. Since MTs can mitigate a large amount of interference with coding, CIA-MAC judges much less severe interferers, e.g. 0.2 on average in our simulation. Hence, CIA-MAC finds no severe interferers in most cases and BSs simply keep on transmitting, as does traditional MAC. However, CIA-MAC still outperforms the traditional MAC because of avoidance of severe interference whenever it exists.

In Figure 16.7, each BS allocates power to maintain a 9 dB received SNR and the coding gains of all MTs are increased from 0 dB to 15 dB. We observe that CIA-MAC always outperforms the traditional MAC. With higher and higher coding gain, interference has less and less impact on frame reception and thus fewer and fewer severe interferers are judged. Hence, with high coding gain, the performance of CIA-MAC and traditional MAC tend to be the same. The static FFR suffers performance loss for low-frequency-reuse efficiency at cell edges in the high-coding-gain region.

Dynamic FFR schemes [132, 20, 21] can be used to further improve frequency-reuse efficiency at the cost of higher network deployment complexity. However, we should note that CIA-MAC improves network performance in the most cost-effective way, and even the static FFR implemented here requires much higher deployment cost than CIA-MAC.

17 Distributed power control

In the previous chapters, we have introduced optimal distributed medium access control (MAC) schemes, which are essentially distributed scheduling approaches that schedule the transmissions of all users on orthogonal resources (time slots) in a decentralized way while exploiting multi-user diversity in both channels and interference environments. Users are usually spatially separated and more than one user may be granted channel access and transmit data at the same time on the same frequency. Because of frequency reuse, the transmissions of different users will interfere with each other. To simplify the MAC designs, we have assumed collision channel models and once a collision, i.e. interference, exists, the transmission fails. In practice, a more realistic SINR channel model can be used. The signal to interference plus noise ratio (SINR) accounts for the cumulative interference level. A signal transmission succeeds if the SINR perceived by the receiver exceeds an SINR threshold. It is a more natural channel model for deciding packet decoding success. SINR is determined by both channel gains as well as transmitter powers of all users in the network. While channel gains are usually fixed depending on user locations, transmitter power control can be used to determine the transmission power of transmitters in wireless networks and thus control network interference to achieve good SINR performance. It is a fundamental component of wireless resource management. It has the benefit of reducing interference, increasing network capacity, and reducing energy consumption. Power control is broadly used in cellular networks, wireless local area networks (WLANs), wireless sensor networks, wireless mobile ad hoc networks, and so on. In this section, we will introduce power control for both real-time and elastic traffic.

17.1   System model

In a wireless system, observe an arbitrary frequency channel where multiple users are communicating at the same time. On this channel, there are K pairs of transmitters and receivers, as shown in Figure 17.1. The transmitters and receivers can be co-located, depending on the network type. For example, in the uplink of a single-cell code division multiple access (CDMA) network, the receivers are co-located at the base station and in the downlink the transmitters are co-located. In the downlink of a multi-cell cellular network, the transmitters are the base stations and the receivers the mobile users in each base station. For the uplink case, their roles are reversed.

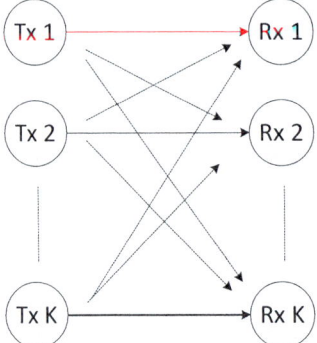

Figure 17.1 Multi-user communications

Define the signal power gain from transmitter i to receiver j to be g_{ij}. In time division multiple access (TDMA) or frequency division multiple access (FDMA) systems, g_{ij} describes the link quality. In a CDMA system, g_{ij} should also capture the spreading effect, i.e.

$$g_{ij} = \widehat{g}_{ij}|\mathbf{c}_i\mathbf{c}_j|^2, \tag{17.1}$$

where $\widehat{g}_{ij}$ is the signal power gain if no spreading is used and $\mathbf{c}_i$ is the spreading code assigned to user i. The power of transmitter i is p_i and let vector $\mathbf{P}$ denote the power values of all transmitters,

$$\mathbf{P} = (p_1, p_2, ..., p_K)^T. \tag{17.2}$$

The SINR of the receiver i is

$$\gamma_i = \frac{p_i g_{ii}}{\sum_{j \neq i} p_j g_{ji} + \sigma_i^2}, \tag{17.3}$$

where σ_i^2 denotes the noise power at receiver i.

17.2   Power control for real-time traffic

For real-time traffic with strict data rate requirements, the SINR of each user should be above a certain threshold, i.e.

$$\gamma_i \geq \overline{\gamma}_i, \tag{17.4}$$

where $\overline{\gamma}_i$ is the target SINR of user i. Assuming fixed power of all other users, the power transmitter i should use is lower bounded by

$$p_i \geq \overline{\gamma}_i \sum_{j \neq i} \left(p_j \frac{g_{ji}}{g_{ii}} + \frac{\sigma_i^2}{g_{ii}} \right), \tag{17.5}$$

which is the same as

$$p_i \geq \sum_{j \neq i} p_j \hat{g}_{ji} + \hat{N}_i, \qquad (17.6)$$

where $\hat{g}_{ji} = \frac{g_{ji} \bar{\gamma}_i}{g_{ii}}$ and $\hat{N}_i = \frac{\sigma_i^2 \bar{\gamma}_i}{g_{ii}}$. Define $\hat{\mathbf{N}} = [\hat{N}_1, \hat{N}_2, ..., \hat{N}_K]^T$ and the $K \times K$ matrix $\hat{\mathbf{G}}$ by

$$\hat{G}_{ij} = \begin{cases} \hat{g}_{ij} & i \neq j, \\ 0 & i = j. \end{cases} \qquad (17.7)$$

Rewriting (17.6), we have

$$(\mathbf{I} - \hat{\mathbf{G}})\mathbf{P} \succeq \hat{\mathbf{N}}, \qquad (17.8)$$

where $\mathbf{I}$ denotes the $K \times K$ identity matrix and $\mathbf{A} \succeq \mathbf{B}$ means each element in $\mathbf{A}$ is no smaller than the corresponding one in $\mathbf{B}$.

The target SINR, $\bar{\gamma} = [\bar{\gamma}_1, \bar{\gamma}_2, ..., \bar{\gamma}_K]$, is achievable if there exists a $\mathbf{P}$ such that for all i,

$$\gamma_i \geq \bar{\gamma}_i.$$

$\bar{\gamma}$ is achievable only if the solution $\mathbf{P}^*$ to $(\mathbf{I} - \hat{\mathbf{G}})\mathbf{P}^* = \hat{\mathbf{N}}$ is non-negative. If $(\mathbf{I} - \hat{\mathbf{G}})^{-1}$ exists,

$$\mathbf{P}^* = (\mathbf{I} - \hat{\mathbf{G}})^{-1} \hat{\mathbf{N}}. \qquad (17.9)$$

Note that

$$(\mathbf{I} - \hat{\mathbf{G}})(\mathbf{I} + \hat{\mathbf{G}} + \hat{\mathbf{G}}^2 + \cdots) = \mathbf{I} - \hat{\mathbf{G}} + \hat{\mathbf{G}} - \hat{\mathbf{G}}^2 + \hat{\mathbf{G}}^2 - \hat{\mathbf{G}}^3 + ... = \mathbf{I}. \qquad (17.10)$$

Therefore,

$$(\mathbf{I} - \hat{\mathbf{G}})^{-1} = \sum_{i=0}^{\infty} \hat{\mathbf{G}}^i. \qquad (17.11)$$

$(\mathbf{I} - \hat{\mathbf{G}})^{-1}$ exists if and only if $\sum_{i=0}^{\infty} \hat{\mathbf{G}}^i$ exists, which happens only when the largest eigenvalue, $\rho(\hat{\mathbf{G}})$, of $\hat{\mathbf{G}}$ satisfies

$$|\rho(\hat{\mathbf{G}})| < 1. \qquad (17.12)$$

Then $(\mathbf{I} - \hat{\mathbf{G}})^{-1} > 0$ and

$$\mathbf{P}^* = (\mathbf{I} - \hat{\mathbf{G}})^{-1} \hat{\mathbf{N}} \succeq 0.$$

When $|\rho(\hat{\mathbf{G}})| = 1$, $\mathbf{I} - \hat{\mathbf{G}}$ is singular and the solution $\mathbf{P}^*$ may exist only when $\hat{\mathbf{N}} = 0$. In this case, there would be multiple non-negative solutions.

Figure 17.2 illustrates the feasible region for a two-user case when $|\rho(\hat{\mathbf{G}})| < 1$. The two lines are the power vectors that satisfy the SINR requirements for the two users, respectively. Regions B and D are feasible for user 1 and C and D for 2. D is the feasible region for both users. The two lines have to cross so that the feasible region is non-empty and the condition is that

$$\hat{g}_{21} \geq \hat{g}_{12}, \qquad (17.13)$$

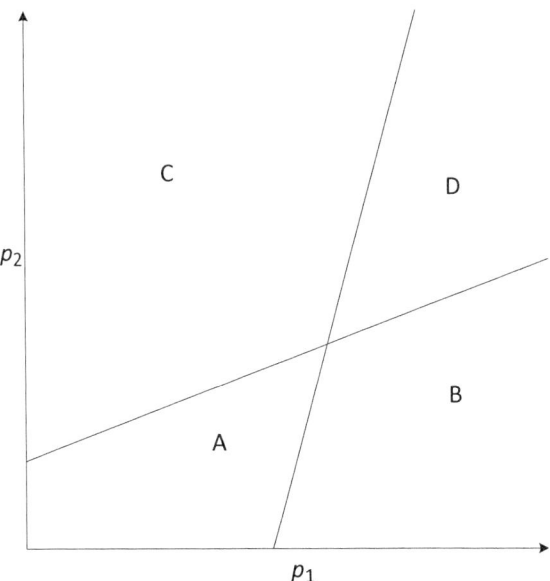

Figure 17.2 Feasible region for a two-user network (B: $p_1 \geq p_2 \hat{g}_{21} + \hat{N}_1$; C: $p_2 \geq p_1 \hat{g}_{12} + \hat{N}_2$;)

indicating

$$\frac{\hat{g}_{12}}{\hat{g}_{21}} \leq 1, \tag{17.14}$$

which can be easily shown to be the same as the eigenvalue condition $|\rho(\hat{\mathbf{G}})| \leq 1$. When $\frac{\hat{g}_{12}}{\hat{g}_{21}} = 1$, the two lines are parallel and only when

$$\hat{N}_1 = \hat{N}_1 = 0$$

will these two lines overlap. In this case, there would be multiple non-negative solutions.

If not all SINR requirements can be met, i.e. no non-negative solutions, there are more users granted channel access than can be supported by the channels. In this case, some users should be removed and the objective of user removal can be to maximize the number of users that can be supported by the channels. This problem is first investigated by Zander in [106] assuming null noise and all users have the same signal-to-interference ratio (SIR) target. It is shown that in the optimum power allocation, some users should be shut off completely and the remaining users use the power levels that meet their SINR requirements. The idea in the removal algorithm is to remove one user in each iteration until the SINRs of all remaining users achieve their targets.

17.2.1   Distributed power control

In the above, we have discussed centralized power control for real-time traffic. While centralized power control algorithms can find the optimal solutions, it is in general difficult to implement them in real wireless systems because estimating all instantaneous channels and exchanging the information requires lots of signaling overhead. Distributed power control is therefore preferred. In the following, we assume the target SINR, $\overline{\gamma}$, is achievable and introduce the distributed power control proposed in [78].

The power can be updated using the following first-order autoregressive filter,

$$\mathbf{P}[k] = \hat{\mathbf{G}}\mathbf{P}[k-1] + \hat{\mathbf{N}}, \tag{17.15}$$

where $\mathbf{P}[k]$ denotes the power vector of the kth iteration. It can be easily seen that

$$\mathbf{P}[k] = \hat{\mathbf{G}}^{k-1}\mathbf{P}[1] + \sum_{k=0}^{n-1} \hat{\mathbf{G}}^k\hat{\mathbf{N}}. \tag{17.16}$$

As k goes to infinity,

$$\lim_{k->\infty} \mathbf{P}[k] = \lim_{k->\infty} \hat{\mathbf{G}}^{k-1}\mathbf{P}[1] + \lim_{k->\infty} \sum_{k=0}^{n-1} \hat{\mathbf{G}}^k\hat{\mathbf{N}} = 0 + (\mathbf{I} - \hat{\mathbf{G}})^{-1}\hat{\mathbf{N}} = \mathbf{P}^*. \tag{17.17}$$

Therefore, this approach converges to the optimal power setting. Furthermore, this power control can be written in a distributed way, as shown below. The power of user i can be written in the following format:

$$p_i[k] = \sum_{j \neq i} p_j[k-1]\hat{g}_{ji} + \hat{N}_i = \left(\sum_{j \neq i} p_j[k-1]g_{ji} + \sigma_i^2\right) \frac{\overline{\gamma}_i}{g_{ii}}. \tag{17.18}$$

$\left(\sum_{j \neq i} p_j[k-1]g_{ji} + \sigma_i^2\right)$ is the total interference plus noise of user i and is measurable locally. Equivalently, the power can be updated by

$$p_i[k] = p_i[k-1] \frac{\overline{\gamma}_i}{\gamma_i[k-1]}. \tag{17.19}$$

Another way to obtain this power update equation is to let each user allocate a power level such that its SINR target is just met in each iteration based on the observation in the previous iteration, i.e.

$$\frac{p_i[k]g_{ii}}{\sum_{j \neq i} p_j[k-1]g_{ji} + \sigma_i^2} = \overline{\gamma}_i. \tag{17.20}$$

Rewriting Equation (17.20), we have the desired update in (17.19).

17.3 Power control for elastic traffic

Elastic traffic, like best-effort data traffic, is more delay tolerant. A higher data rate is also desired as it will help improve network performance. One intuitive objective is to control the power values to maximize the network throughput and the optimization problem is

$$\max_{\{p_i\}} \sum_i r_i(p_i) = \sum_i \log_2\left(1 + \frac{p_i g_{ii}}{\sum_{j \neq i} p_j g_{ji} + \sigma_i^2}\right). \tag{17.21}$$

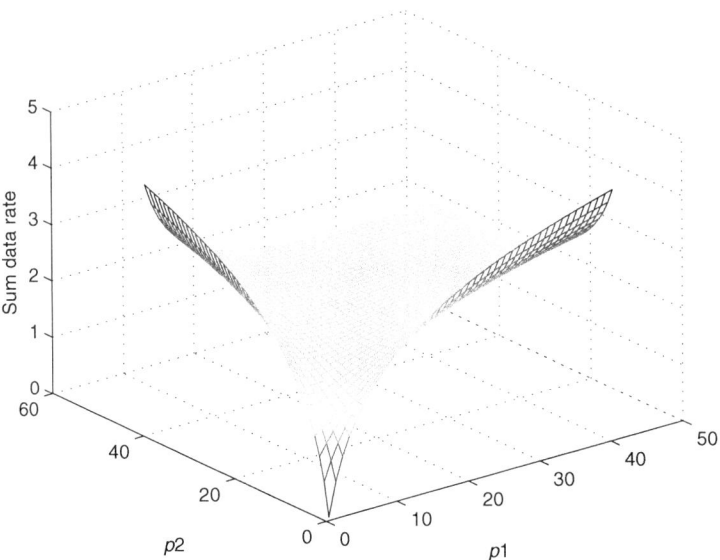

Figure 17.3 Sum network throughput for various transmitter power values ($a_{12} = 0.5$ and $a_{21} = 0.8$ and others are 1)

s.t.

$$r_i \geq r_{i,min}, \forall i.$$

Unfortunately, this objective function is non-concave and many local maxima may exist depending on channel conditions. To illustrate this, consider the two-user case,

$$r_1(p_1) + r_2(p_2) = \log_2\left(1 + \frac{p_1 g_{11}}{p_2 g_{12} + \sigma_1^2}\right) + \log_2\left(1 + \frac{p_2 g_{22}}{p_2 g_{12} + \sigma_2^2}\right). \qquad (17.22)$$

Figures 17.3 and 17.4 illustrate the corresponding relationship between the sum data rate and the two power values. The results in Figure 17.3 show that when inter-user interference is the dominating source of disturbance (e.g. shorter ranges like WLANs), the maximal total (sum) data rates by single transmissions are higher. This holds true in many cases, but not always. When inter-user interference is not dominating, e.g. longer ranges like inter-cell interference in cellular networks, maximal sum rates are mostly reached for simultaneous transmission schemes, as demonstrated in Figure 17.4.

The characterizations are general in the sense that they are not sensitive to the choice of modulation schemes. As a practical consequence, the results demonstrate that for short range, interference-limited systems, e.g. indoor systems, coordinated orthogonal multiplexing provides better performance than simultaneous transmissions. In Part II and the first four sections of this part, the MAC and scheduling protocols and algorithms are designed to coordinate the orthogonal transmissions of all users in these interference-limited systems. For longer ranges when noise comes into play,

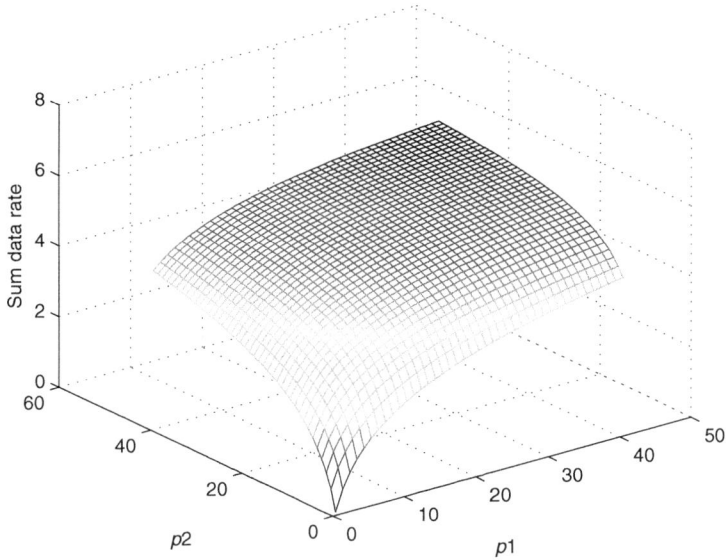

Figure 17.4 Sum network throughput for various transmitter power values ($a_{12} = 0.5$ and $a_{21} = 0.8$ and others are 1)

non-orthogonal schemes have advantages and power control algorithms are needed to determine the transmission power values.

Generally speaking, the non-concave power control problem in (17.21) is hard to solve. In [121], a centralized algorithm called MAPEL is proposed to find an optimal solution based on increasingly accurate approximation of the feasible SINR region. However, the algorithm has exponential complexity and cannot be used in practice. Furthermore, a centralized algorithm requires complete channel knowledge, including the interference channels, $\{g_{ij}\}$. In the following, we introduce distributed heuristic power control algorithms for elastic data traffic.

With distributed power control, each transmitter decides its power level based on local observations. The power of transmitter i is

$$p_i[k + 1] = f_i(A_i^{(k)}), \tag{17.23}$$

where $A_i^{(n)}$ is the parameter set observed locally in the previous iteration. For example, it can be the previous transmission power $p_i[k + 1]$, or the effective interference of other users plus noise that user i must overcome or both. f_i is called the response function of user i. Some examples of f_i can be found in the above section. Another example of the response function is that each user selfishly chooses power allocation to maximize its own utility. Let $u_k(p_k, \mathbf{p}_{-k})$ denote the utility of user k where $\mathbf{p}_{-k}$ is the vector of transmission power of all users other than k. The response function can be written as

$$f_k = \arg \max_{p_k} u_k(p_k, \mathbf{p}_{-k}). \tag{17.24}$$

For a whole network perspective, the network response function can be defined as

$$f = [f_1, f_2, ..., f_K]^T. \tag{17.25}$$

As shown above, each user allocates its power independently. The variation in power allocation of one user impacts that of all others. Equilibrium is the condition of a network in which competing influences are balanced assuming invariant channel conditions. Its properties are important to network performance. In the following, we introduce the existence condition and sufficient conditions of a unique network equilibrium.

17.3.1   Existence of equilibrium

A set of strategies is said to be at a Nash equilibrium, referred to as equilibrium in the following, if no user can gain individually by unilaterally altering its own strategy.

In this and the following two sections, to make the analysis generic, we will consider multi-channel power control, where each user will transmit data on N channels simultaneously. Let $\mathbf{P}_k^* = [p_{k1}, p_{k2}, ..., p_{kn}]^T$ denote the transmission power vector of user k, where p_{kn} is the respective power used on channel n. Define the equilibrium as

$$\mathbf{P}^* = (\mathbf{P}_1^*, \mathbf{P}_2^*, \dots, \mathbf{P}_K^*). \tag{17.26}$$

Nash equilibrium can be described by the following definition.

DEFINITION 17.1   *In a distributed power control scheme, an equilibrium is a set of power allocations that no user has the incentive to choose a different set of power allocations, i.e.*

$$\mathbf{P}^* = f(\mathbf{P}^*) = [f_1(\mathbf{P}_{-1}^*), f_2(\mathbf{P}_{-2}^*), \dots, f_K(\mathbf{P}_{-K}^*)]^T, \tag{17.27}$$

where $f(\mathbf{P})$ is the network response function and $\mathbf{P}_{-k}^*$ is the transmission power of all users other than user k. The equilibrium can also be called the fixed point.

The network response relies on the response functions of all users, which determines the equilibrium properties. To facilitate our discussion, we first introduce the concept of quasiconcavity.

DEFINITION 17.2   *As defined in [66], a function z, which maps from a convex set of real n-dimensional vectors, $\mathcal{D}$, to a real number, is called strictly quasiconcave if for any $\mathbf{x}_1, \mathbf{x}_2 \in \mathcal{D}$ and $\mathbf{x}_1 \neq \mathbf{x}_2$,*

$$z(\lambda \mathbf{x}_1 + (1 - \lambda)\mathbf{x}_2) > \min\{z(\mathbf{x}_1), z(\mathbf{x}_2)\}, \tag{17.28}$$

for any $0 < \lambda < 1$.

An equivalent definition can be based on the convex contour sets.

DEFINITION 17.3   *A function z, which maps from a convex set of real n-dimensional vectors, $\mathcal{D}$, to a real number, is called strictly quasiconcave if every upper level set of z is convex, i.e. $P_\alpha = x\mathcal{D} : f(x)\alpha$ is convex for all α.*

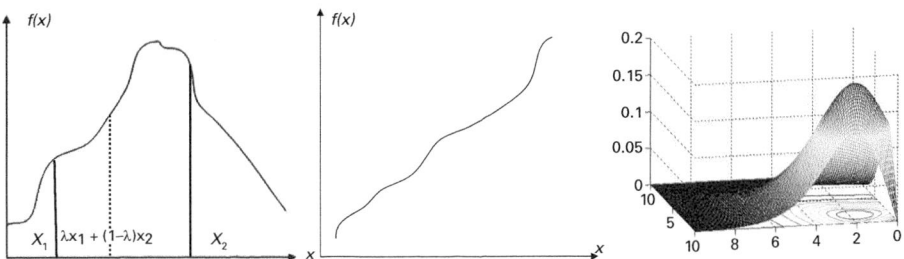

Figure 17.5 Some examples of quasiconcave functions

The above definitions apply to all functions, including single-variable functions. For single-variable functions, the following proposition can be used to determine if a function is quasiconcave or not.

PROPOSITION 17.4   *A function z of a single variable is quasiconcave if and only if either*

- *it is non-decreasing,*
- *it is non-increasing,*
- *or there exists x^* such that z is non-decreasing for $x < x^*$ and non-increasing for $x > x^*$.*

All concave functions are quasiconcave but not every quasiconcave function is a concave function. Some examples of quasiconcave functions are given in Figure 17.5.

In [105], it has been shown that a Nash equilibrium exists if for any k,

- $\mathbf{P}_k$ is a non-empty, convex, and compact subset of some Euclidean space $\Re^L$.
- and $u_n(\mathbf{p}_n, \mathbf{p}_{-n})$ is continuous and quasiconcave in $\mathbf{p}_n$.

17.3.2   Uniqueness of equilibrium in single-channel systems

In this section, we discuss the uniqueness of the equilibrium in a single-channel system and introduce the framework discovered by Yates [159].

DEFINITION 17.5   *A network response function is standard if it satisfies three properties for all power values:*

(i)   *Positivity: $f(\mathbf{P}) > 0$.*
(ii)   *Monotonicity: If $\mathbf{P} \succeq \mathbf{Q}, f(\mathbf{P}) \succeq f(\mathbf{Q})$, where $\succeq$ denotes vector inequality and each element of the vector satisfies the inequality.*
(iii)   *Scalability: For all $\alpha > 1, \alpha f(\mathbf{P}) > f(\alpha \mathbf{P})$.*

The positivity indicates that the background noise is non-zero. The monotonicity indicates that if some users increase their transmitter power, the interference to other users will also increase and at least one user will increase its transmitter power to overcome the increasing interference power. The scalability indicates that if some users increase their transmission power to $\alpha\mathbf{P}$, the variation in the network response will be smaller than scaling up the network response directly; therefore, that variation of transmit power is always smaller than that of the interference power. Furthermore, if $p_j \geq f_j(\mathbf{P})$, then $\alpha p_j \geq \alpha f_j(\mathbf{P}) > f_j(\alpha\mathbf{P})$ for $\alpha > 1$. So if user j has an acceptable connection with $\mathbf{P}$, then the user will have a more acceptable connection when all users are scaled up uniformly. These assure the convergence to a unique equilibrium.

It can be easily proven that the response function in (17.19) is standard. In the following, we give several other standard interference functions in a multi-cell cellular network.

- Assume user k is connected to the base station with the maximum SIR. The user desires a hard SIR requirement, γ_k. Therefore, the power allocation is given by

$$\max_j p_k \beta_{kj}(\mathbf{P}) \geq \gamma_k, \tag{17.29}$$

where $\beta_{kj}(\mathbf{P}) = \frac{g_{kj}}{\sum_{i \neq k} p_i g_{ij} + \sigma_j^2}$ is the carrier to interference plus noise ratio if the user is connected to base station j. The response function is

$$p_k \geq \min_j \frac{\gamma_k}{\beta_{kj}(\mathbf{P})}. \tag{17.30}$$

- Assume a subset, $\mathcal{B}$, of base stations will combine all signals received from user k, i.e. macro diversity is used and the user desires a SIR higher than γ_k. The power allocation is given by

$$p_k \sum_{j \in \mathcal{B}} \beta_{kj}(\mathbf{P}) \geq \gamma_k, \tag{17.31}$$

The response function is

$$p_k \geq \frac{\gamma_k}{\sum_{j \in \mathcal{B}} \beta_{kj}(\mathbf{P})}. \tag{17.32}$$

Distributed power control algorithms that have standard network response functions have the following convergence properties [159].

THEOREM 17.6 *Uniqueness of equilibrium: there is a unique equilibrium, i.e. one power vector* $\mathbf{P}$ *such that* $\mathbf{P}^* = f(\mathbf{P}^*)$.

Proof Assume two different equilibria, $\mathbf{P}$ and $\mathbf{P}'$. Since $f(\mathbf{P}) > 0$, $p_k > 0$ and $p_k' > 0$ for all users. Assume there exists a k such that $p_k < p_k'$. Therefore, there exists $\alpha > 1$ such that $\alpha\mathbf{P} \geq \mathbf{P}'$ and that for some k, $\alpha p_k = p_k'$. The monotonicity and scalability properties imply

$$p_k' = f_k(\mathbf{P}') \leq f_k(\alpha\mathbf{P}) < \alpha f_k(\mathbf{P}) = \alpha p_k.$$

Since $p_k' = \alpha p_k$, we find a contradiction. The uniqueness follows immediately.

THEOREM 17.7   *Convergence to equilibrium: if the equilibrium is achievable, then from any initial value of power vector, the distributed power control algorithm will converge to the unique equilibrium.*

Proof   Since the equilibrium is achievable, there exists a unique equilibrium $\mathbf{P}^*$. Since for all k, $p_k^* > 0$, for any initial $\mathbf{P}$, we can find $\alpha \geq 1$ such that $\alpha \mathbf{P} \succeq \mathbf{P}$. $\alpha \mathbf{P}^*$ is also feasible according to the scalability property. Let $\mathbf{z}$ denote the all zero vector. The monotonicity implies that

$$f^n(\mathbf{z}) \leq f^n(\mathbf{P}) \leq f^n(\alpha \mathbf{P}^*).$$

In the following, we can prove that $\lim_{n \to \infty} f^n(\mathbf{z}) = \mathbf{P}^*$ and $\lim_{n \to \infty} f^n(\alpha \mathbf{P}^*) = \mathbf{P}^*$. Then it is straightforward that $\lim_{n \to \infty} f^n(\mathbf{P}) = \mathbf{P}^*$.

$\lim_{n \to \infty} f^n(\mathbf{z}) = \mathbf{P}^*$: Let $\mathbf{z}[n] = f^n(\mathbf{z})$. $\mathbf{z}(0) = \mathbf{z} \succ \mathbf{P}^*$ and $\mathbf{z}(1) = f(\mathbf{z}) \succeq \mathbf{z}$.

The monotonicity implies that

$$\mathbf{z} \preceq \mathbf{z}(1) \preceq \ldots \preceq \mathbf{z}(n) \preceq \mathbf{P}^*.$$

Therefore,

$$\mathbf{P}^* = f(\mathbf{P}^*) \succeq f(\mathbf{z}(n)) \succeq f(\mathbf{z}(n-1)) = \mathbf{z}(n).$$

Therefore, $\mathbf{P}^* \succeq \mathbf{z}(n+1) \succeq \mathbf{z}(n)$. Hence, the sequence of $\mathbf{z}(n)$ is non-decreasing and upperbounded by $\mathbf{P}^*$. The uniqueness property further leads to $\lim_{n \to \infty} f^n(\mathbf{z}) = \mathbf{P}^*$.

$\lim_{n \to \infty} f^n(\alpha \mathbf{P}^* = \mathbf{P}^*)$: $f(\alpha \mathbf{P}^*) \succeq f(\mathbf{P}^*) = \mathbf{P}^*$. With scalability, we have $f(\alpha \mathbf{P}^*) \preceq \alpha f(\mathbf{P}^*) = \alpha \mathbf{P}^*$. Then we have

$$f^{n-1}(\alpha \mathbf{P}^*) \succeq f^n(\alpha \mathbf{P}^*) \succeq \mathbf{P}^*.$$

Therefore, $f^n(\alpha \mathbf{P}^*)$ is non-increasing and lower bounded by $\mathbf{P}^*$. It also converges to $\mathbf{P}^*$.

The above properties ensure that a distributed power control with standard response functions will converge to a unique equilibrium from any initial settings if the equilibrium exists. In addition, it is shown in [159] that these properties are still maintained when maximum and minimum constraints are applied; i.e., if $f(\mathbf{P})$ is standard, the following response functions

$$f'(\mathbf{P}) = \max(f(\mathbf{P}), \mathbf{P}_{\min})$$

and

$$f'(\mathbf{P}) = \min(f(\mathbf{P}), \mathbf{P}_{\max})$$

are also standard. In addition, suppose $f(\mathbf{P})$ and $\hat{f}(\mathbf{P})$ are two standard response functions and define component-wise maximum and minimum response functions as below

$$f_1'(\mathbf{P}) = \max(f(\mathbf{P}), \hat{f}(\mathbf{P}))$$

and

$$f_2'(\mathbf{P}) = \min(f(\mathbf{P}), \hat{f}(\mathbf{P})).$$

Then both $f_1'(\mathbf{P})$ and $f_2'(\mathbf{P})$ are standard. This result can be easily extended to the case of multiple interference functions. This indicates that a user can run two distributed power control algorithms at the same time and choose the maximum or minimum power value and the resulting power control will still be standard. One example would be letting a user connect to two base stations at the same time and apply the power control in (17.19) for the connection to each base station. In each time instant, the user will choose the base station that requires lower power values and the resulting power control will still be standard.

Note that the above discussion only provides a sufficient condition for a distributed power control to converge to a unique equilibrium and a distributed power control with a standard response function is guaranteed to converge. There exist non-standard network response functions such that the distributed power control can still converge.

17.3.3   Uniqueness of equilibrium in multi-channel systems

In this section, we discuss generic sufficient conditions that assure a unique equilibrium in multi-channel wireless systems. Especially we will show that the number of equilibria in the network will depend on the overall interference and signal channel gains in the whole network. A general distributed power control over multiple subchannels will be considered.

Consider a system with K subchannels. There are N users, each consisting of a pair of transmitter and receiver and operating on these subchannels. All users interfere with each other. Accurate channel state information (CSI) is available to any pair of transmitter and receiver. Define the signal power attenuation of user i at subchannel k to be $g_{ii}^{(k)}$ and the interference power gain from the transmitter of user i to the receiver of user j at subchannel k to be $g_{ij}^{(k)}$. The noise power on each subchannel is σ^2. The power allocation of user n on all subchannels is denoted by vector

$$\mathbf{p}_n = [p_n^{(1)} p_n^{(2)} \cdots p_n^{(K)}].$$

The interference on all subchannels of user n is denoted by vector

$$\mathbf{I}_n = [I_n^{(1)} I_n^{(2)} \cdots I_n^{(K)}],$$

where

$$I_n^{(k)} = \sum_{i=1, i \neq n}^{N} p_i^{(k)} g_{in}^{(k)}. \tag{17.33}$$

Consequently, the SINR, $\eta_n^{(k)}$, of user n at subchannel k can be expressed as

$$\eta_n^{(k)} = \frac{p_n^{(k)} g_{nn}^{(k)}}{\sum_{i=1, i \neq n}^{N} p_i^{(k)} g_{in}^{(k)} + \sigma^2}. \tag{17.34}$$

The data rate at subchannel k of user n, $r_n^{(k)}$, is assumed to be a function of $\eta_n^{(k)}$ and can be expressed as

$$r_n^{(k)} = R(\eta_n^{(k)}), \tag{17.35}$$

where $R()$ is assumed to be strictly concave and increasing in SINR with $R(0) = 0$. For capacity approaching coding [161],

$$r_n^{(k)} = w \log(1 + \eta_n^{(k)}), \tag{17.36}$$

where w is the bandwidth of each subchannel.

Let the data rate vector of user n across the K subchannels be

$$\mathbf{r}_n = [r_n^{(1)}, r_n^{(2)}, \ldots, r_n^{(K)}],$$

then the overall data rate is

$$r_n = \sum_{k=1}^{K} r_n^{(k)}. \tag{17.37}$$

The total transmission power is

$$p_n = \sum_{k=1}^{K} p_n^{(k)}. \tag{17.38}$$

Each user selfishly chooses power allocation to maximize its own utility in an interference-limited environment. The utility of user n, denoted by

$$U_n(\mathbf{p}_n, \mathbf{I}_n(\mathbf{p}_{-n})),$$

is assumed to be quasiconcave in $\mathbf{p}_n$ given $\mathbf{I}_n$, interference on all subchannels. $\mathbf{I}_n$ is a function of $\mathbf{p}_{-n}$ and is determined by (21.31). The best response of power allocation of user n is denoted to be

$$\mathbf{p}_n^o = F_n(\mathbf{p}_{-n}) = \tilde{F}_n(\mathbf{I}_n(\mathbf{p}_{-n})) = \arg\max_{\mathbf{p}_n} U_n(\mathbf{p}_n, \mathbf{I}_n(\mathbf{p}_{-n})). \tag{17.39}$$

Define the Jacobian matrix of $\tilde{F}_n$ at $\mathbf{I}_n$ to be $\frac{\partial \tilde{F}_n}{\partial \mathbf{I}_n}$ and the Jacobian matrix of $\mathbf{I}_n$ at $\mathbf{p}_{-n}$ to be $\frac{\partial \mathbf{I}_n}{\partial \mathbf{p}_{-n}}$. Then

$$\frac{\partial \tilde{F}_n}{\partial \mathbf{I}_n} = \begin{pmatrix} \frac{\partial p_n^{(1)o}}{\partial I_n^{(1)}} & \cdots & \frac{\partial p_n^{(K)o}}{\partial I_n^{(1)}} \\ \vdots & \ddots & \vdots \\ \frac{\partial p_n^{(1)o}}{\partial I_n^{(K)}} & \cdots & \frac{\partial p_n^{(K)o}}{\partial I_n^{(K)}} \end{pmatrix} \tag{17.40}$$

and

$$\frac{\partial \mathbf{I}_n}{\partial \mathbf{p}_{-n}} = \begin{pmatrix} g_{1n}^{(1)} & & & & \mathbf{0} \\ & \ddots & & & \\ \mathbf{0} & & & & g_{1n}^{(K)} \\ & & \vdots & & \\ g_{(n-1)n}^{(1)} & & & & \mathbf{0} \\ & \ddots & & & \\ \mathbf{0} & & & & g_{(n-1)n}^{(K)} \\ g_{(n+1)n}^{(1)} & & & & \mathbf{0} \\ & \ddots & & & \\ \mathbf{0} & & & & g_{(n+1)n}^{(K)} \\ & & \vdots & & \\ g_{Nn}^{(1)} & & & & \mathbf{0} \\ & \ddots & & & \\ \mathbf{0} & & & & g_{Nn}^{(K)} \end{pmatrix}. \tag{17.41}$$

Define $||A||$ to be the Frobenius norm of matrix $A = (a_{ij})$, i.e. $||A|| = \sqrt{\sum_{i,j} a_{ij}^2}$. We know that when a contraction mapping has a fixed point, the fixed point is unique [160]. Readily, we have the following sufficient condition, which comes from [105], that assures a unique equilibrium.

THEOREM 17.8 (Uniqueness) *In frequency-selective channels, if for any user n,* $||F_n(\mathbf{p}_{-n}) - F_n(\check{\mathbf{p}}_{-n})|| < ||\mathbf{p}_{-n} - \check{\mathbf{p}}_{-n}||$ *for any different* $\mathbf{p}_{-n}$ *and* $\check{\mathbf{p}}_{-n}$*, there exists one and only one equilibrium* $\mathbf{p}^*$ *in the non-cooperative power control game defined by (17.39).*

Intuitively, Theorem 17.8 says that if other users change their transmit powers by some amount, the best power allocation of the user is altered by a lesser amount, then the equilibrium is unique. Note that the transmit powers of other users and the best response $F_n(\mathbf{p}_{-n})$ in (17.39) are related through interference channel gains, which therefore determines the variation of the best response and whether the sufficient condition can be guaranteed. Stronger interference channel gains result in higher correlation and vice versa. The above two-user network illustrates an example where one subchannel has extremely strong interference channel gains. In this case, the sufficient condition is violated and there are multiple equilibria.

Based on Theorem 17.8, Theorem 17.9 explicitly shows the impact of interference channel gains on the number of equilibria and is proved in Appendix A.12.

THEOREM 17.9 (Uniqueness) *In frequency-selective channels, if for any user n,*

$$\left|\left|\frac{\partial \mathbf{I}_n}{\partial \mathbf{p}_{-n}}\right|\right| < \frac{1}{\sup_{\mathbf{I}_n} \left|\left|\frac{\partial \tilde{F}_n}{\partial \mathbf{I}_n}\right|\right|}, \tag{17.42}$$

where $\sup_{\mathbf{I}_n}$ is the supremum on all feasible $\mathbf{I}_n$, there exists one and only one equilibrium $\mathbf{p}^*$ in the non-cooperative power control game defined by (17.39).

After examining the Jacobian matrices, we see that the left-hand side of (17.42) depends on interference channel gains only, while the right-hand side is independent of interference channel gains. Hence, interference channel gains directly impacts the number of equilibria. Consider an example where different users are sufficiently far away and all interference channel gains are close to zero. It is easy to see that transmit powers of other users have almost no effect on the best response of the user and there is a unique equilibrium.

The sufficient conditions of a unique equilibrium for distributed multi-channel power controls in Theorems 17.8 and 17.9 can be applied in different kinds of distributed multiple-input multiple-output (MIMO) and orthogonal frequency division multiplexing (OFDM) systems.

17.3.4 Distributed power control with pricing

There are many types of elastic data services. For these delay-tolerant services, e.g. email, web browsing, and file transferring, it is not desirable to use the distributed power control algorithms that are developed for supporting strict-SIR requirements. These delay-tolerant services usually desire high data rate and higher data rates will always result in higher user satisfaction. One simple approach is for each user to allocate power to selfishly maximize its own data rate, i.e. $u_k = r_k$ and

$$p_k[n+1] = \arg\max_{p_k} r_k(p_k) = \arg\max_{p_k} \log\left(1 + \frac{p_k g_{kk}}{\sum_{i \neq k} p_i g_{ik} + \sigma_k^2}\right). \tag{17.43}$$

This may not work because all users will simply use the highest power possible and network performance might be poor because of excessive interference. The problem can be resolved by introducing the concept of pricing. The pricing mechanism introduces a cost for each user to use a certain amount of power and thus regulates the aggressive power control behavior to reduce interference. A linear pricing example is

$$u_k = r_k - \mu_k p_k = \log\left(1 + \frac{p_k g_{kk}}{\sum_{i \neq k} p_i g_{ik} + \sigma_k^2}\right) - \mu_k p_k, \tag{17.44}$$

where μ_k is the pricing parameter and

$$p_k[n+1] = \arg\max_{p_k} u_k. \tag{17.45}$$

Some examples of the utility function, u_k, are illustrated in Figure 17.6. We can see different pricing parameters will result in different power values and using a proper μ_k will help control the network interference and improve network throughput.

u_k is concave in p_k and its maximum can be found by setting its first-order derivative to be zero as follows,

$$\frac{du_k}{dp_k} = \frac{1}{\frac{I_k}{g_{kk}} + p_k} - \mu_k = 0, \tag{17.46}$$

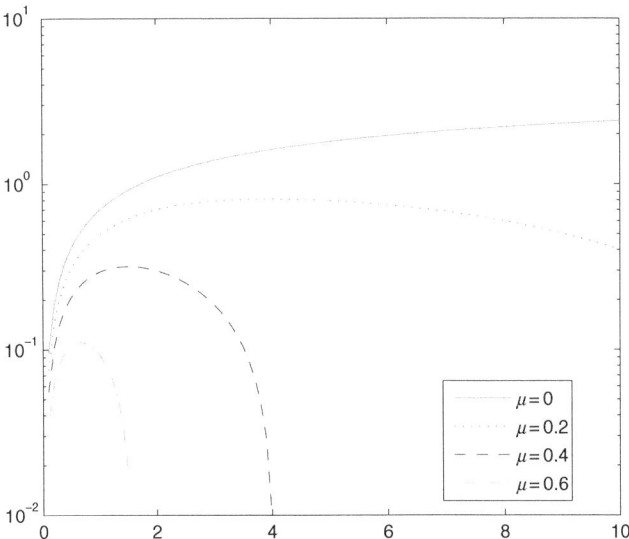

Figure 17.6 Utility function (the interference plus noise power is 1, all parameters, if not specified, are 1)

where $I_k = \sum_{i \neq k} p_i g_{ik} + \sigma_k^2$. Based on the above equation, we have the power control formula

$$p_k = \max\left(\frac{1}{\mu_k} - \frac{I_k}{g_{kk}}, 0\right). \tag{17.47}$$

Compare the power control in (17.18) and (17.47). In (17.18), the transmission power will be increased if the transmitter observes increasing interference. So the power control mechanism is more aggressive and allocates power actively to achieve the SIR requirements. On the other hand, with (17.47), the transmission power will be reduced if increasing interference is observed. This is a passive way of avoiding interference and the power control behavior is more socially favorable.

This power control function is not a standard interference function and there might be multiple equilibria. According to Theorem 17.9, if for each user k,

$$\sqrt{\sum_{i \neq k} g_{ik}^2} < \delta(g_{kk}), \tag{17.48}$$

where $\delta(g_{kk})$ is a function of only g_{kk} and independent of interference channels, there will be a unique equilibrium for the above distributed power control. Hence, the number of equilibria is determined by how strong interference channel gains and signal channel gains are. The power control results of different users are related through interference channels, which determine the variation of the power control results when other users change their power values. Stronger interference channels result in higher correlation and vice versa. When all interference channels are weak, the above sufficient condition will be satisfied and there will be a unique equilibrium. Similarly, in [190], it is shown that if the spreading gain in a CDMA system is sufficiently large, i.e. interference channels are sufficiently weak, the algorithm in (17.47) will converge to a unique equilibrium. Please refer to [190] for more detailed discussions of the global stability and convergence of the power control in (17.47).

There are many other ways of defining the utility function u_k as well as the pricing mechanism for distributed power control. The properties of the power-control algorithms depend on the characteristics of the utility functions and pricing mechanisms. Readers are referred to [133] for more discussions.

Part IV

Cross-layer optimization for energy-efficient networks

18 Overview

The exponential growth in both mobile data traffic and the number of terminals connected to cellular networks calls for a future-ultra dense deployment of cellular networks. This growth will inevitably be limited by energy consumption of mobile terminals and, predominantly, the radio access networks.

While semiconductor processing speed has been increasing exponentially, doubling almost every two years according to the Moore's law, processor power consumption also continues to grow by 150% every two years [114]. By contrast, advances in battery technology have not kept pace, with capacity increasing at a modest rate of 10% every two years [114]. This leads to an increasingly large gap between power thirst and battery capacity. On the other hand, information and communication technology (ICT) plays an important role in global greenhouse gas emissions since the amount of energy consumed by ICT increases dramatically to meet the explosive growing service requirements. It is shown that nowadays the total energy used by the infrastructure of cellular networks, wired networks, and Internet takes up more than 3% of worldwide electric energy consumption [80]. Its CO_2 emissions have increased to about 2% of worldwide CO_2 emissions over the past decade [74]. More importantly, a large portion of energy consumption originates from the operation of wireless access networks. Indeed, base station energy consumption accounts for about 70% of the total energy used by mobile operators, which is still growing fast to accommodate the increasing need for mobile communication [91]. Based on [91], while the number of base stations grows linearly with approximately 120 000 more each year, the energy consumption increases at an exponential rate. As shown in [91], average energy consumption maintaining one base station was already as high as 14 000 kWh in 2011. The above scenario calls for an essentially different mobile access, exhibiting unprecedented sustainability and efficiency in the usage of all resources, especially energy.

To address the rising concern about energy consumption, there have been several ongoing industry efforts to improve wireless wide-band networks, such as long-term evolution (LTE) and LTE-A. Examples include the GreenTouch [3] and EARTH (Energy Aware Radio and NeTwork TecHnologies) projects [2], whose goals are to drastically reduce energy wastage of wireless broadband networks without compromising the quality of service provided to users. While these projects focus more on industrial solutions, there is a lack of understanding of energy efficiency from a more fundamental perspective, e.g. what would be the minimum amount of energy needed to enable wireless communications, what are the tightest energy efficiency upper bounds of different systems, and what are the corresponding optimal designs. These are the focus of this chapter.

Similar to the previous two parts, we emphasize cross-layer design as an integrative cross-layer approach which exploits the interactions between different layers and will significantly improve *energy efficiency* and adaptability to service, traffic, and environmental dynamics. Since wireless networks utilize a shared medium, network energy consumption is affected not only by the layers responsible for the point-to-point link, but also by the interaction among links in the entire network. Thus, a systematic approach is essential in achieving the most energy-efficient design.

18.1   Lighting analogy

Lighting lamps are quite similar to radio transmitters, except the lamps are working on higher electromagnetic frequencies. Lamps are deployed everywhere to illuminate human beings while wireless networks are deployed ubiquitously to "illuminate" mobile devices. In outdoor areas, high-power floodlights or street lights can be used to provide lighting while in indoor environments, ceiling lights, table and floor lamps can be used for illumination. Similarly, in mobile communications, sparsely distributed macro base stations (BSs) with 20W or higher radiation power are used to serve large areas and hotspot BSs like pico and femto BSs serve indoors. The development of lighting technologies are inspiring for us to look into the future development of wireless technologies.

Figure 18.1 briefly describes the timeline of modern lighting technologies. In 1780, Aime Argand invented the central draught fixed oil lamp, which was much brighter than a candle by a factor of five to ten. William Murdoch was the first to use gas for the commercial application of lighting. Early in the 19th century, gas lights were used in the streets of most cities in the United States and Europe. The brighter lighting provided by gas enabled people to read more easily and far longer. Gas lighting gave way to electric lighting at the turn of the 19th century. Throughout the 19th century and early 20th century, scientists worked on improving the brightness and lifetime of electric lamps, from platinum filament, arc lamp to neon lighting. With Edison's design of the first electric incandescent lamps in the 1870s, electric lighting became viable for the first time.

Incandescent bulbs work by using electricity and a filament. In incandescent bulbs, electricity is forced through a filament with high resistance to electricity, which heats the

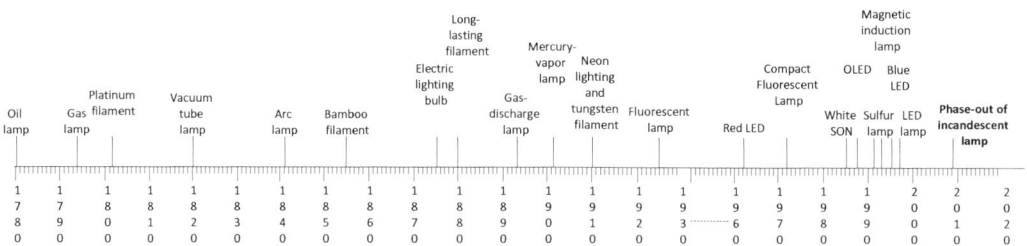

Figure 18.1 Timeline of lighting technology [205]

filament to an extremely high temperature. The filament then radiates electromagnetic waves with a frequency spectrum determined by the temperature. At extremely high temperatures, the electromagnetic waves are partly in the visible light spectrum. The first incandescant bulbs made by Edison used a carbon filament and were 1% efficient in turning electricity into light. Later, in 1904, tungsten replaced carbon as the filament material. Also some other techniques such as adding halogen-element gas into the bulb, were discovered to enable filaments to last longer and even to run hotter. Nowadays, the efficiency of incandescent bulbs has been improved from 1% to about 10% and the remaining 90% of the electrical energy is converted mainly into heat. Therefore, a 40-watt halogen bulb only radiates 4 watts of visible light and the remaining 36 watts is lost, primarily as heat.

It took about 130 years, from 1780 to 1910, to improve the brightness of light bulbs for higher lumens. Interestingly, the trend changed after 1910. While lifetime extension of the bulb was still one main driver, the other focus shifted towards more energy-efficient designs. Fluorescent lamps were invented in 1926. Compared to incandescent bulbs, fluorescent lamps consume only one fifth of the energy consumption for the same lumens. While incandescent and halogen bulbs range in efficacy from two to 30 lumens per watt, fluorescent ones range from 25 to 105 lumens per watt. Nowadays, the compact fluorescent lamp (CFL) is available almost everywhere and used as an energy-saving alternative. In 1962, the first practical visible-spectrum light-emitting diode (LED) was invented. They are solid-state devices with lifetime many times longer than incandescent bulbs. Many more solid-state light (SSL) sources, e.g. OLED, blue LED, light-emitting polymer (LEP), were invented in the following decades. These SSL devices are among the most efficient devices for turning electricity into light. Compared to fluorescent devices, it is projected that SSL bulbs hold the promise to reduce electric energy use by 50% [145]. SSL devices are evolving to displace some of the traditional light sources in some applications. From the timeline in Figure 18.1, the energy-efficient designs for lighting technologies have lasted more than a hundred years and are still ongoing.

Besides continuing the development of more energy-efficient lighting devices, countries worldwide have started phasing out incandescent bulbs. For example, on 1 September 2009, the European Union began phasing out incandescent bulbs and banning stores from buying new stock. CFLs are more expensive than incandescent bulbs, but they pay for themselves in energy savings within several months. In addition, these energy-efficient long-lasting CFLs also have the benefit of reducing environmental impact. According to the US Department of Energy, replacing incandescent bulbs with CFL bulbs in all USA families would be the environmental equivalent of removing 7.5 million cars from the road.

Now let us take a look at the development of wireless technologies, which has been ongoing for a bit more than 100 years. Figure 18.2 illustrates the main history of wireless technology. Especially in the recent years, cellular technologies have outstripped every other form of wireless technology as two-way communication systems. While we spent almost a hundred years developing analog communications, it took just over 30 years to migrate from the first-generation analogue networks to 4G systems capable

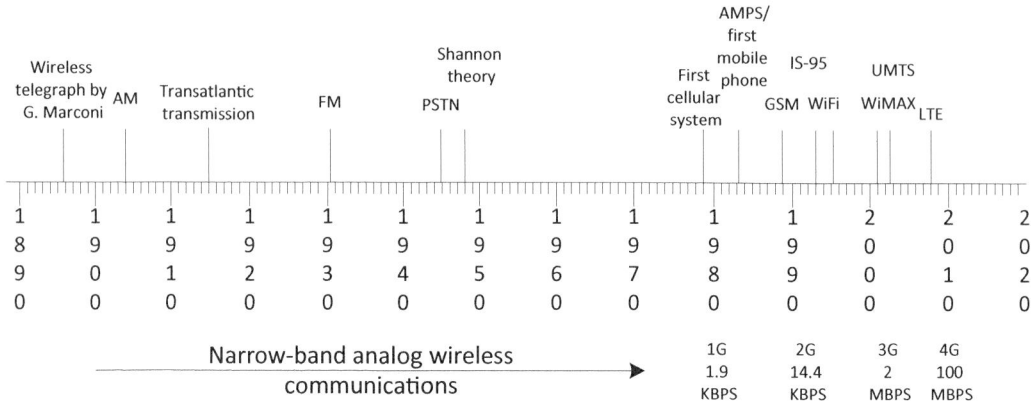

Figure 18.2 Timeline of wireless technology

of multimedia content transfers. The first 4G standard is already capable of data rates of over 100 Mbps. In 3GPP Rel 10, the peak downlink data rate can be higher than 1 Gbps and the following releases support even higher speed. With numerous developments still occurring, the outlook for wireless technology would be particularly interesting. Comparing Figure 18.2 with Figure 18.1, we can see tremendous similarity between the timelines for wireless and lighting technologies. Both technologies took around one hundred years for capacity improvement, one in data rate and the other in luminous intensity. In the latter one hundred years of development, the focus of lighting technologies shifted toward energy-efficient designs. Analogously, how about the future development of wireless technologies? Is it also time for us to take the turn? After meeting demand in performance, we can confidently project that in the following decades, energy-efficient designs would be one main trend.

18.2   Methodology

Radio resource management (RRM) controls radio transmission characteristics and system level co-channel interference in wireless networks. Figure 18.3 summarizes the basic methodology of conventional RRM technologies. It involves network architecture, protocol, and algorithm designs and the design parameters can be anything, such as cellular deployment, frequency planning, handover criteria, medium access control (MAC) policy, modulation order, coding rate, and so on. The objective is to control the design parameters either statically or dynamically such that the quality of service (QoS) of all users can be assured. Efficient RRM schemes may increase the system spectral efficiency by an order of magnitude, which is typically not possible by using solely advanced coding schemes.

Let us take a look at Figure 18.3 from a different angle and we will see energy conservation is almost completely ignored in the conventional RRM designs. In the past, because of technology limitations, we have focused solely on improving the spectral efficiency of wireless networks. Now wireless networks are almost able to

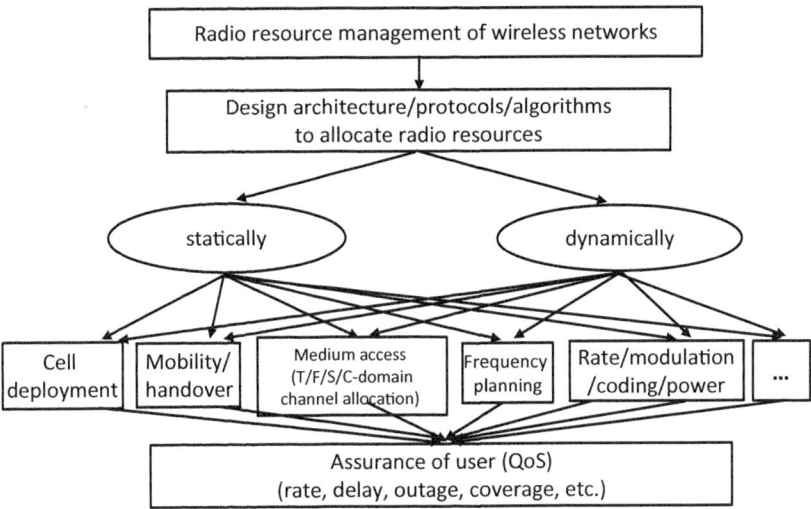

Figure 18.3 Radio resource management in wireless networks

meet most of our demands and it is time to integrate energy efficiency into the system designs. This indeed means huge opportunities ahead of us, as all the conventional design philosophies of existing network architectures, protocols, and algorithms should be revisited. More importantly, we need to come up with new ones.

The goal of this part is to investigate thoroughly the relation between energy efficiency, spectral efficiency, implementation, and network resource management for various wireless networks and their corresponding optimal designs. We target understanding of the highest energy efficiency wireless systems can achieve; equivalently, how much energy is the least necessary to drive the desired data communications. We will introduce a set of the tightest energy efficiency upper bounds in wireless networks and their corresponding optimal designs. The investigations will take into account not only radio transmission, but also other sources of energy consumption so that the theoretical framework can be easily used in practice.

Energy is consumed by all layers of the implementation and network protocols, from the radio frequency (RF) front to the operating system, from the physical and MAC layer to the whole network, etc. In general, communication devices have two main categories of energy consumption.

- Communication: energy spent by reliable communications. Communication energy depends on communication states, such as signal-to-noise ratio (SNR) requirements, wireless channel (cell radius, fading, shadowing), network load, and so on. Communication energy consumption models all the power used for reliable data transmission. In the following, we will call this portion of energy consumption transmission energy and power is called transmission power.
- Computation: energy spent by signal processing, computing, cooling, and so on to maintain the operation of the device. Computation energy depends on hardware and software used for running the operating system, compressing data,

coding, decoding, filtering, etc. This portion of energy consumption is usually independent of communication states, like modulation order and data rate. We will call it circuit energy and the corresponding power consumption circuit power. The circuit power captures the power consumption that is independent of transmission states.

The communication energy consumption dominates for long-distance communications such as in macro cellular or satellite networks. On the other hand, the computation energy consumption may dominate for short-distance communications, e.g. in femto cells or WiFi. Generally speaking, minimizing total energy consumption requires minimizing the contributions from both communication and computation and finding the best tradeoff between them. With the development of semiconductor technologies, which has been shown to follow Moore's Law, the computation energy will get smaller and smaller. However, communication energy is determined by wireless channels and user requirements, and is increasing constantly to meet user demand. The portion of energy consumed for reliable communications is therefore increasingly bigger and more important.

Intuitively, there are several ways to improve the energy efficiency of communication devices. From the implementation perspective, low-power techniques can be used in semiconductor technology, circuit design, system architecture, operating system, cooling system, and application designs. The consideration of energy efficiency will affect the whole design and implementation process. However, these will not be discussed in detail in this book as we will focus on the communications perspective for energy-efficient network designs.

In this part, we will take a bottom-up approach to study the fundamental limits and corresponding optimal designs. This means we will first study simple point-to-

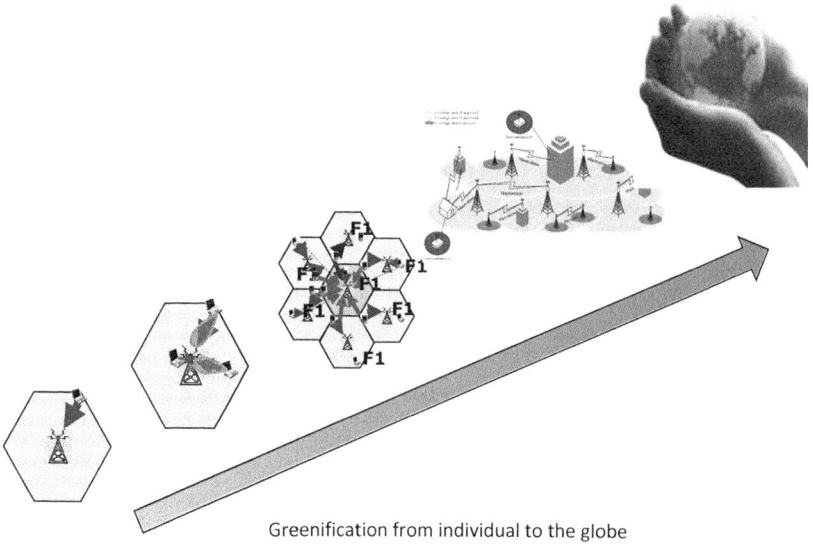

Greenification from individual to the globe

Figure 18.4 Green designs from individual links to the global network

point energy efficient communication and then more complicated multi-user networks. Figure 18.4 illustrates the framework of this part. We will first introduce single-user energy-efficient transmission technologies and then energy-efficient orthogonal resource management schemes for a multi-user single-cell network. After that we will study distributed energy-efficient cross-layer designs, including both distributed random access and power control. Then we introduce the fundamental tradeoffs in wireless resource allocation. Finally, we move on to the whole network level and investigate system-level energy-efficient designs for both homogeneous and heterogeneous cellular networks. Implementation issues in practice will also be discussed at the end.

19 Energy-efficient transmission

In this chapter, we will introduce the concept of energy efficiency capacity, i.e. the highest energy efficiency a wireless link can achieve, and the corresponding optimal designs for point-to-point link communications across time, frequency, and spatial domains.

19.1 Energy efficiency capacity

While channel capacity is the maximum data rate at which information can be transmitted through a communication channel, we define energy efficiency capacity, U, of a transceiver link as the maximum energy efficiency at which information can be transmitted through its channel, i.e.

$$C_E = \max_{R \text{ and } P} u = \max_{R \text{ and } P} \frac{R}{P}, \tag{19.1}$$

where the maximum is taken over all possible choices of coding, and thus rate, and power allocation. Note that

$$C_E = \max_{R \text{ and } P} \frac{R}{P} = \max_{P} \max_{R \text{ given } P} \frac{R}{P} = \max_{P} \frac{\max_{R \text{ given } P} R}{P} = \max_{P} \frac{C(P)}{P}, \tag{19.2}$$

where $C(P)$ is the conventional channel capacity when the transceiver power consumption is P. We can see that one necessary condition to achieve energy efficiency capacity is that the coding scheme should achieve the channel capacity. Equivalently, the problem is the same as

$$C_E = \max_{R} \max_{P \text{ given } R} \frac{R}{P} = \max_{R} \frac{R}{\min_{P \text{ given } R} P} = \max_{R} \frac{R}{C^{-1}(R)}, \tag{19.3}$$

where $C^{-1}()$ is the inverse of $C()$. Again, the channel capacity has to be achieved for the energy efficiency capacity. Equations (19.2) and (19.3) represent two equivalent ways of achieving the energy efficiency capacity. Always using capacity-approaching coding, either the appropriate transmission power or the data rate can be chosen to maximize the energy efficiency. Choosing the power is the same as choosing the data rate as they are uniquely related through the power-rate function, e.g. the Shannon capacity formula here. In this book, we will use these two methods interchangeably.

19.2   Ideal transmission

With ideal capacity-approaching channel codes, such as turbo or low density parity check (LDPC) codes, the reliable data rate, r, of a band-limited additive white Gaussian noise (AWGN) channel is given by

$$r = W \log_2 \left(1 + \frac{pg}{WN_0} \right),$$

(19.4)

where p is the transmission power, g the channel power gain, W the signal bandwidth, and N_0 the noise spectral density. The time to deliver one bit is t, which equals

$$t = \frac{1}{r} = \frac{1}{W \log_2 \left(1 + \frac{pg}{WN_0} \right)},$$

(19.5)

and the power–time relationship is

$$p = \left(2^{\frac{1}{Wt}} - 1 \right) WN_0/g.$$

(19.6)

Thus, the relationship between the energy and time used in sending this one bit is

$$E = pt = \left(2^{\frac{1}{Wt}} - 1 \right) WtN_0/g,$$

(19.7)

and the energy efficiency is

$$u = \frac{1}{E} = \frac{g}{\left(2^{\frac{1}{Wt}} - 1 \right) WtN_0}.$$

(19.8)

The energy is monotonically decreasing and convex in the transmission time t, as illustrated in Figure 19.1. Achieving the highest energy efficiency $u = \frac{1}{E}$ is the same as minimizing the energy consumption E. We can see that energy is minimized when t approaches infinity, i.e. using an infinite amount of time to send one bit. In this case, an infinitely small amount of power will be used according to (19.6). To ensure reliable transmission, we need to use a coding scheme with an infinitely small coding rate. The minimum energy needed to send one bit is

$$E_{min} = \lim_{t->\infty} \left(2^{\frac{1}{Wt}} - 1 \right) WtN_0/g = \frac{N_0 \ln 2}{g}.$$

(19.9)

We can also look at the energy consumption from the receiver perspective.

$$\frac{Eg}{N_0} \geq \frac{E_{min}g}{N_0} = \ln 2 \geq \ln 2 = -1.59\text{dB},$$

(19.10)

indicating that the minimum received signal energy in total can be 1.59 dB smaller than the noise spectral density.

The energy efficiency capacity is

$$C_E = \max \frac{W \log_2(1 + \frac{pg}{WN_0})}{p} = \lim_{p->0} \frac{W \log_2(1 + \frac{pg}{WN_0})}{p} = \frac{g}{N_0 \log 2},$$

(19.11)

which can also be obtained by letting t in (19.8) approach infinity. Note this energy efficiency capacity is for AWGN channels. The energy efficiency capacity for other types of channels can be derived in a similar way.

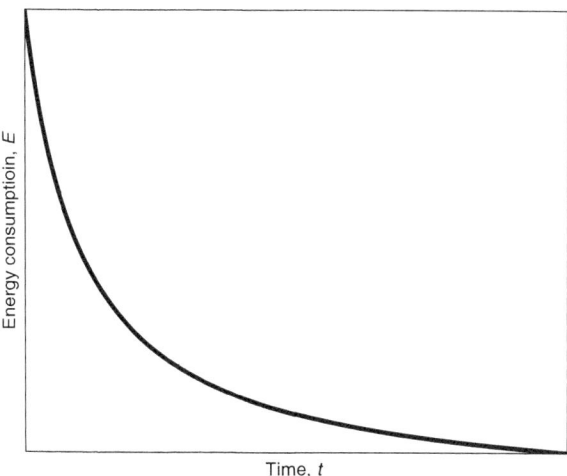

Figure 19.1 Energy consumption and time to send one bit

19.3   Energy-efficient transmission in practice

In addition to transmission power, a device will also incur additional circuit power that is relatively independent of the transmission rate. Thus, a fixed energy cost of transmission is incurred, which must also be accounted for when designing energy-efficient transmission systems. We first consider the case of flat-fading wireless channels before addressing the more complex frequency-selective channels.

The overall energy used in sending one bit is

$$E = pt + P_c t = \left(2^{\frac{1}{Wt}} - 1\right) WN_0 t/g + P_c t, \qquad (19.12)$$

where P_c is the average circuit power, including all electronic power consumption except transmit power for reliable data transmission. Figure 19.2 shows the circuit energy and transmit energy tradeoff for overall energy efficiency. We can see that when circuit power is taken into account, the method to transmit with the longest duration is no longer the best since circuit energy consumption increases with transmission duration.

Considering the circuit power, system energy efficiency is

$$u(r) = \frac{r}{P(r)} = \frac{r}{P_C + P_T(r)}, \qquad (19.13)$$

where $P_T(r)$ is a generic description of the relationship between the achieved data rate and transmission power. For an AWGN channel with capacity-approaching coding,

$$P_T(r) = (e^{\frac{r}{W}} - 1)N_o W/g, \qquad (19.14)$$

which is monotonically increasing and strictly convex in r. In general, we assume $P_T(r)$ to be monotonically increasing and strictly convex and $P_T(0) = 0$. For a channel other than AWGN ones, the following theories and algorithms can be used to find system energy efficiency capacity and the optimal design after figuring out the corresponding

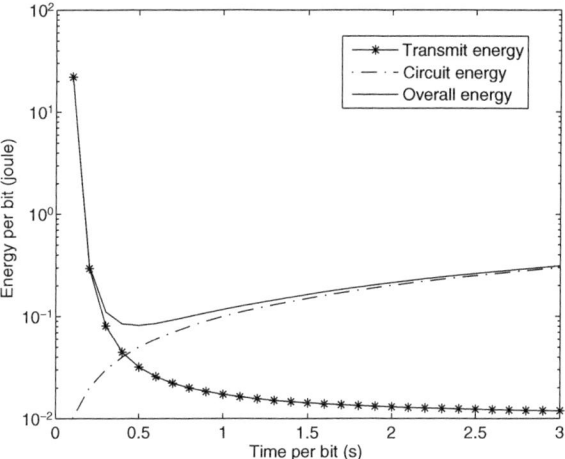

Figure 19.2 Relationship between energy consumption and symbol duration

$P_T(r)$ of the channel, where r should be the sum rate and $P_T(r)$ the sum power of all potential subchannels. In Section 22.1, we will give a thorough discussion on the power and rate relationship, $P_T(r)$, and its inverse function for different channels.

The optimal transmission data rate is given by the following theorem.

THEOREM 19.1   *If $P_T(r)$ is monotonically increasing and strictly convex in r, there exists a unique globally optimal transmission data rate given by*

$$r^* = \frac{P_C + P_T(r^*)}{P'_T(r^*)},$$ (19.15)

where $P'_T(\cdot)$ is the first derivative of the function $P_T(\cdot)$. The energy efficiency capacity is

$$C_E = \frac{r^*}{P_C + P_T(r^*)}.$$ (19.16)

Proof $r^* = \arg\max_r \frac{r}{P_C+P_T(r)} = \arg\min_r \frac{P_C+P_T(r)}{r}$. Define $f(r) = \frac{P_C+P_T(r)}{r}$. Let $r = \frac{1}{t} > 0$, and $g(t) = f(\frac{1}{t}) = P_C t + P_T(\frac{1}{t})t$. Then $r^* = \frac{1}{t^*}$ and $t^* = \arg\min_t g(t)$. Since $\frac{\partial^2 g(t)}{\partial t^2} = \frac{1}{t^3} P''_T(\frac{1}{t}) > 0$, $g(t)$ is strictly convex in t. Since $P_T(r)$ is monotonically increasing and strictly convex in r, the derivative satisfies $\lim_{r->\infty} P'_T(r) = \infty$. According to the L'Hopital's rule, $\lim_{t->0} g(t) = \lim_{t->0} P_T(\frac{1}{t})t = \lim_{r->\infty} \frac{P_T(r)}{r} = \lim_{r->\infty} \frac{P'_T(r)}{1} = \infty$. Besides, $\lim_{t->\infty} g(t) = \infty$. Since $g(t) < \infty$ for $0 < t < \infty$, t^* uniquely exists and is globally optimal. By letting $\frac{\partial g(t)}{\partial t} = 0$ and $R = \frac{1}{t}$, we have the solution in Equation (19.15).   □

It can be easily verified that C_E is tightly upper bounded by $\frac{1}{P'_T(0)}$, which is achieved when $P_C = 0$. When Shannon capacity is achieved in AWGN channels, the tightest upper bound is $\frac{g}{N_o}$.

In the following, Propositions 19.2 and 19.3 summarize the impact of channel gain and circuit power on the optimal energy-efficient transmission.

PROPOSITION 19.2   *Both the optimal data rate and energy efficiency increase with channel gain.*

PROPOSITION 19.3   *The optimal data rate increases with circuit power, while energy efficiency decreases with it. With zero circuit power, the highest energy efficiency,* $\frac{1}{P_T'(0)}$, *is obtained by transmitting with an infinite small data rate.*

Proof   Define $P_R(r)$ to be the received power on a subchannel for reliable detection when the data rate on the subchannel is r. We have $P_T(r) = \frac{P_R(r)}{g} = \frac{P_R(r)}{g}$, where g is the channel power gain. It is easy to see that $P_R(r)$ is monotonically increasing and strictly convex, and $P_T(0) = P_R(0) = 0$. According to Theorem 19.1, we have $r^* P_T'(r^*) = P_C + P_T(r^*)$, which is equivalent to $R^* P_R'(r^*) - P_R(r^*) = P_C g$. By differentiating the left-hand side with respect to r^*, $\frac{\partial \left(r^* P_R'(r^*) - cP_R(r^*) \right)}{\partial r^*} = r^* P_R''(r^*) > 0$. Hence, the left-hand side is strictly increasing in r^*. Therefore, a higher data rate should be used when the channel has higher power gain. Suppose $g_1 > g_2$, and the corresponding optimal modulation and coding result in data rates r_1^* and r_2^*, respectively. Hence, $U_1(r_1^*) > U_1(r_2^*)$. Besides, $U_1(r_2^*) = \dfrac{r_2^*}{P_C + \frac{P_R(r_2^*)}{g_1}} > \dfrac{r_2^*}{P_C + \frac{P_R(r_2^*)}{g_2}} = U_2(r_2^*)$. Hence, energy efficiency increases with channel gain.

According to Theorem 19.1, $r^* P_T'(r^*) - P_T(r^*) = P_C$. The derivative of the left-hand side is $r^* P_T''(r^*) > 0$. Hence, r^* increases with P_C. The proof that energy efficiency decreases with circuit power is similar to the proof that energy efficiency increases with channel gain. When $P_C = 0$, $U(r)$ is maximized when r approaches zero, i.e. $U_{max} = \lim_{r \to 0} \frac{r}{P_T(r)} = \frac{1}{P_T'(0)}$.   □

From Proposition 19.3, when circuit power dominates power consumption, which is usually true with short-range communication, the highest data rate should be used to finish transmission as soon as possible, which has been commonly assumed by most MAC layer energy-efficient optimization schemes as described in the introduction to this chapter. However, when the circuit power is negligible, which is usually true with long-range communications such as satellite communications, the lowest data rate should be used, which coincides with the results in [71] and [42].

To illustrate the application of Theorem 19.1, we derive the optimal link settings for uncoded *multiple quadrature amplitude modulation* (M-QAM) in an AWGN channel. The frame structure of the system is shown in Figure 19.3. Each transmission slot consists of a data interval, T_s, and a signaling interval, τ. Assume block-fading, i.e. the channel state remains constant during each data interval and is independent from one to another. For M-QAM, the number of bits transmitted per symbol is $b = \log_2 M$, where M is the modulation order. In each data interval, l symbols are transmitted on

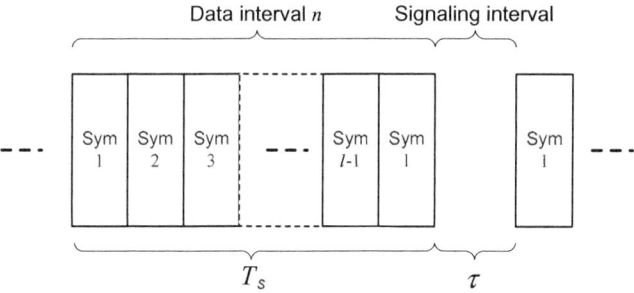

Figure 19.3 Frame structure

each subchannel. The data rate on each subchannel is $r = \frac{bl}{T_s+\tau}$, and the overall data rate is

$$R = cr = \frac{cbl}{T_s + \tau}. \tag{19.17}$$

Consequently, for a given data transmission rate, the number of bits transmitted per symbol will be $b = \frac{R(T_s+\tau)}{cl}$.

The *bit error rate* (BER) for coherently detected M-QAM with Gray mapping over an AWGN channel is approximated by [4]

$$P_e(\gamma) \approx 0.2 \exp\left(-\frac{1.5\gamma}{M-1}\right), \tag{19.18}$$

where γ is the *signal-to-noise ratio* (SNR). Define g to be the power gain of the channel. The SNR on each subchannel will be

$$\gamma = \frac{P_T(R)g}{cN_oW}, \tag{19.19}$$

where N_o is the power spectral density and W is the signal bandwidth in each subchannel. For a given BER, P_e, the required SNR can be determined by (19.18). Consequently, the required transmit power will be

$$P_T(R) = \frac{\gamma cN_oW}{g} = A(1 - 2^{BR}), \tag{19.20}$$

where $A = \frac{2c \ln(5P_e)N_oW}{3g}$ and $B = \frac{T_s+\tau}{cl}$. Usually, $5P_e < 1$ for effective transmission; therefore, $A < 0$. It can be seen that $P_T(R)$ is monotonically increasing and strictly convex in R. P_C characterizes circuit power consumption in both the data and signaling intervals. According to Theorem 19.1, the desired data rate is

$$R^* = \frac{A(2^{BR^*} - 1) - P_c}{AB2^{BR^*} \ln 2}. \tag{19.21}$$

Correspondingly, $b^* = \frac{R^*(T_s+\tau)}{cl}$ and $M^* = 2^{\frac{R^*(T_s+\tau)}{cl}}$.

To demonstrate Theorem 19.2, we present energy-efficient link transmission for an uncoded M-QAM system with the frame structure as in Figure 19.3. System parameters

Table 19.1 System parameters

Carrier frequency	1.5 GHz
Subchannel bandwidth	10 kHz
BER	10^{-6}
Symbol number of data interval, l	100
Time duration of data interval, T_s	0.01s
Time duration of signaling interval, τ	0.001s
Thermal noise power, N_o	-141 dBW/MHz
User antenna height	1.6 m
BS antenna height	40 m
Environment	Macro cell in urban area
Circuit power, P_C	100 mW
Maximum transmit power	33 dBm
Propagation model	Okumura–Hata model
Fading	Flat-fading
Modulation	Uncoded M-QAM

are listed in Table 19.1. Each user is assigned ten subchannels, unless otherwise specified.

Figure 19.4(a) shows the energy efficiency of users located at different distances from the base station (BS). The lower axis shows the data rate achieved given the modulation order indicated by the top axis. The figure shows that by selecting an optimal modulation scheme, energy efficiency increases as the user moves closer to the BS. Furthermore, the optimal modulation for energy-efficient transmission varies with the distance between the user and BS. In general, for transmission with maximum energy efficiency, the closer the user is to the BS, the higher the modulation order should be. Figure 19.4(b) shows the energy efficiency of a user located 1 km away from the BS with different numbers of assigned subchannels. From Figure 19.4(b), the more the number of subchannels assigned to a user, the higher the maximum energy efficiency is, and the more sensitive to modulation order the energy efficiency is.

Figure 19.5 shows the energy consumed for sending one megabit. Figure 19.5(a) compares energy consumption for a system with fixed modulation and with an optimal modulation order determined by the energy-efficient transmission. For fixed modulation, the transmit power is allocated such that the BER is 10^{-6}. Figure 19.5(b) compares the optimal energy-efficient scheme with traditional adaptive modulation. In traditional adaptive modulation, the transmit power is fixed to be 15 dBm, 20 dBm, 25 dBm, or 30 dBm. The energy values are normalized with those of the optimal energy-efficient scheme. From the figures, the optimal scheme always achieves the lowest energy consumption.

19.4   Energy-efficient link adaptation in frequency-selective channels

In wide-band applications, different frequency bands usually experience different levels of fading. Link adaptation can be used to improve transmission performance. With link adaptation, the modulation order, coding rate, and transmission power can be selected

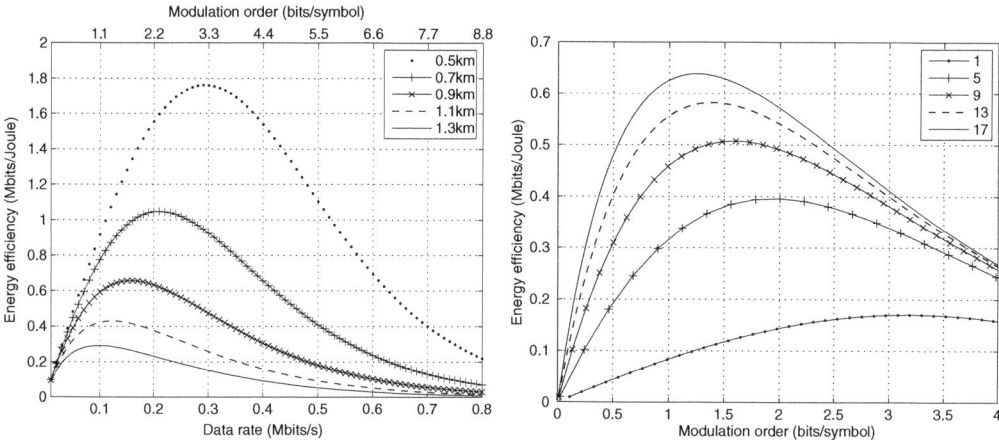

(a) Relationship of energy efficiency, distance, modulation, and transmission data rate

(b) Relationship of energy efficiency, modulation, and subchannel assignment

Figure 19.4 Energy efficiency relationship of per link transmission

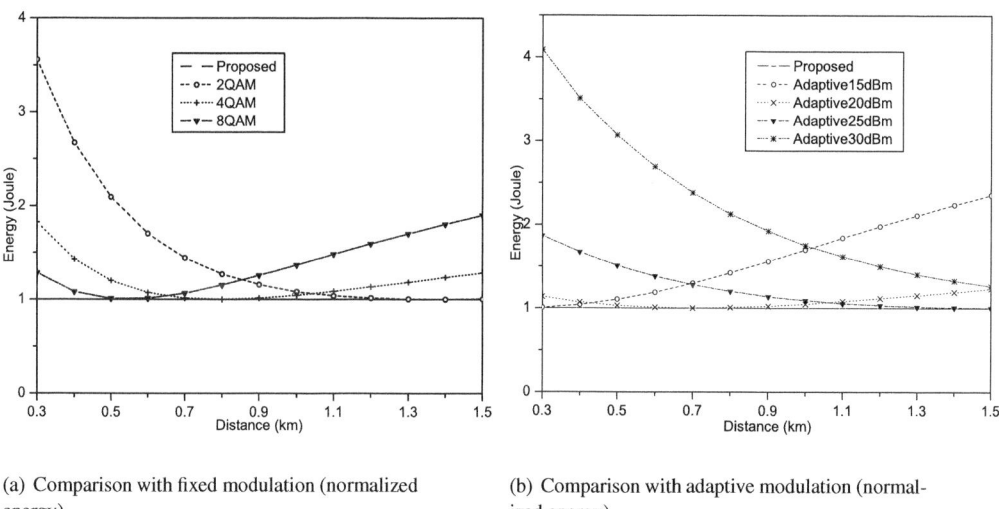

(a) Comparison with fixed modulation (normalized energy)

(b) Comparison with adaptive modulation (normalized energy)

Figure 19.5 Energy consumed for transmitting one megabit

according to *channel state information* (CSI). In addition to throughput improvement, energy efficiency can also be increased using link adaptation, as will be shown in this chapter. With sufficient battery power, link adaptation can be geared toward peak performance delivery. However, with limited battery capacity, link adaptation could be adapted toward energy conservation to minimize battery drain. Orthogonal frequency division multiplexing (OFDM) and Multiple-input multiple-output (MIMO) are primary physical (PHY) techniques for next-generation broadband wireless standards [63, 18]. While extensive research has been conducted to improve their throughput [213, 81], limited work has been done to address their energy-efficient communications. In this chapter, we discuss energy-efficient link adaptation for these types of systems in frequency-selective channels.

19.4.1   Modeling of energy-efficient link adaptation

Current communication systems design deals with frequency selectivity through subdividing the bandwidth into small segments, where the channel can be assumed to be flat. So for an ideal channel-orthogonalization technology, such as MIMO or OFDM, the channel may be divided into K subchannels, each experiencing flat-fading. Consider static channels to gain insight, and the results for time-varying channels can be derived similarly. Assume that K subchannels are used for transmission, each with a different channel gain. An example of this scenario is OFDM transmission over frequency-selective channels and MIMO transmission over several spatial channels. Another example is the downlink transmission of a BS serving multiple users at the same time using orthogonal frequency division multiple access (OFDMA) or multiple-user MIMO (MU-MIMO). After scheduling and subchannel assignment, the BS needs to determine modulation order and coding on all the subchannels. Assume block-fading [163, 59], i.e. the channel state remains constant during each data frame and is independent from one to another. Define the data rate on subchannel i as r_i and the data rate vector on all subchannels as

$$\mathbf{R} = [r_1, r_2, \ldots, r_K]^T, \tag{19.22}$$

where $[]^T$ is the transpose of a vector. The data rate vector, $\mathbf{R}$, depends on the channel state, coding, and power allocation. Correspondingly, the overall data rate is

$$R = \sum_{i=1}^{K} r_i. \tag{19.23}$$

For a given channel state, the transmit power on each subchannel is determined by the requirement of reliable data transmission. If we define W as the subchannel bandwidth, N_o the power spectral density, g_i the power gain, and P_{T_i} the allocated transmit power on subchannel i, the channel output SNR will be

$$\eta_i = \frac{P_{T_i} g_i}{N_o W} \tag{19.24}$$

and the achievable data transmission rate r_i is determined by [102]

$$r_i = W \log\left(1 + \frac{\eta_i}{\Gamma}\right), \tag{19.25}$$

where Γ is the SNR gap that defines the gap between channel capacity and a practical coding and modulation scheme. The SNR gap depends on the coding and modulation scheme used and on the target probability of error. Define the overall transmit power as $P_T(\mathbf{R})$ and

$$P_T(\mathbf{R}) = \frac{\sum_{i=1}^{K} P_{T_i}}{\zeta} = \sum_{i=1}^{K} (e^{\frac{r_i}{W}} - 1) \frac{N_o W \Gamma}{g_i \zeta}, \tag{19.26}$$

where $\zeta \in [0,1]$ is the power amplifier efficiency and depends on the design and implementation of the transmitter. $P_T(\mathbf{R})$ is strictly convex and monotonically increasing in $\mathbf{R}$. In fact, the developed theory and approaches taken can be used for any $P_T(\mathbf{R})$ that is strictly convex and monotonically increasing in $\mathbf{R}$ with $P_T(\mathbf{0}) = 0$, where $\mathbf{0} = [0, 0, \ldots, 0]^T$.

Defining the circuit power as P_C, the overall power consumption given a data rate vector will be

$$P(\mathbf{R}) = P_C + P_T(\mathbf{R}). \tag{19.27}$$

It is desirable to maximize the amount of data sent with a given amount of energy, which is equivalent to maximizing

$$U(\mathbf{R}) = \frac{R}{P_C + P_T(\mathbf{R})}. \tag{19.28}$$

$U(\mathbf{R})$ is called energy efficiency. As stated in Section 19.1, the energy efficiency can also be written as a function of $\mathbf{P}_T = [P_{T_1}, P_{T_2}, ..., P_{T_K}]^T$ and the following results will be the same. The unit of the energy efficiency is bits per Joule, which has been frequently used in literature for energy-efficient communications [162, 178, 58, 142, 71]. The optimal energy-efficient link adaptation achieves maximum energy efficiency, i.e.

$$\mathbf{R}^* = \arg\max_{\mathbf{R}} U(\mathbf{R}) = \arg\max_{\mathbf{R}} \frac{R}{P_C + P_T(\mathbf{R})}. \tag{19.29}$$

Note that if we fix the overall transmit power, the objective of Equation (19.29) is equivalent to maximizing the overall throughput, and the existing water-filling power allocation approach [161] gives the solution. However, besides adapting the power distributions on all subchannels, the overall transmit power can also be adapted according to the states of all subchannels to maximize the energy efficiency. Hence, the solution to Equation (19.29) is in general different from existing power allocation schemes that maximize throughput with power constraints.

19.4.2 Design principles

In the following, we demonstrate that a unique globally optimal data rate vector always exists and give the necessary and sufficient conditions for a data rate vector to be globally optimal.

Conditions of optimality

The concept of quasiconcavity will be used in our discussion and is defined as [66]

DEFINITION 19.4   *A function f, which maps from a convex set of real n-dimensional vectors, $\mathcal{D}$, to a real number, is called strictly quasiconcave if for any $\mathbf{x}_1, \mathbf{x}_2 \in \mathcal{D}$ and $\mathbf{x}_1 \neq \mathbf{x}_2$,*

$$f(\lambda \mathbf{x}_1 + (1 - \lambda)\mathbf{x}_2) > \min\{f(\mathbf{x}_1), f(\mathbf{x}_2)\}, \tag{19.30}$$

for any $0 < \lambda < 1$.

Any strictly monotonic function is quasiconcave. Besides, any strictly concave function is also strictly quasiconcave but the reverse is not generally true. An example is the Gaussian function, which is strictly quasiconcave but not concave. Figure 19.6 illustrates an example of a strictly quasiconcave function that is not concave.

$U(\mathbf{R})$ has the following properties, which are proved in Appendix A.13.

LEMMA 19.5   *If $P_T(\mathbf{R})$ is strictly convex in $\mathbf{R}$, $U(\mathbf{R})$ is strictly quasiconcave. Furthermore, $U(\mathbf{R})$ is either strictly decreasing or first strictly increasing and then strictly decreasing in any r_i of $\mathbf{R}$, i.e. the local maximum of $U(\mathbf{R})$ for each r_i exists at either 0 or a positive finite value.*

For strictly quasiconcave functions, if a local maximum exists, it is also globally optimal [66]. The existence of the local maximum has been proved in Lemma 19.5. Hence, a unique globally optimal transmission rate vector always exists and its characteristics are summarized in Theorem 19.6.

Note that in the rest of the book, the property that "if a local maximum of a strictly quasiconcave function exists, it is also globally optimal" will be frequently used. However, we will skip the discussions about the existence of the local maximums for the functions that will be studied, as their proofs can be done in the same way as the one for Lemma 19.5. We will simply state that if a function is strictly quasiconcave, it has a unique globally optimal value.

THEOREM 19.6   *If $P_T(\mathbf{R})$ is strictly convex, there exists a unique globally optimal transmission data rate vector $\mathbf{R}^* = [r_1^*, r_2^*, \ldots, r_K^*]^T$ for (19.29), where r_i^* is given by:*

$$\text{(i)} \quad \text{when } \frac{P_C + P_T(\mathbf{R}_i^{(0)})}{R_i^{(0)}} \geq \left. \frac{\partial P_T(\mathbf{R})}{\partial r_i} \right|_{\mathbf{R}=\mathbf{R}_i^{(0)}}, \left. \frac{\partial U(\mathbf{R})}{\partial r_i} \right|_{\mathbf{R}=\mathbf{R}^*} = 0, \text{ i.e. } \frac{1}{\frac{\partial P_T(\mathbf{R}^*)}{\partial r_i^*}} = \frac{R^*}{P_C + P_T(\mathbf{R}^*)}$$

$$= U(\mathbf{R}^*);$$

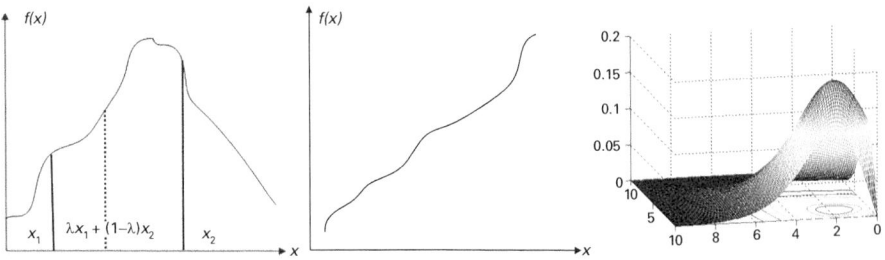

Figure 19.6 An example of strictly quasiconcave function

(ii)   when $\frac{P_C + P_T(\mathbf{R}_i^{(0)})}{R_i^{(0)}} < \frac{\partial P_T(\mathbf{R})}{\partial r_i}\bigg|_{\mathbf{R}=\mathbf{R}_i^{(0)}}, \; r_i^* = 0,$

where $\mathbf{R}_i^{(0)} = [r_1^*, r_2^*, \ldots, r_{i-1}^*, 0, r_{i+1}^*, \ldots, r_K^*]$ and $R_i^{(0)} = \sum_{j \neq i} r_j^*$, *i.e. the overall data rate on all other subchannels except i. The energy efficiency capacity is*

$$C_E = \frac{R^*}{P_C + P_T(\mathbf{R}^*)}. \tag{19.31}$$

Theorem 19.6 has clear physical insights. $P_C + P_T(\mathbf{R}_i^{(0)})$ is the power consumption of both circuit and all other subchannels when subchannel i is not used. $\frac{P_C + P_T(\mathbf{R}_i^{(0)})}{R_i^{(0)}}$ is the per-bit energy consumption when subchannel i is not used and the overall per-bit energy consumption needs to be minimized for energy-efficient communications. $\frac{\partial P_T(\mathbf{R})}{\partial r_i}\big|_{\mathbf{R}=\mathbf{R}_i^{(0)}}$ is the per-bit energy consumption transmitting an infinitely small data rate on subchannel i conditioned on the optimal status of all other subchannels. Hence, subchannel i should not transmit anything when $\frac{P_C + P_T(\mathbf{R}_i^{(0)})}{R_i^{(0)}} < \frac{\partial P_T(\mathbf{R})}{\partial r_i}\big|_{\mathbf{R}=\mathbf{R}_i^{(0)}}$. Otherwise, there should be a tradeoff between the desired data rate on subchannel i and the incurred power consumption. The tradeoff closely depends on the power consumption of both circuits and transmission on all other subchannels, and can be found through the unique zero derivative of $U(\mathbf{R})$ with respect to r_i.

To further understand Theorem 19.6, we consider an example when each subchannel achieves the Shannon capacity and the transmit power on each subchannel is given in (19.26) with $\Gamma = 0$ dB and $\zeta = 1$. The overall transmit power is

$$P_T(\mathbf{R}) = \sum_{k=1}^{K} (e^{\frac{r_k}{W}} - 1)\frac{N_o W}{g_k}. \tag{19.32}$$

According to condition (*i*) of Theorem 19.6, when $r_k > 0$, we have

$$\frac{1}{\frac{\partial P_T(\mathbf{R})}{\partial r_k}} = \frac{1}{e^{\frac{r_k}{W}}\frac{N_o}{g_k}} = U(\mathbf{R}^*). \tag{19.33}$$

Hence, the transmit power on subchannel k is

$$P_{T_n} = (e^{\frac{r_k}{W}} - 1)\frac{N_o W}{g_k} = \frac{W}{U(\mathbf{R}^*)} - \frac{N_o W}{g_k}, \tag{19.34}$$

which is a water-filling to level $\frac{W}{U(\mathbf{R}^*)}$. Since the water level is determined by the optimal energy efficiency, we refer to our scheme as *dynamic energy-efficient water-filling*. Note that while the absolute value of power allocation is determined by the maximum energy efficiency $U(\mathbf{R}^*)$, which relies on both the circuit power and channel state, the relative differences of power allocation on different subchannels depend only on the channel gains on those subchannels.

19.4.3 Constrained energy-efficient link adaptation

When the link adaptation has physical requirements, e.g. peak or average power limits, data rate requirement, and so on, it should achieve the highest energy efficiency while meeting these requirements. If the globally optimal link adaptation meets the requirements, it can be used. Otherwise, the transmission should be adapted right on boundary conditions. In the following, we give examples for energy-efficient link adaptation with constraints. In practice, there might be many other constraints of QoS requirements and resources. In those cases, optimization techniques can be used to search for the solutions.

With a data rate requirement
The energy-efficient link adaptation is given by

$$\widehat{\mathbf{R}}^* = \arg \max_{\mathbf{R}} \frac{R}{P_C + P_T(\mathbf{R})}, \tag{19.35a}$$

subject to

$$R \geq \bar{R}, \tag{19.35b}$$

where $\bar{R}$ is the rate requirement. If the optimal data rate vector without constraint in (19.29) satisfies $R^* \geq \bar{R}$, it is also the solution to Problem (19.35), i.e. $\widehat{\mathbf{R}}^* = \mathbf{R}^*$. Otherwise, Problem (19.35) is equivalent to

$$\widehat{\mathbf{R}}^* = \arg \max_{\bar{R}} \frac{\bar{R}}{P_C + P_T(\mathbf{R})} = \arg \min_{\mathbf{R}} P_T(\mathbf{R}), \tag{19.36a}$$

subject to

$$R = \bar{R}. \tag{19.36b}$$

Since $P_T(\mathbf{R})$ is strictly convex, a unique globally optimal $\widehat{\mathbf{R}}^*$ exists. Define

$$f_k(r_k) = \frac{\partial P_T(\mathbf{R})}{\partial r_k} \tag{19.37}$$

and its inverse function to be $f_k^{-1}()$. Then $\widehat{\mathbf{R}}^*$ can be easily obtained via the Lagrangian technique [171] and is

$$\widehat{r}_k^* = \max \left\{ f_k^{-1}(\lambda), 0 \right\} \tag{19.38}$$

for $k = 1, \ldots, K$, where λ is determined by

$$\sum_{k=1}^{K} \widehat{r}_k^* = \bar{R}. \tag{19.39}$$

When the channel capacity is achieved on each subchannel, the corresponding optimal power allocation is a water-filling allocation, which achieves the sum channel capacity $\bar{R}$.

With a power limit

Similarly, with a maximum transmission power constraint, the problem is to find

$$\widetilde{\mathbf{R}}^* = \arg\max_{\mathbf{R}} \frac{R}{P_C + P_T(\mathbf{R})}, \tag{19.40a}$$

subject to

$$P_T(\mathbf{R}) \leq P_m. \tag{19.40b}$$

If the optimal data rate vector without constraint in (19.29) satisfies $P_T(\mathbf{R}^*) \leq P_m$, it is also the solution to Problem (19.40), i.e. $\widetilde{\mathbf{R}}^* = \mathbf{R}^*$. Otherwise, via the Lagrangian technique again, we have the unique optimal solution as follows

$$\widetilde{r}_k^* = \max\left\{f_k^{-1}(\lambda), 0\right\}, k = 1, \ldots, K, \tag{19.41}$$

where λ is determined by

$$P_T(\widetilde{\mathbf{R}}^*) = P_m. \tag{19.42}$$

When channel capacity is achieved on each subchannel, the power allocation is the classical water-filling where the water level is determined by P_m [161].

19.4.4   Energy-efficient downlink OFDMA transmission

The downlink transmission of a base station serving multiple users at the same time using OFDMA is a special case of wireless transmission in frequency-selective channels, except that, before the transmission, the BS needs to assign the subcarriers to users according to certain rules. After scheduling and subcarrier assignment, the BS needs to determine modulation order and coding on all the subchannels. In the following, we demonstrate how to apply Theorem 19.6 in energy-efficient BS designs. Note that there are many ways of formulating the BS downlink energy-efficient transmission, e.g. the methods in [5, 6], and the key idea stays the same, finding out $P_T(\mathbf{R})$ for the system and then using the strict quasiconcave property of the energy efficiency function to determine the optimal transmission setting.

Downlink OFDMA transmission with diverse QoS requirements

Consider the downlink transmission of a single-cell OFDMA cellular network that serves N users. The subcarrier assignment is given by $\rho = [\rho_{k,n}]_{K \times N}$, where $\rho_{k,n} = 1$ indicates subcarrier k is allocated to user n and 0 otherwise. Here we assume the subcarrier assignment has been scheduled by the BS and ρ is fixed. In addition, each subcarrier is assigned to at most one user and

$$\sum_n \rho_{k,n} \leq 1, \forall k. \tag{19.43}$$

Now we need to determine the power and modulation order on each subcarrier so that the downlink energy efficiency is maximized.

We consider several constraints. The BS has a total power limit given by

$$\sum_{k,n} p_{k,n} \rho_{k,n} \leq P_m. \tag{19.44}$$

In addition, users with diverse quality of service requirements are considered. Let $\mathcal{K}_1$ denote the set of users demanding real-time services and these users have fixed data rate requirements. Let $\mathcal{K}_2$ denote the set of users asking for non-real-time services and they have minimum data rate requirements. In addition, the system has a minimum sum throughput requirement. As shown in [5], the energy efficiency of the BS is given by

$$U = \frac{\sum_{k,n} \rho_{k,n} r_{k,n}}{\sum_{k,n} \rho_{k,n} (\zeta p_{k,n} + \xi r_{k,n}) + P_C}, \tag{19.45a}$$

subject to

$$\sum_{k,n} p_{kn} \rho_{k,n} \leq P_m, \tag{19.45b}$$

$$\sum_{k,n} r_{kn} \rho_{k,n} \geq \bar{R}, \tag{19.45c}$$

$$\sum_{k} r_{kn} \rho_{k,n} = \bar{R}_n, \forall n \in \mathcal{K}_1, \tag{19.45d}$$

and

$$\sum_{k} r_{kn} \rho_{k,n} \geq \bar{R}_n, \forall n \in \mathcal{K}_2, \tag{19.45e}$$

where $\bar{R}$ and $\bar{R}_n$ are the corresponding data rate requirements, ζ is the inverse of power amplifier efficiency, and ξ models the additional power consumption that is proportional to the data rates. The data rate on each subcarrier is given in (19.25). It can be easily found that the power–rate relationship is given by

$$P_T(\mathbf{R}) = \sum_{k,n} \rho_{k,n} (\zeta p_{k,n} + \xi r_{k,n}) = \zeta \sum_{k,n} \rho_{k,n} \left(e^{\frac{r_{k,n}}{W}} - 1 \right) \frac{N_o W \Gamma}{g_{k,n} \zeta} + \xi \sum_{k,n} \rho_{k,n} r_{k,n}. \tag{19.46}$$

Note that here $\mathbf{R}$ is a vector of rates on all subcarriers and the data rates on different subcarriers can be for different users.

Obviously $P_T(\mathbf{R})$ is strictly convex and monotonically increasing in $\mathbf{R}$ with $P_T(\mathbf{0}) = 0$. Therefore, Theorem 19.6 and the algorithms discussed in this chapter can be used to determine the optimal R^*. Furthermore, we need to check if R^* meets the limiting conditions in (19.45). If R^* satisfies the constraints, it is also the optimal solution to maximize energy efficiency in (19.45). Otherwise, optimization techniques using iterative numerical algorithms are needed to find the optimal transmission scheme.

19.4.5   Iterative algorithm design

Theorem 19.6 provides the necessary and sufficient conditions for a rate vector to be the unique and globally optimum one. However, it is usually difficult to directly solve the joint non-linear equations according to Theorem 19.6 to obtain the optimal vector $\mathbf{R}^*$. Therefore, numerical algorithms are needed to search for the optimal $\mathbf{R}$ for maximizing $U(\mathbf{R})$. In the following, we show two intuitive gradient-based iterative methods. The global optimality of the methods is guaranteed by the strict quasiconcavity of $U(\mathbf{R})$.

Gradient-assisted binary search

When there is only one subchannel, Lemma 19.5 shows that function $U(r)$ has a unique r^* such that for any $r < r^*$, $\frac{dU(r)}{dr} > 0$, and for any $r > r^*$, $\frac{dU(r)}{dr} < 0$. Hence, we have the following lemma to seek two points r_1 and r_2 such that $r_1 \le r^* \le r_2$.

PROPOSITION 19.7 *Let the initial setting $r^{[0]} > 0$ and set $\alpha > 1$. For any $i \ge 0$, let*

$$r^{[i+1]} = \begin{cases} \frac{r^{[i]}}{\alpha} & \frac{dU(r)}{dr}\Big|_{r^{[0]}} < 0 \\ \alpha r^{[i]} & \text{otherwise} \end{cases}. \tag{19.47}$$

Repeat (19.47) until $r^{[I]}$ such that $\frac{dU(r)}{dr}\Big|_{r^{[I]}}$ has a different sign from $\frac{dU(r)}{dr}\Big|_{r^{[0]}}$. Then r^* must be between $r^{[I]}$ and $r^{[I-1]}$.

To locate r^* between r_1 and r_2, let $\widehat{r} = \frac{r_1+r_2}{2}$. If $\frac{dU(r)}{dr}\Big|_{\widehat{r}} = 0$, r^* is found. If $\frac{dU(r)}{dr}\Big|_{\widehat{r}} < 0$, $r_1 < r^* < \widehat{r}$ and replace r_2 with $\widehat{r}$; otherwise, replace r_1 with $\widehat{r}$. This leads to the *gradient-assisted binary search* (GABS) for maximizing $U(r)$, which is summarized in Algorithm 19.1.

Algorithm 19.1: *GABS(r_o)*
(∗ algorithm for single-subchannel transmission. ∗)
Input: initial guess: $r_o > 0$
Output: optimal transmission rate: r^*

1. $r_1 = r_o$, $h_1 \leftarrow \frac{dU(r)}{dr}\Big|_{r_1}$, initialize $\alpha > 1$ (e.g.10)
2.   **if** $h_1 < 0$
        (∗ seek r_1 and r_2 such that $r_1 < r^* < r_2$ ∗)
3.   **then** $r_2 \leftarrow r_1$, $r_1 \leftarrow \frac{r_1}{\alpha}$, and $h_1 \leftarrow \frac{dU(r)}{dr}\Big|_{r_1}$
4.   **while** $h_1 < 0$
5.   **do** $r_2 \leftarrow r_1$, $r_1 \leftarrow \frac{r_1}{\alpha}$, and $h_1 \leftarrow \frac{dU(r)}{dr}\Big|_{r_1}$
6.   **else** $r_2 \leftarrow r_1 * \alpha$ and $h_2 \leftarrow \frac{dU(r)}{dr}\Big|_{r_2}$
7.   **while** $h_2 > 0$
8.   **do** $r_1 \leftarrow r_2$, $r_2 \leftarrow r_2 * \alpha$, and $h_2 \leftarrow \frac{dU(r)}{dr}\Big|_{r_2}$
9.   **while** no convergence
        (∗ seek r^* between r_1 and r_2 ∗)
10.   **do** $\widehat{r} \leftarrow \frac{r_2+r_1}{2}$; $\widehat{h} \leftarrow \frac{dU(r)}{dr}\Big|_{\widehat{r}}$
11.   **if** $\widehat{h} > 0$
12.   **then** $r_1 = \widehat{r}$;
13.   **else** $r_2 = \widehat{r}$
14. **return** $\widehat{r}$

Binary search assisted ascent

To find the optimal data rate vector for the multiple subchannel case, we design a gradient ascent method to produce a maximizing sequence $\mathbf{R}^{[i]}$, $n = 0, 1, \ldots$, and

$$\mathbf{R}^{[i+1]} = \left[\mathbf{R}^{[i]} + \mu \nabla U(\mathbf{R}^{[i]})\right]^+, \tag{19.48}$$

where $[\mathbf{R}]^+$ sets the negative part of the vector $\mathbf{R}$ to be zero, $\mu > 0$ is the search step size, and $\nabla U(\mathbf{R}^{[i]})$ is the gradient at iteration i. With sufficiently small step size, $U(\mathbf{R}^{[i+1]})$ will be always bigger than $U(\mathbf{R}^{[i]})$ except when $\nabla U(\mathbf{R}^{[i]}) = 0$, which indicates the optimality of $\mathbf{R}^{[i]}$ [171]. However, small step size leads to slow convergence. Besides, each element of the gradient depends on the corresponding subchannel power gain, which potentially differ from each other by orders of magnitude. Hence, a line search of the optimal step size needs to cover a large range to ensure global convergence on all subchannels, which is computationally expensive. Therefore, at each $\mathbf{R}^{[i]}$, an efficient algorithm is needed to find the optimal step size. Define

$$f_i(\mu) = U\left(\left[\mathbf{R}^{[i]} + \mu \nabla U(\mathbf{R}^{[i]})\right]^+\right). \tag{19.49}$$

Similar to the proof of Lemma 19.5, it is easy to show that $g_i(\mu)$ is also strictly quasi-concave in μ and has a unique globally maximum μ^* such that for any $\mu < \mu^*$, $\frac{df_i(\mu)}{d\mu} > 0$, and for any $\mu > \mu^*$, $\frac{df_i(\mu)}{d\mu} < 0$. Let $\nabla U(\mathbf{R}^{[i]}) = [\widehat{g}_1, \widehat{g}_2, \ldots, \widehat{g}_K]$. Replace $\frac{dU(r)}{dr}$ in GABS to be

$$\frac{df_i(\mu)}{d\mu} = [\nabla U(\mathbf{R}^{[i+1]})]^T \widetilde{\mathbf{G}}[i], \tag{19.50}$$

where $\widetilde{\mathbf{G}}[i] = \frac{d[\mathbf{R}^{[i]} + \mu \nabla U(\mathbf{R}^{[i]})]^+}{d\mu} = [\widetilde{g}_1, \widetilde{g}_2, \ldots, \widetilde{g}_K]$, in which $\widetilde{g}_k = \widehat{g}_k$ if the kth component of $\mathbf{R}^{[i]} + \mu \nabla U(\mathbf{R}^{[i]})$ is positive and $\widetilde{g}_k = 0$ otherwise. Then GABS can be used for quick location of the optimal step size. This leads to the *binary search assisted ascent* (BSAA) algorithm in Algorithm 19.2.

While the global convergence of both GABS and BSAA is guaranteed by the strict quasiconcavity of $U(\mathbf{R})$ [110], we further study the convergence rate in this section.

Theorem 19.8 characterizes the convergence of GABS and can be easily proved.

Algorithm 19.2: *BSAA($\mathbf{R}_o$)*

($*$ algorithm for multi-subchannel transmission. $*$)

Input: initial guess: $\mathbf{R}_o$(default transmission rate can be used)

Output: optimal transmission rate vector: $\mathbf{R}^*$

1. $\mathbf{R} = \mathbf{R}_o$,
2.   **while** no convergence
3.   **do** use GABS to find the optimal step size μ^*;
4. $\mathbf{R} = \left[\mathbf{R} + \mu^* \nabla U(\mathbf{R})\right]^+$
5.   **return** $\mathbf{R}$

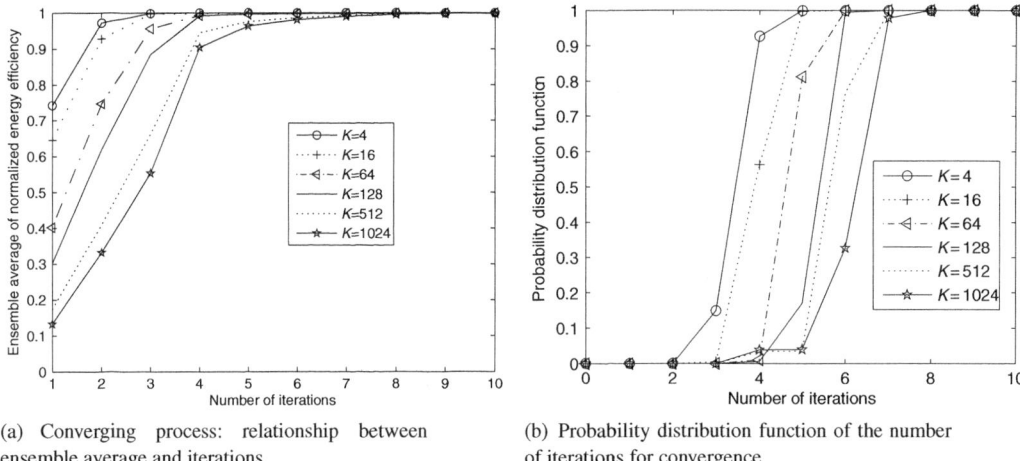

(a) Converging process: relationship between ensemble average and iterations

(b) Probability distribution function of the number of iterations for convergence

Figure 19.7 Convergence rate of BSAA

THEOREM 19.8   *GABS converges to the globally optimal transmission data rate r^*. A rate r, which satisfies $|r - r^*| \leq \epsilon$, can be found within at most M iterations, where M is the minimum integer such that $M \geq \log_2(\frac{(\alpha-1)r^*}{\epsilon} - 1)$.*

It is difficult to theoretically analyze the global convergence rate of BSAA because of the non-concavity of $U(\mathbf{R})$. Instead, we run numerical simulations and observe the convergence. Figure 19.7(a) illustrates the improvement in energy efficiency with iterations. Here we assume the channel gain of each subchannel has Rayleigh distribution with a unit average. The circuit power is 5. The noise power on each subchannel is 0.01. The transmit power is given by Equation (19.26) with $\Gamma = 0$ dB. The energy efficiency is normalized by the optimal value and the curves are the ensemble averages of 5000 channel instances. Figure 19.7(b) shows the corresponding probability distribution functions of the numbers of iterations necessary for convergence. In both figures, we vary the number of subchannels to verify the impact on the convergence rate. We can see that BSAA converges very fast to the global optimum, even with 1024 subchannels.

Other advanced searching algorithms

There are many other more advanced optimization methods that can be used to search for $\mathbf{R}^*$. For example, the optimization problem to maximize the energy efficiency $U(\mathbf{R})$ is a type of concave fractional program. These programs share some important properties with concave programs and many standard methods for concave programs can be used to solve the energy efficiency maximization problem. For example, Dinkelbach's algorithm [16] can be adopted for this purpose with slight modifications. Dinkelbach's algorithm is indeed Newton's method for non-linear optimization and converges superlinearly. Dinkelbach's algorithms for finding $\mathbf{R}^*$ in both flat-fading and frequency-selective channels have been thoroughly studied in [47]. As this book focuses on the communication aspect, the discussion of detailed optimization techniques is beyond its scope. Readers interested are referred to [47] or other optimization books for more information.

19.4.6   Energy efficiency gain

The energy-efficient link adaptation discussed in this chapter is generic and can be applied in different OFDM, MIMO, and MIMO-OFDM systems. To apply it, we only need to find the transmit power relationship $P_T(\mathbf{R})$ of those systems. In this section, we discuss the optimal energy-efficient link adaptation for OFDM with subchannelization as an example.

We compare the performance of energy-efficient OFDM transmission with that of traditional transmission schemes. The system parameters are listed in Table 19.2. The *International Telecommunication Union* (ITU) pedestrian channel model B [95] is used to implement multi-path frequency-selective fading. Each subchannel consists of a block of ten contiguous subcarriers. Figures 19.8(a) and 19.8(b) compare energy efficiency and throughput of different transmission schemes. Two energy-efficient OFDM transmission schemes are implemented: Frequency selection energy efficiency (FS EE), i.e. the optimal energy-efficient transmission developed in this chapter, and flat energy efficiency (Flat EE) that treats the channel as flat-fading. Transmissions with both fixed and adaptive QAM modulations are also implemented for comparison. For fixed modulation, the transmit power is adapted to meet the BER requirement while not exceeding the 15 dBm maximum power constraint. For adaptive modulation, transmit power is equally distributed over all subchannels and the modulation is adapted to meet the BER requirement. From Figures 19.8(a) and 19.8(b), fixed and adaptive modulations

Table 19.2 System parameters

Carrier frequency	1.5 GHz
Subcarrier number	256
Subcarrier bandwidth	10 kHz
BER requirement	10^{-3}
Symbol number of data interval, l	100
Time duration of data interval, T_s	0.01s
Time duration of signaling interval, τ	0.001s
Thermal noise power, N_o	-141 dBW/MHz
User antenna height	1.6 m
BS antenna height	40 m
Environment	Macro cell in urban area
Circuit power, P_C	100 mW
Modulation	Uncoded M-QAM
Subchannelization	Fixed-interval and contiguous
Propagation model	Okumura–Hata model
Shadowing	Log-normal with standard deviation of 10 dB
Frequency-selective fading	ITU pedestrian channel B
User speed	3 km/h

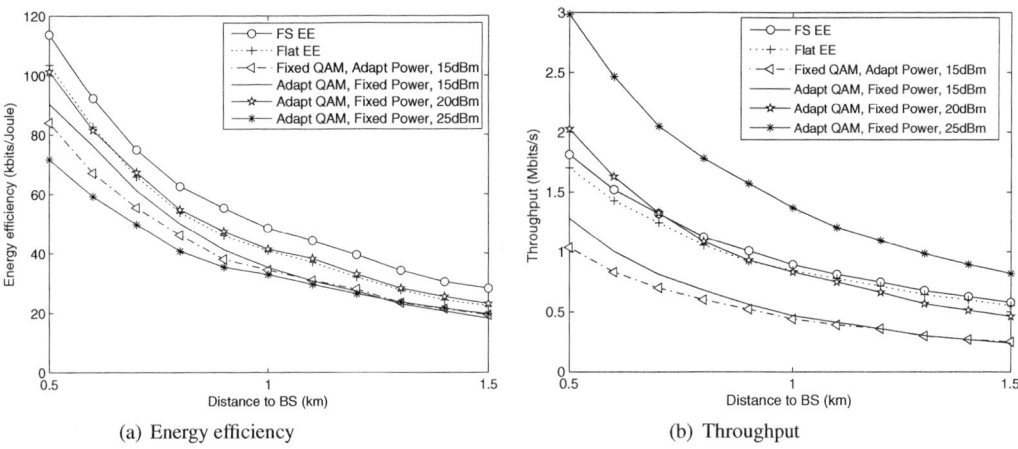

(a) Energy efficiency   (b) Throughput

Figure 19.8 Performance comparison for contiguous subchannelization

perform closely to each other, especially when far away from the BS, for both energy efficiency and throughput, when the maximum transmit power is 15 dBm. By increasing the transmit power from 15 dBm to 25 dBm, the throughput of adaptive modulation increases; however, the energy efficiency first increases and then decreases. Due to global optimality, the energy-efficient transmission for frequency-selective channels always achieves the highest energy efficiency, and outperforms the others by at least 15%. However, the throughput is not necessarily maximum; the other schemes, especially the adaptive QAM modulation with 25 dBm transmit power, sacrifice power to obtain higher throughput. Therefore, we see a tradeoff between spectral and energy efficiency.

19.5   Low-complexity energy-efficient link adaptation

In this section, we introduce closed-form energy-efficient link adaptation techniques.

Define the index set of all subchannels as $\mathcal{K} = \{1, 2, \ldots, K\}$. The data rate of user n at frame t is

$$r[t] = \sum_{k \in \mathcal{K}} r_k[t], \qquad (19.51)$$

where $r_k[t]$ is the data rate at subchannel k and depends on the frequency-selective fading. If we use an exponentially weighted low-pass filter to get the average data rate at frame t, it can be expressed as

$$T[t] = \left(1 - \frac{1}{w}\right) T[t-1] + \frac{1}{w} r[t], \qquad (19.52)$$

where $w \gg 1$.

Define the SNR for reliable reception of $r_k[t]$ to be

$$\eta_k = S(r_k[t]). \tag{19.53}$$

For example,

$$r_i = B \log_2 \left(1 + \frac{\eta_i}{\Gamma}\right), \tag{19.54}$$

and correspondingly $S()$ is given by

$$\eta_k = \left(2^{\frac{r_i}{B}} - 1\right)\Gamma, \tag{19.55}$$

where Γ is the SNR gap. $S(r)$ is strictly convex in r and $S(0) = 0$. Hence, in general, we do not specify the exact form of $S(r)$ and only assume $S(r)$ to be strictly convex in r and $S(0) = 0$. Define the signal power attenuation on subchannel k at frame t to be $g_k[t]$, then the required power on subchannel k to transmit at a rate of $r_k[t]$ will be

$$p_k[t] = \frac{\eta_k \sigma^2}{g_k[t]} = \frac{S(r_k[t])\sigma^2}{g_k[t]}, \tag{19.56}$$

where σ^2 is the noise power. The overall transmit power is

$$p[t] = \sum_{k \in \mathcal{C}[t]} p_k[t]. \tag{19.57}$$

The overall weighted moving average power consumption, $P[t]$, is also obtained using an exponentially weighted moving average low-pass filter, i.e.

$$P[t] = \left(1 - \frac{1}{w}\right)P[t-1] + \frac{1}{w}(p[t] + p_c[t]). \tag{19.58}$$

The circuit power, $p_c[t]$, is measured at frame t. Here, the power consumption of each user is divided into two parts; $p_c[t]$ models the circuit power, while $p[t]$ the transmission power.

For energy-efficient communications, users want to send as much data as possible with a given amount of energy. User n wants to maximize

$$u[t] = \frac{T[t]}{P[t]}. \tag{19.59}$$

u is called average energy efficiency of user n.

We need to determine the data rates at all subchannels to maximize

$$u[t] = \frac{T[t]}{P[t]} = \frac{(1 - \frac{1}{w})T[t-1] + \frac{1}{w}\sum_k r_k[t]}{(1 - \frac{1}{w})P[t-1] + \frac{1}{w}(\sum_k p_k[t] + p_c[t])}, \tag{19.60}$$

where $p_k[t+1]$ is given by (19.56). Define the data rate vector on all assigned subchannels to be $\mathbf{r}[t]$. Then $u[t]$ is a function of $\mathbf{r}[t]$ and is strictly quasiconcave in $\mathbf{r}[t]$. Hence, a unique globally optimal rate vector, $\mathbf{r}^*[t]$, always exists [66] and every element in $\mathbf{r}^*[t]$ satisfies

$$\frac{\partial u[t]}{\partial r_k[t]} = 0 \tag{19.61}$$

if $r_k[t] > 0$. Note that in (19.60), only $r_k[t]$ and $p_k[t]$ are functions of $r_k[t]$ and other terms are independent of $r_k[t]$. Then solving (19.61) yields the following optimal rate condition:

$$\frac{\partial p_k[t]}{\partial r_k[t]} = \frac{P[t]}{T[t]} = \frac{1}{u[t]}, \forall k. \tag{19.62}$$

If $w \gg 1$, as assumed, $P[t] \approx P[t-1]$ and $T[t] \approx T[t-1]$,

$$\frac{\partial p_k[t]}{\partial r_k[t]} = \frac{P[t-1]}{T[t-1]} = \frac{1}{u[t-1]}, \forall k. \tag{19.63}$$

Together with (19.56), we have

$$S'(r_k[t]) = \frac{1}{u[t-1]} \frac{g_k[t]}{\sigma^2}, \forall k. \tag{19.64}$$

where $S'(\cdot)$ is the derivative of the function $S(\cdot)$. Consequently, the optimal data rate follows immediately,

$$r_k^*[t] = \max\left(S'^{-1}\left(\frac{1}{u[t-1]} \frac{g_k[t]}{\sigma^2}\right), 0\right) \forall k \in \mathcal{C}[t], \tag{19.65}$$

where $S'^{-1}()$ is the inverse function of S'. The corresponding optimal power allocation is

$$p_k^*[t] = \frac{S(r_k^*[t])\sigma^2}{g_k[t]}, \forall k \in \mathcal{C}[t]. \tag{19.66}$$

Note that, in the above derivation, approximation is only used in (19.63). If w is sufficiently large, the approximation error is close to zero, and (19.65) and (19.66) are almost globally optimal.

If each subchannel is experiencing AWGN and the Shannon capacity (p.373 of [161]) is achieved on each subchannel, $r = B \log_2(1 + \eta)$, then $S(r) = 2^{\frac{r}{B}} - 1$, where B is the subchannel bandwidth. The optimal data rate on subchannel n is

$$r_k^*[t] = \max\left(B \log_2\left(\frac{B g_k[t]}{u[t-1]\sigma^2 \log 2}\right), 0\right) \forall k \in \mathcal{C}[t]. \tag{19.67}$$

The corresponding optimal power allocation on subchannel n is

$$p_k^*[t] = \max\left(\frac{B}{u[t-1]\log 2} - \frac{\sigma^2}{g_k[t]}, 0\right) \forall k \in \mathcal{C}[t], \tag{19.68}$$

which is a water-filling power allocation with a water level of $\frac{B}{u[t-1]\log 2}$, as in Figure 19.9. We can see that the energy-efficient link adaptation in (19.65), (19.66), (19.67), and (19.68) is determined by $u[t-1]$ and $g_k[t]$, and is expressed in closed form. This significantly reduces the complexity associated with the iterative solutions developed earlier in Section 19.4.5. From Figure 19.9, we can see that every shadowed part corresponds to the power allocated on each subchannel.

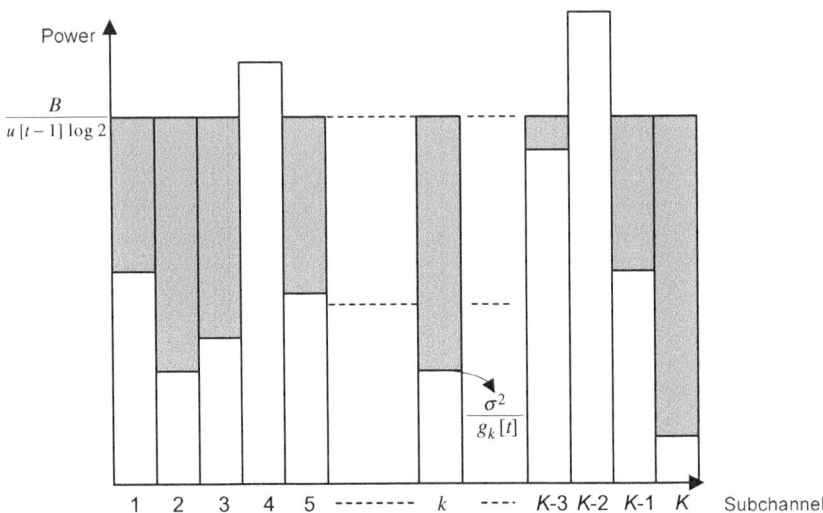

Figure 19.9 Low-complexity energy-efficient water-filling power allocation

19.6 Energy-efficient MIMO and MU-MIMO link adaptation

Multiple-input and multiple-output (MIMO) techniques have been shown to be effective in improving wireless system spectral efficiency, diversity, and interference suppression [26]. However, the advantage of the MIMO technique comes with an overhead in circuit implementation due to duplicated transmitter and receiver radio front ends. The exploitation of multiple antennas requires more active circuit components, which increases both transmission power and circuit power. Hence, characterizing how multiplexing gain, diversity gain, and circuit cost impact overall system energy efficiency is very important.

In a multi-user scenario, MU-MIMO systems can provide substantial gain in networks by allowing multiple users to communicate in the same frequency and time slots [157, 30]. MU-MIMO takes advantage of both the high capacity achieved with MIMO processing and the benefits of space-division multiple access, and has been accepted by major wireless standards such as IEEE 802.16m [156] and 3GPP long-term evolution (LTE) [99]. Recently, there has been some research interest in energy-efficient communications for MIMO systems [87, 172, 143, 67]. For example, in [172], MIMO systems based on Alamouti diversity, schemes are studied to improve the energy efficiency of sensor networks. A mechanism to switch between MIMO with two transmitter antennas and single-input multiple-output (SIMO) to conserve mobile terminal energy is proposed in [87]. A low-complexity energy-efficient and reconfigurable reduced dimension maximum likelihood MIMO detector is proposed in [67]. However, there is very limited research studying energy-efficient MU-MIMO and its optimal power allocation. In this chapter, we will study the relationships between power allocated on each antenna, channel states, antenna circuit power consumption, and antennas that should be turned on for energy-efficient MIMO and MU-MIMO communications.

In the following, we first assume that all users consume a fixed amount of circuit power in addition to the radio frequency (RF) power and demonstrate the existence of a unique globally optimal power allocation that achieves energy efficiency capacity. In practice, users may have improved circuit management capability and turn off part of the circuit operations when some antennas are not used to reduce circuit power consumption. Therefore, we will further introduce energy-efficient UL MU-MIMO with improved circuit management. In this case, the problem is non-concave and multiple local maximums may exist. However, we will introduce algorithms that converge to the global optimum.

19.6.1   Energy-efficient MU-MIMO modeling

In this section, we introduce energy-efficient MU-MIMO. Consider an MU-MIMO system, as illustrated in Figure 19.10, where one access point (AP) is serving K users that desire best-effort data service, e.g. file transfer and email, and have no data rate requirements. Both the AP and all users desire energy-efficient communications. The AP has N antennas. User i has k_i antennas and $\sum_{i=1}^{K} k_i \leq N$. Assume block-fading and that the channel state remains constant during each data frame. The CSI between the AP and users is predetermined earlier through either training pilots, as in a time-division duplex system, or a feedback channel, as in a frequency-division duplex system. Each user has its CSI, while the AP has CSI of all users. Signaling overhead and incomplete channel state information will result in performance loss and the discussion on its impact is beyond the scope of this chapter. Besides we focus on narrow-band flat-fading channels. However, the solutions, especially the methodology, can be easily adapted in broadband channels. For example, in LTE, MU-MIMO can be applied in each resource block (RB) to multiplex the transmission of multiple users in the network [155]. Here each RB has a small number of, e.g. 12, adjacent subcarriers and therefore experiences flat-fading. We can apply power allocations schemes in each RB to improve

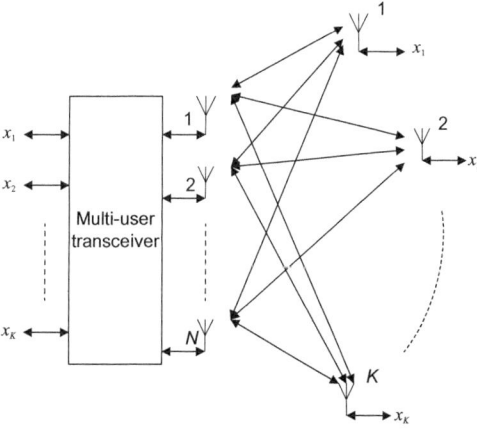

Figure 19.10 System diagram of a multi-user system

network energy efficiency. Some improvements may be needed. For example, as the data transmission of a user may occur in multiple RBs, circuit power in the algorithms should be replaced by total circuit power divided by the number of RBs used, i.e. the average circuit power per RB.

In a flat-fading propagation environment, the received signal at the AP is given by

$$y = H \cdot Q \cdot P \cdot x + n = \sum_{i=1}^{K} H_i \cdot Q_i \cdot P_i \cdot x_i + n, \qquad (19.69)$$

where $y = [y_1, y_2, ..., y_N]^T$. $x_i = [x_{i1}, x_{i2}, ..., x_{ik_i}]^T$ consists of transmitted signals of user i and $E[|x_{ij}|^2] = 1$, where E is the expectation. Here $[]^T$ is the transpose of a vector. $P_i = \text{diag}\{\sqrt{p_{i1}}, \sqrt{p_{i2}}, ..., \sqrt{p_{ik_i}}\}$ is the power allocation matrix of user i. Q_i is the precoding matrix of user i. H_i is the $N \times k_i$ channel matrix of user i and is assumed to have rank k_i, which is generally true in a rich-scattering environment. n is the length-N noise vector, which is Gaussian distributed with a zero mean and a covariance matrix $\sigma^2 I_N$, where I_N is the identity matrix of size N.

$$x = [x_1, x_2, ..., x_K]^T,$$

$$P = \text{diag}\{P_1, P_2, ..., P_K\},$$

$$Q = \text{diag}\{Q_1, Q_2, ..., Q_K\},$$

and

$$H = [H_1, H_2, ..., H_K].$$

With a linear detector, the decision vector for the transmitted symbols is

$$\hat{x} = w \cdot y = w \cdot H \cdot Q \cdot P \cdot x + w \cdot n. \qquad (19.70)$$

Using singular value decomposition (SVD),

$$H_i = U_i \begin{bmatrix} \Lambda_i \\ 0 \end{bmatrix} V_i^H = [\dot{U}_i \ddot{U}_i] \begin{bmatrix} \Lambda_i \\ 0 \end{bmatrix} V_i^H = \dot{U}_i \Lambda_i V_i^H, \qquad (19.71)$$

where U_i and V_i are $N \times N$ and $k_i \times k_i$ unitary matrices and $[]^H$ is the Hermitian transpose. $\dot{U}_i$ consists of the first k_i columns of U_i.

$$\Lambda_i = \text{diag}\{\lambda_{i1}, \lambda_{i2}, ..., \lambda_{ik_i}\},$$

where $\lambda_{ij} \geq 0$.

With local channel knowledge H_i, user i sets the precoding matrix $Q_i = V_i$. Define

$$U = [\dot{U}_1, \dot{U}_2, ..., \dot{U}_K]$$

and

$$\Lambda = \text{diag}\{\Lambda_1, \Lambda_2, ..., \Lambda_K\}.$$

It is easy to see the decision vector at the AP is

$$\hat{x} = w \cdot U \cdot \Lambda \cdot P \cdot x + w \cdot n. \qquad (19.72)$$

There are many ways of designing the linear receiver $\mathbf{w}$. Since receiver design is not the focus of this chapter, we use the zero-forcing (ZF) receiver [177], i.e.,

$$\mathbf{w} = (\mathbf{U}^H \mathbf{U})^{-1} \mathbf{U}^H, \tag{19.73}$$

in its simplicity. Note that the restriction on $\sum_{i=1}^{K} k_i \leq N$ is needed for the existence of the ZF receiver. The decision vector is

$$\hat{\mathbf{x}} = \mathbf{\Lambda} \cdot \mathbf{P} \cdot \mathbf{x} + \hat{\mathbf{n}}, \tag{19.74}$$

where $\hat{\mathbf{n}} = (\mathbf{U}^H \mathbf{U})^{-1} \mathbf{U}^H \cdot \mathbf{n}$, which is also Gaussian distributed with a zero mean and a covariance matrix

$$E[\hat{\mathbf{n}}\hat{\mathbf{n}}^H] = \sigma^2 [\left(\mathbf{U}^H \mathbf{U}\right)^{-1}]^H, \tag{19.75}$$

with all elements in the diagonal being σ^2.

From (19.74), the transmissions of different users are uncoupled. The AP can detect each symbol independently and the achieved SNR of all the symbols for user i is

$$\eta_i = \left[\frac{p_{i1}\lambda_{i1}^2}{\sigma^2}, \frac{p_{i2}\lambda_{i2}^2}{\sigma^2}, ..., \frac{p_{ik_i}\lambda_{ik_i}^2}{\sigma^2} \right]^T. \tag{19.76}$$

Given the transceiver structure and the channel state, each user determines the optimal data rate and power on each antenna. Define the data rate vector of user i to be, $\mathbf{R}_i = [r_{i1}, r_{i2}, ..., r_{ik_i}]$. Correspondingly, the overall data rate is

$$R_i = \sum_{k=1}^{k_i} r_{ik}, \tag{19.77}$$

where the achievable data rate r_{ik} is

$$r_{ik} = B \log_2 \left(1 + \frac{\eta_{ik}}{\Gamma}\right), \tag{19.78}$$

where B is the system bandwidth, $\eta_{ik} = \frac{p_{ik}\lambda_{ik}^2}{\sigma^2}$ and Γ is the SNR gap. The SNR gap may also capture effects like the several dBs performance loss because of using the simple ZF receiver. Define the overall transmission power of user i to be P_{Ti} such that

$$P_{Ti} = \frac{\sum_{k=1}^{k_i} p_{ik}}{\zeta}, \tag{19.79}$$

where $\zeta \in [0, 1]$ is the power amplifier efficiency. We assume all users are operating in the linear ranges of their power amplifiers and do not assume power constraints for any user. This is because energy-efficient communications are different from spectral-efficient communications. With spectral-efficient communications, users want to use as much power as possible to achieve high throughput and it is necessary to consider power constraints for power allocation [203]. But with energy-efficient communications, all users tend to use as little power as possible. Therefore, the chance that the power allocated exceeds the amplifier linear range is very limited and usually there is no need to consider power constraints particularly.

The overall power consumption of user i is

$$P_i = P_{Ci} + P_{Ti}. \tag{19.80}$$

The AP also consumes electronic circuit energy to receive and decode signals. Define the receiver circuit power as P_r. Similar to the circuit power, P_r models the average energy consumption of AP device electronics, such as mixers, filters, and analog-to-digital converters. For readers not interested in the receiver circuit power, they can assume $P_r = 0$. The MU-MIMO system wants to send a maximum amount of data by choosing the optimal transmission power allocation to maximize energy efficiency

$$U(\mathbf{P}) = \frac{\sum_i R_i}{\alpha \sum_i P_i + \beta P_r}, \tag{19.81}$$

where the weights $\alpha \in [0, 1]$ and $\beta \in [0, 1]$ characterize the priorities of transmitter and receiver power consumption. For example, $\alpha = 1$ and $\beta = 0$ indicate that receiver power consumption is not considered.

The energy efficiency capacity of MU-MIMO is defined as

$$C_E = \max_{\mathbf{P}} \frac{\sum_i R_i}{\alpha \sum_i (P_{Ti} + P_{Ci}) + \beta P_r}, \tag{19.82}$$

and the optimal energy-efficient power allocation achieving energy efficiency capacity is

$$\mathbf{P}^* = \arg\max_{\mathbf{P}} U = \arg\max_{\mathbf{P}} \frac{\sum_i R_i}{\alpha \sum_i (P_{Ti} + P_{Ci}) + \beta P_r}. \tag{19.83}$$

When $K = 1$, (19.82) and (19.83) give the energy efficiency capacity and the optimal power allocation for a point-to-point MIMO system. Therefore, the results in this chapter are also applicable to MIMO systems. For downlink MU-MIMO, similar approaches can be derived and the discussions will be skipped.

19.6.2   Principles of energy-efficient MU-MIMO power allocation

We can easily see that the modeling of MU-MIMO energy efficiency also fits the model in Section 19.6.1 and, based on Theorem 19.6, we have the following principles of energy-efficient MU-MIMO power allocation.

THEOREM 19.9   *There exists a unique globally optimal energy-efficient power allocation* $\mathbf{P}^*$ *that achieves energy efficiency capacity, where* p_{ik}^* *is given by*

$$p_{ik}^* = \begin{cases} \dfrac{B\zeta}{\alpha U^* \ln 2} - \dfrac{\Gamma\sigma^2}{\lambda_{ik}^2} & \text{if } \dfrac{B\zeta\lambda_{ik}^2}{\alpha\Gamma\sigma^2 \ln 2} > U^*, \\ 0 & \text{otherwise,} \end{cases} \tag{19.84}$$

Correspondingly, the energy efficiency capacity is

$$C_E = U(\mathbf{P}^*). \tag{19.85}$$

Theorem 19.9 says that the kth antenna of user i should be used only when the corresponding spatial channel, characterized by λ_{ik}^2, is sufficiently good such that using it improves overall network energy efficiency.

Similar to Propositions 19.2 and 19.3, we have the following properties of power allocation.

PROPOSITION 19.10   *The energy efficiency capacity decreases strictly, while the optimal allocated power on each spatial channel, if non-zero, increases strictly with the circuit power of any user. When receiving power is considered ($\beta > 0$), the energy efficiency capacity decreases strictly, while the optimal allocated power on each spatial channel, if non-zero, increases strictly with the receiving power.*

The algorithms introduced in Section 19.4.5 can also be used to find the optimal power allocation. In the following, this algorithm is named energy-efficient MU-MIMO power allocation (EMMPA). The EMMPA algorithm should be implemented at the AP. Each user needs to report its circuit power to the AP before the communication. This is a one-time report and the signaling overhead is negligible. After running EMMPA, the AP only needs to broadcast $\frac{B\zeta}{\alpha U^* \ln 2}$ and all users can determine their optimal power allocations according to (19.84).

19.6.3   Energy-efficient MU-MIMO with improved circuit management

In the above, we have assumed user i consumes a fixed amount of circuit power, P_{C_i}, regardless of how many antennas are used. However, according to Theorem 19.9, the power allocated on some antennas may be zero. User i can turn off these antennas to reduce circuit energy consumption. With the improved circuit management, circuit power is a function of the set of antennas that are turned on, e.g. a function of the number of antennas turned on. In the following, for notation simplicity, assign the circuit power of user i to $P_{C_i}(k_i^o)$, where k_i^o is the number of antennas that have positive power allocation. $P_{C_i}(k_i^o)$ is increasing in k_i^o. Please note the result in this section also applies in generic cases such that P_{C_i} is a function of the set of antennas that are turned on. A simple example is

$$P_{C_i}(k_i^o) = k_i^o P_\alpha + I(k_i^o)P_\beta, \qquad (19.86)$$

where P_α is the extra antenna-related circuit power consumption when one more antenna is turned on and P_β is the power consumption of circuit components that are independent of the number of antennas turned on. When $k_i^o = 0$, the user can be turned off completely to avoid any circuit power consumption and $P_{C_i}(0) = 0$. The indicator function $I(A)$ is defined as

$$I(A) = \begin{cases} 1 & \text{if } A > 0, \\ 0 & \text{otherwise.} \end{cases} \qquad (19.87)$$

In this scenario, the energy efficiency capacity is given by

$$U^* = \max_{\mathbf{P}} \tilde{U}(\mathbf{P}) = \max_{\mathbf{P}} \frac{\sum_i R_i}{\alpha \sum_i (P_{Ti} + P_{Ci}(k_i^o)) + \beta P_r}, \qquad (19.88)$$

and the optimal energy-efficient power allocation achieving energy efficiency capacity is

$$\mathbf{P}^* = \arg\max_{\mathbf{P}} \tilde{U}(\mathbf{P}) = \arg\max_{\mathbf{P}} \frac{\sum_i R_i}{\alpha \sum_i (P_{Ti} + P_{Ci}(k_i^o)) + \beta P_r}, \qquad (19.89)$$

where $k_i^o = \sum_k I(p_{ik})$.

Principles of energy-efficient power allocation

With improved circuit management, the energy efficiency function $\tilde{U}$ is no longer continuous or quasiconcave in $\mathbf{P}$. Theorem 19.6 is not appropriate in characterizing the globally optimal power allocation. Observe antenna j of user i and define it to be antenna (i,j). Assume the power on all other antennas have been optimally allocated and define $\mathbf{P}_{ij}^{(o)}(p_{ij})$ to be the power allocation that equals the optimal power allocation except that the power on antenna (i,j) is p_{ij}. Antenna (i,j) may have two states, on or off. If it is on, the energy efficiency is

$$\tilde{U}\left(\mathbf{P}_{ij}^{(o)}(p_{ij})\right) = \frac{R_{ij}^o + B\log_2\left(1 + \frac{p_{ij}\lambda_{ij}^2}{\Gamma\sigma^2}\right)}{P_{ij}^o + p_{ij}\frac{\alpha}{\zeta} + P_{Ci}(k_i^o)}, \qquad (19.90)$$

where $R_{ij}^o = \sum_{\{u,k:u\neq i,k\neq j\}} B\log_2\left(1 + \frac{p_{uk}^*\lambda_{uk}^2}{\Gamma\sigma^2}\right)$ and $P_{ij}^o = \alpha\sum_{u\neq i} P_{Ci}(k_u^o) + \beta P_r + \frac{\alpha}{\zeta}\sum_{\{u,k:u\neq i,k\neq j\}} p_{uk}^*$. $\tilde{U}$ is also strictly quasiconcave in p_{ij}. Hence, there is a unique p_{ij} that maximizes $\tilde{U}$ if $p_{ij} > 0$, i.e. antenna (i,j) should be turned on. In the following, we study the condition that antenna (i,j) should be turned on. The partial derivative of $\tilde{U}$ with respect to p_{ij} is

$$\frac{\partial\tilde{U}}{\partial p_{ij}} = \frac{f(p_{ij})}{\left(1 + \frac{p_{ij}\lambda_{ij}^2}{\Gamma\sigma^2}\right)(\frac{\alpha}{\zeta}p_{ij} + P_{ij}^o + P_{Ci}(k_i^o))^2}, \qquad (19.91)$$

where $f(p_{ij}) = \frac{B}{\ln 2}\frac{\lambda_{ij}^2}{\Gamma\sigma^2}(\frac{\alpha}{\zeta}p_{ij} + P_{ij}^o + P_{Ci}(k_i^o)) - \frac{\alpha}{\zeta}R_{ij}^o(1 + \frac{p_{ij}\lambda_{ij}^2}{\Gamma\sigma^2}) - B\frac{\alpha}{\zeta}\left(1 + \frac{p_{ij}\lambda_{ij}^2}{\Gamma\sigma^2}\right)\log_2$ $\left(1 + \frac{p_{ij}\lambda_{ij}^2}{\Gamma\sigma^2}\right)$. Because $\tilde{U}$ is strictly quasiconcave, if there exists a $p_{ij} > 0$ such that $\frac{\partial\tilde{U}}{\partial p_{ij}} = 0$, it is unique. This further indicates that, if $p_{ij} > 0$, there is only one p_{ij} such that $f(p_{ij}) = 0$. It can be easily proven that only when

$$\frac{\lambda_{ij}^2}{\sigma^2} > \frac{R_{ij}^o \alpha \Gamma \ln 2}{(P_{ij}^o + P_{Ci}(k_i^o))B\zeta}, \qquad (19.92)$$

where k_i^o is the number of antennas when antenna (i,j) is turned on, will there exist a $p_{ij} > 0$ such that $f(p_{ij}) = 0$. This is also the condition for antenna (i,j) to be turned on.

Readily, by setting $\frac{\partial\tilde{U}}{\partial p_{ij}}$ to be zero, we have the following necessary condition of globally optimal energy-efficient power allocation.

THEOREM 19.11 *With improved circuit management, the optimal energy-efficient power allocation* $\mathbf{P}^*$ *achieving energy efficiency capacity satisfies, for antennas that are turned on,*

$$p_{ij}^* = \frac{B\zeta}{\alpha U^* \ln 2} - \frac{\Gamma \sigma^2}{\lambda_{ij}^2} \tag{19.93}$$

and these antennas have channel conditions

$$\frac{\lambda_{ij}^2}{\sigma^2 \Gamma} > \frac{R_{ij}^o \alpha \ln 2}{(P_{ij}^o + P_{Ci}(k_i^o))B\zeta}, \tag{19.94}$$

where k_i^o is the number of antennas of user i when antenna (i,j) is turned on. Correspondingly, the energy efficiency capacity is

$$C_E = \tilde{U}(\mathbf{P}^*). \tag{19.95}$$

According to Theorem 19.11, whether or not antenna (i,j) should be turned on is determined by multiple factors. $\frac{\lambda_{ij}^2}{\sigma^2 \Gamma}$ characterizes the channel condition of antenna (i,j) and determines the effective receiver SNR once the power is allocated. If it is above the threshold $\frac{R_{ij}^o \alpha \ln 2}{(P_{ij}^o + P_{Ci}(k_i^o))B\zeta}$, antenna (i,j) should be used since using it improves overall network energy efficiency. The threshold decreases with power amplifier efficiency, ζ, indicating that with improved power amplifier efficiency, poorer channel conditions can be used for data transmission on antenna (i,j) to achieve energy efficiency capacity. The threshold is also determined by the states of all other antennas. The energy efficiency achieved by all other antennas is $\frac{R_{ij}^o}{P_{ij}^o + P_{Ci}(k_i^o)}$ assuming antenna (i,j) has an infinite small amount of power allocated. The threshold increases with $\frac{R_{ij}^o}{P_{ij}^o + P_{Ci}(k_i^o)}$, which indicates that if the energy efficiency achieved by all other antennas increases, the channel state for antenna (i,j) should also be better such that turning it on improves the overall network energy efficiency.

Algorithm development

Unlike Theorem 19.6, Theorem 19.11 only gives the necessary conditions of globally optimal energy-efficient power allocation. With improved circuit power management, there may be multiple power allocation schemes that satisfy Theorem 19.11 because there may be multiple local maximums of the energy efficiency function $\tilde{U}$. An example is given in Figure 19.11, where we assume one user with two antennas is communicating to the AP, i.e. a MIMO system. The circuit power of the user is assumed to be $P_{C_1} = k_1^o + I(k_1^o)$. Observing Figure 19.11, $\tilde{U}$ has three local maximums, each of which satisfies Theorem 19.11. When both antennas are turned on, there is a unique power allocation that maximizes $\tilde{U}$. When the state of one antenna switches from on to off, i.e. p_{11} or p_{12} goes to zero, energy efficiency $\tilde{U}$ increases abruptly because of the reduction in circuit power. In this case, we need to compare the three local maximums to determine the energy efficiency capacity and the optimal power allocation. Therefore, to find the optimal power allocation, we can use EMMPA in Algorithm 19.1 to determine the

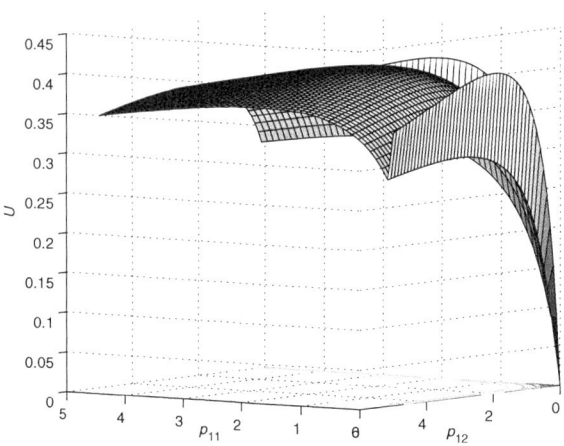

Figure 19.11 An example of non-quasiconcave energy efficiency function $\tilde{U}$

Algorithm 19.3: *Exhaustive search power allocation*
1. $U_{max} \leftarrow 0$; $\mathbf{P}^* \leftarrow \mathbf{0}$;
2.   **for** all antenna configurations
3.   Calculate the circuit power for each user;
4.   Use EMMPA to find the optimal μ, power allocation $\mathbf{P}$, and energy effi-
              ciency U;
5.   **if** $U > U_{max}$ and all antennas turned on have positive power allocation
6. $\mathbf{c} \leftarrow$ current antenna configuration;
7. $\mu^* = \mu$, $U_{max} \leftarrow U$, and $\mathbf{P}^* \leftarrow \mathbf{P}$;
8.   **return c**, μ^*, $\mathbf{P}^*$, and U_{max}.

corresponding optimal power allocation and the maximum energy efficiency achieved
when Antenna $(1, 1)$ is on, $(1, 2)$ is on, or both are on. The antenna configuration and
power allocation that achieve the highest energy efficiency is the optimal one.

An Exhaustive search algorithm: For a generic case, define the antenna configuration
to be a binary vector, $\mathbf{c}$, of length $\sum_i k_i$, in which 1 indicates the corresponding antenna
is on and 0 otherwise. Given a certain antenna configuration, the circuit power of all
users is determined and the power allocation is the same as that in Section 19.6.2.
So EMMPA can be used to determine the corresponding optimal power allocation
for that antenna configuration. Therefore, we need to determine the globally optimal
antenna configuration and use EMMPA to find the corresponding power allocation. One
simple approach is that we can exhaustively search all antenna configurations and use
EMMPA to determine the maximum energy efficiency achieved for each configuration.
The antenna configuration and corresponding power allocation that achieve the highest
energy efficiency is the globally optimal one. We call this the exhaustive search power
allocation (ESPA) algorithm and summarize it in Algorithm 19.3.

The complexity of ESPA grows exponentially with the total number of antennas of all users and, based on Proposition 19.8, it can be easily shown that the convergence rate is characterized by Proposition 19.12.

PROPOSITION 19.12   *ESPA converges to the globally optimal power allocation. The optimal antenna configuration, as well as the power allocation for antennas turned on,*
$$p_{ik} = \left[\mu - \frac{\Gamma \sigma^2}{\lambda_{ik}^2} \right]^+, \text{ where } \mu \text{ satisfies } \left| \mu - \frac{B\zeta}{\alpha U^*} \right| \leq \epsilon, \text{ can be found within at most}$$
$$\left\lceil \log_2 \left(\frac{(\alpha-1)\mu^*}{\epsilon} - 1 \right) \right\rceil (2^{\sum_i k_i} - 1) \text{ iterations.}$$

A quadratic-complexity algorithm: For a small number of users and antennas, ESPA is effective in finding the globally optimal solution. When there are many users in the system and each has many antennas, ESPA has high complexity. In the following, we further develop a low-complexity algorithm. This algorithm consists of two steps. In the first step, the linear-complexity EMMPA is used to filter out antennas that are a subset of antennas that should be turned off in the globally optimal antenna configuration. These antennas are turned off because of their poor channel conditions. In the second step, the remaining antennas are examined and some may be turned off to achieve higher energy efficiency because they consume relatively large amounts of circuit power. We will show that, while the algorithm has quadratic complexity, it is also globally optimal.

In the first step, assume all antennas are turned on and the circuit power, $P_{C_i}(k_i^o)$, of all users can be determined. According to Theorem 19.6, antennas with channel conditions $\frac{B\zeta \lambda_{ik}^2}{\alpha \Gamma \sigma^2 \ln 2} \leq U^*$ will have zero power allocation and we define these antennas by set $S^{(1)}$. If any antenna is turned off, the circuit power of the corresponding user is reduced. According to Proposition 19.10, the optimal power allocated on each spatial channel, if non-zero, will be reduced. Therefore, any antenna in $S^{(1)}$ will not be allocated positive power if any antenna is turned off. This indicates that $S^{(1)}$ belongs to the set of antennas that should be turned off in the globally optimal antenna configuration. $S^{(1)}$ can be determined by EMMPA. In the second round, we turn off all antennas in $S^{(1)}$ and calculate the circuit power, $P_{C_i}(k_i^o)$, of all users. Then we use EMMPA again to determine $S^{(2)}$, the set of antennas that should be turned off in this round. Similarly, $S^{(2)}$ also belongs to the set of antennas that should be turned off in the globally optimal antenna configuration. In addition, $S^{(2)}$ is a super set of $S^{(1)}$. We iterate this process until reaching a round when all antennas turned on have positive power allocation. This iterative EMMPA algorithm is summarized in Algorithm 19.4, whose property is given in Proposition 19.13 based on the above analysis.

PROPOSITION 19.13   *The output of the iterative EMMPA algorithm, $S^{(m)}$, is a subset of antennas that should be turned off in the globally optimal antenna configuration. $S^{(m)}$ can be found within at most $\sum_i k_i - 1$ rounds.*

We can look at the iterative EMMPA algorithm from another perspective. In each round, some additional antennas are turned off to achieve higher energy efficiency. Therefore, U^* increases in each round. More antennas may be turned off in each round because, although in the first several rounds, their channel conditions may not fulfill

Algorithm 19.4: *Iterative EMMPA*

1. Let $S^{(0)}$ be an empty set.
2. Assume all antennas are turned on and calculate the circuit power, $P_{C_i}(k_i^o)$, of all users.
3. Use EMMPA to determine $S^{(1)}$ and the corresponding optimal μ^* and $\mathbf{P}^*$. $m \leftarrow 1$.
4. **while** $S^{(m)}$ differs from $S^{(m-1)}$
5.   **do** Turn off all antennas in $S^{(m)}$ and calculate the circuit power, $P_{C_i}(k_i^o)$, of all users.
6. $m \leftarrow m + 1$.
7.   Use EMMPA to determine $S^{(m)}$ and the corresponding optimal μ^* and $\mathbf{P}^*$.
8. **return** $S^{(m)}$, μ^*, and $\mathbf{P}^*$.

$\frac{B\zeta\lambda_{ik}^2}{\alpha\Gamma\sigma^2\ln 2} \leq U^*$, they fulfill this condition in a later round owing to the increase of U^*. Finally, only the antennas with the best channel conditions are kept on for data transmission.

The iterative EMMPA algorithm finds a subset of antennas that should be turned off in the globally optimal antenna configuration. However, some more antennas may still need to be turned off to achieve energy efficiency capacity. An example has been given in Figure 19.11. In this example, both antennas have good channel states and the iterative EMMPA algorithm determines that both antennas should be turned on. However, turning both antennas on may not be globally optimal because one of them may consume too much circuit power. Therefore, the remaining antennas that are kept on by the iterative EMMPA algorithm should be further examined.

In the second step, we determine which remaining antennas should be turned off. Define $\overline{S}^{(0)}$ to be the set of the remaining antennas and $U^{(0)}$ to be the highest energy efficiency achieved when all the remaining antennas are turned on. Define $U_{i,j}^{(0)}$ to be the highest energy efficiency achieved when only antenna (i,j) in $\overline{S}^{(0)}$ is turned off. If turning off any antenna in $\overline{S}^{(0)}$ will not improve the energy efficiency, i.e.

$$\max_{(i,j)\in\overline{S}^{(0)}} U_{i,j}^{(0)} \leq U^{(0)}, \tag{19.96}$$

no antennas in $\overline{S}^{(0)}$ should be turned off and the selection process is done. Otherwise, we turn off the antenna that results in the highest energy efficiency, i.e. antenna (k,l), where

$$(k,l) = \arg \max_{(i,j)\in\overline{S}^{(0)}} U_{i,j}^{(0)}. \tag{19.97}$$

Then higher energy efficiency $U_{k,l}^{(0)}$ is achieved. In the second round, let $U^{(1)} = U_{k,l}^{(0)}$. Define the set of remaining antennas that are still on to be $\overline{S}^{(1)}$. The above selection process can be repeated until in round m, no antennas should be turned off, i.e.

$$\max_{(i,j)\in\overline{S}^{(m)}} U_{i,j}^{(m)} \leq U^{(m)}. \tag{19.98}$$

Note that m is always smaller than the number of antennas in $\overline{S}^{(0)}$ because in each round, one antenna will be turned off and, finally, at least one antenna in $\overline{S}^{(0)}$ must be kept on to achieve non-zero energy efficiency.

The whole algorithm for energy-efficient MU-MIMO with improved circuit management is named improved EMMPA and summarized in Algorithm 19.5. The global optimality of improved EMMPA is verified later in Section 19.6.4.

The complexity of the improved EMMPA grows quadratically with the total number of antennas on the user side and, based on Proposition 19.8, it can be easily shown that the convergence rate is characterized by Proposition 19.14.

Algorithm 19.5: *Improved EMMPA*
1.   Use iterative EMMPA to determine $\overline{S}^{(0)}$ and $U^{(0)}$.
2. $m \leftarrow 0$ and $U_{max} \leftarrow 0$.
3.   **repeat**
4.   **for** Antenna (i,j) in $\overline{S}^{(m)}$
5.   **do** Turn on only antennas in $\overline{S}^{(m)}$ excluding (i,j) and calculate circuit power of all users.
6.   Use EMMPA to determine $U_{i,j}^{(m)}$ and the corresponding μ and $\mathbf{P}$.
7.   **if** $U_{i,j}^{(m)} > U_{max}$
8.   **then** $U_{max} \leftarrow U_{i,j}^{(m)}$ and $(k,l) \leftarrow (i,j)$. $\mu^* \leftarrow \mu$, $\mathbf{P}^* \leftarrow \mathbf{P}$.
9.   **if** $U_{max} > U^{(m)}$
10.   **then** $\overline{S}^{(m+1)} \leftarrow \overline{S}^{(m)}$ excluding (k,l)
11. $U^{(m+1)} \leftarrow U_{max}$.
12. $m \leftarrow m + 1$;
13. **until** $U_{max} \leq U^{(m-1)}$.
14. **return** $\overline{S}^{(m-1)}$, μ^*, and $\mathbf{P}^*$.

PROPOSITION 19.14   *The output of the improved EMMPA is obtained within at most*

$$\left\lceil \log_2(\frac{(\alpha-1)\mu^*}{\epsilon} - 1) \right\rceil \frac{1}{2}\left(\sum_i k_i - 1\right)\sum_i k_i$$

iterations.

Similar to EMMPA, the algorithms in this section should also be implemented at the AP. The AP needs to broadcast both the antenna configuration and the μ^*, and all users can determine their optimal power allocations according to (19.93).

19.6.4   Energy efficiency gain

In this section, we provide simulation results for a single-cell cellular network to demonstrate the performance of energy-efficient MU-MIMO. System parameters are listed in Table 19.3. In each trial, users are dropped uniformly within 250 meters from the AP. The performance below is the average over all trials.

Table 19.3 Simulation parameters

Carrier frequency	1.5 GHz
System bandwidth	10 kHz
Thermal noise power, N_o	−141 dBW/MHz
User antenna height	1.6 m
BS antenna height	40 m
Environment	Macro cell in urban area
Receiver power, P_r	1000 mW
Propagation model	Okumura–Hata model
Shadowing	10 dB lognormal
Fading	Rayleigh flat-fading
Power amplifier efficiency, ζ	0.5
SNR gap, Γ	0 dB
α	1
β	1

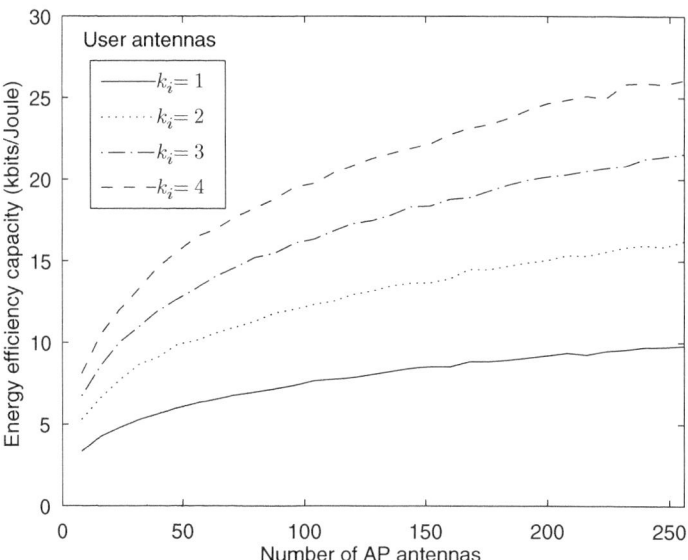

Figure 19.12 Relationship between energy efficiency capacity, transmitter antennas, and receive antennas

Without improved circuit management

First we consider the case that no circuit management is used and each user consumes a fixed amount, $P_{C_i} = 100$ mW, of circuit power. Figure 19.12 gives the average energy efficiency capacity when there are two users in the network and each user has 1, 2, 3, or 4 antennas. The average energy efficiency capacity is an average of multi-user droppings and channel realizations. The number of AP antennas is varied to observe its impact on energy efficiency capacity. Here we assume the same P_r regardless of the number of AP antennas so that only the impact of receiver diversity on the system energy efficiency can be observed. On the other hand, Figure 19.13 compares the average energy efficiency

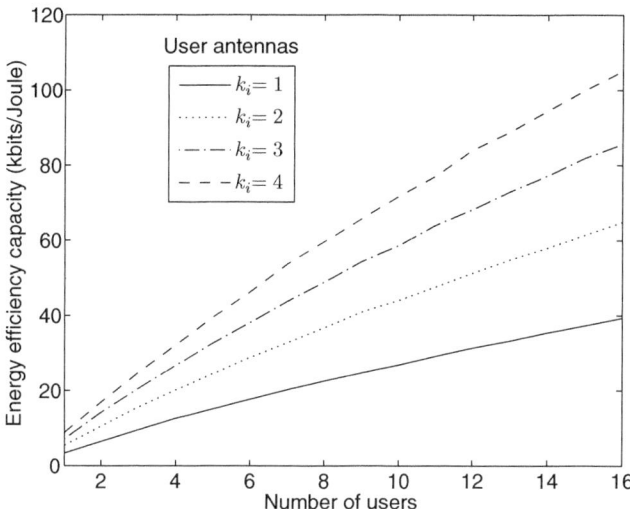

Figure 19.13 Relationship between energy efficiency capacity, users, and transmitter antennas

capacity when the AP has 64 antennas. We can see that without circuit management, more users and more antennas always help improve the energy efficiency capacity of MU-MIMO.

With improved circuit management

In the following, we consider energy-efficient MU-MIMO with improved circuit management and assume $P_{C_i}(k_i^o) = P_\alpha k_i^o + P_\beta I(k_i^o)$ mW for all users.

First, we verify the global optimality of improved EMMPA. Assume $P_\beta = 20$ mW and P_α is varied to observe the impact of circuit power consumption of individual antennas on the suboptimality gap of improved EMMPA and iterative EMMPA algorithms. Figure 19.14 gives the normalized energy efficiency of improved and iterative EMMPA when the AP has 16 antennas and each user has four antennas. The average energy efficiency achieved by improved and iterative EMMPA is normalized by that of ESPA, which is globally optimal. From Figure 19.14, improved EMMPA performs exactly the same as ESPA. In addition to the same average energy efficiency, improved EMMPA always obtains the same instantaneous power allocation and energy efficiency as ESPA according to our observations of simulation results. Therefore, improved EMMPA is also globally optimal. When P_α is small, iterative EMMPA performs very closely to the global optimum because the channel state is the main factor that determines which antennas should be turned on. When P_α is larger, the impact of individual antenna circuit power consumption grows and since iterative EMMPA does not consider this impact, it has greater performance loss. Therefore, improved EMMPA is needed to take this impact into account to achieve the globally optimal performance.

Figure 19.15 compares the average computing time of improved EMMPA, iterative EMMPA, and ESPA with the same simulation setting as that in Figure 19.14. When K

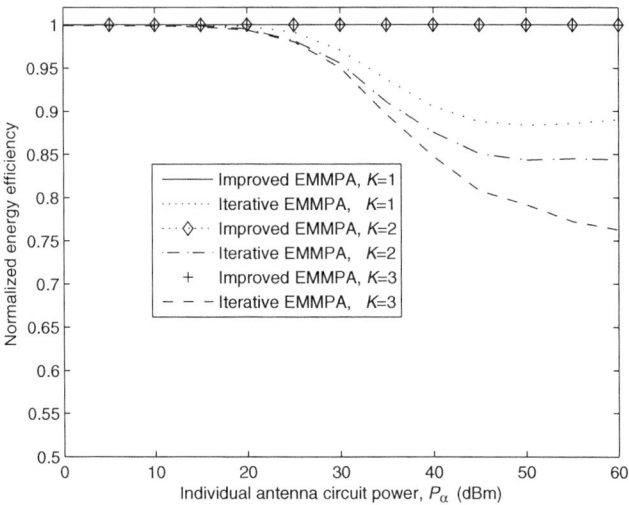

Figure 19.14 Suboptimality gap of improved EMMPA and iterative EMMPA ($N = 16$ and $k_i = 4$)

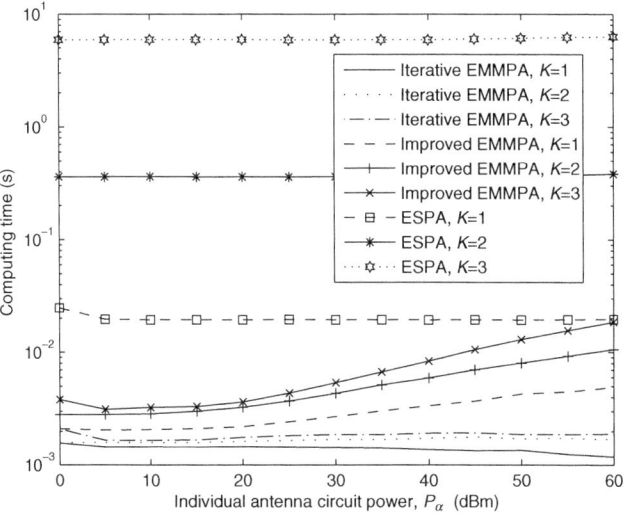

Figure 19.15 Complexity comparison of improved EMMPA, iterative EMMPA, and ESPA ($N = 16$ and $k_i = 4$)

increases, the computing time of ESPA grows exponentially, more than ten times when K is increased by one. On the other hand, the computing time of iterative and improved EMMPA grows very slightly. The individual antenna circuit power consumption P_α has impact on the computing time of improved EMMPA because it determines on average how many additional rounds in the second step are needed to identify antennas that should be turned off because of high antenna circuit power consumption.

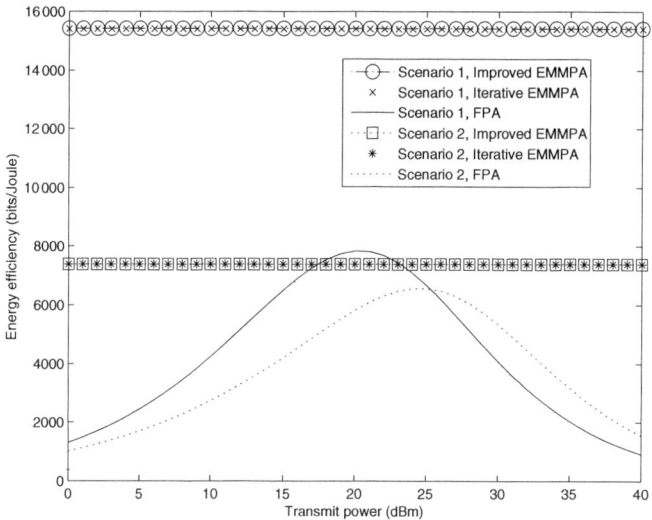

Figure 19.16 Comparison between Improved EMMPA, Iterative EMMPA, and FPA (Scenario 1: $N = 8$, $K = 4$, $k_i = 2$; Scenario 2: $N = 8$, $K = 1$, $k_i = 4$)

Figure 19.16 compares the average energy efficiency of improved EMMPA, iterative EMMPA, and fixed power allocation (FPA). Two scenarios are considered. In one scenario, there are four users in the network, each with two antennas, and the AP has eight antennas. In the second scenario, there is only one user with four antennas in the network and the AP has eight antennas. We can see iterative EMMPA achieves very close performance to that of improved EMMPA. This indicates that with $P_{C_i}(k_i^o) = 20k_i^o + 40I(k_i^o)$, the impact of circuit power consumed by different antennas on energy-efficient MU-MIMO is negligible and the channel states mainly determine which antennas should be turned on. However, when the amount of circuit power consumed by each antenna increases, the impact would grow and the performance difference between improved EMMPA and iterative EMMPA may not be negligible, as shown in Figure 19.14. Compared to FPA, significant gain of energy efficiency can be observed. For example, in scenario 1, more than 100% gain can be observed, indicating half energy is needed to transmit the same amount of information bits using energy-efficient MU-MIMO.

20 Centralized energy-efficient wireless resource management

Multi-user networks can support a large number of users at the same time and the system resources need to be shared among different users. The system performance can be improved significantly by exploiting multi-user diversity in either centralized or distributed ways, which has been demonstrated in Parts II and III. Similarly, multi-user diversity can also be exploited to improve network energy efficiency. This will be the focus of this chapter and the next one. In this chapter, we will first provide an overview of centralized energy-efficient wireless resource management. To better understand the topic, we will later discuss in detail how to manage resources in orthogonal frequency division multiple access (OFDMA) networks so that the overall network achieves the highest energy efficiency.

20.1 Overview

Because of limited wireless resources, there exist intricate tradeoffs between individual performance and the whole network. For example, in a time division multiple access (TDMA) system, all users share a common frequency band. Lowering the rate of one user requires longer transmission duration and thus reduces the available time of other delay-sensitive users. This forces other users to increase modulation order to support higher data rate and consume more energy while potentially the transmission may still suffer from a higher bit error rate (BER). Flexible cross-layer optimization allowing each user to adapt to its environment will enable huge energy savings. Furthermore, the exploitation of diversity across all users will further reduce overall network energy consumption. In this section, we will briefly summarize energy-efficient wireless resource management policies considering time, frequency, and spatial resources. Note that in practice, resources in different domains should be managed together to further enhance network energy efficiency as performances in different domains may affect each other. Also covered are conventional circuit component management techniques, which target reducing circuit power consumption through maximizing idle and sleep time for each device.

20.1.1 Circuit component management

As circuit power occupies a large portion of total energy consumption, effective circuit power management policies are very important. In order to save energy, components should be shut down as fast as possible when a user is inactive and medium access control (MAC) schedules shutdown intervals according to system states, quality of service (QoS) requirements, and channel states. Coordination between users will help power management at the MAC layer to further reduce standby power consumption. For example, users can wake up precisely when they need to transmit or receive data. These have been extensively studied in the conventional energy-efficient wireless network designs.

For example, the energy consumption of a Lucent WaveLAN IEEE 802.11 wireless network interface is measured in [120] to obtain knowledge of energy consumption behavior of actual wireless devices. In [96], different MAC protocols are investigated and their energy consumptions compared. From [96], collisions should be reduced as much as possible to save energy in contention-based MAC protocols. Many wireless standards have integrated components to support energy-efficient communication capabilities. For example, IEEE 802.11 [94] recommends that a mobile be switched to sleep mode, while the access point buffers packets and periodically sends beacons with information about the buffered packets. Based on the beacon information, the mobile decides whether to receive the buffered packets upon waking up and informs the base station when it is ready. In this way, the mobile stays in sleep mode as long as possible and reduces power consumption. Similar schemes to extend sleep durations are supported in the IEEE 802.16e standard [93] and analyzed in [216]. When in sleep mode, mobiles decide whether to wake up or not after periodically checking whether there is downlink traffic or not. The sleep interval is increased exponentially when no arrival traffic is notified.

20.1.2   Time-domain resource management

Most energy-efficient transmission techniques assume that the buffer always has data to transmit. This is not true in general. Due to random and bursty packet arrivals and varying physical (PHY) transmission states, buffers may be occupied or emptied, or may even overflow. Hence, to further enhance energy efficiency, traffic characteristics must be considered and scheduling is necessary to determine the transmission of each arriving packet while satisfying delay constraints. The problem is to minimize the energy used to transmit packets within a given period. From [65], packet transmission times and power levels can be varied to optimize energy efficiency.

When the transmission energy dominates the energy consumption, it is desirable to transmit a packet over a longer period of time to conserve energy. Since all packets are delay constrained, the transmission time of any one packet cannot be arbitrarily long. With knowledge of the arrival time of each packet, it is shown in [65] that the optimal schedule is to have equal transmission times for each packet under feasibility constraint. A lazy scheduling that trades off delay for energy has been developed in [65]. In this scheduling, packets must be buffered. With a small buffer, it is shown that energy consumption can be significantly reduced as compared with a zero-buffer scheme. Based on this observation, transmission periods can be varied according to buffer states and the statistics of the packet arrival process to save energy for practical applications where the future packet arrival time is unknown. It has been demonstrated that lazy schedulers can achieve over 40% energy savings compared to a deterministic scheduler. Similar approaches are also proposed in [42, 22, 150].

In a TDMA network, the channel medium is shared through time division. Each user tends to extend their transmission time to save energy and contradicts the interests of energy savings of other users. Thus, the allocation of time duration among all users is critical in determining network energy efficiency. Consider an energy-efficient variable-length TDMA scheme [173]. As the modulation order determines data rate and thus time for transmitting a certain amount of information, finding the optimal slot length for each user is thus equivalent to determining its corresponding constellation size. Therefore, the MAC layer and PHY layer should be jointly designed for overall energy efficiency. Consequently, the modulation orders of all users should be jointly optimized. A wireless network with central MAC is considered in [164]. The resource allocation scheme within the *access point* (AP) assigns time slots of the channel to all users and specifies transmission parameters of each user for energy-efficient communications. To make the resource management scheme applicable, the scheduling is partitioned into a design-phase and a run-time phase. In the design-time phase, energy-performance representation can be derived for each user to capture the relevant energy and performance tradeoffs. In the run-time phase, a fast greedy algorithm is used to fine tune the operating points to further improve energy efficiency.

20.1.3   Frequency-domain resource management

Increasing transmission bandwidth helps improve energy efficiency. However, the entire system bandwidth cannot be allocated exclusively to one user in a multi-user system since this may hurt the energy efficiency of other users as well as that of the overall network. Hence, the allocation of frequency-domain resources is critical in determining overall network energy efficiency. We will expand the discussions in detail in Sections 20.2 and 20.3.

20.1.4   Spatial-domain resource management

Transmission of one user may interfere with neighboring users and reduce their energy efficiency. However, users can gain in energy efficiency if cooperation among neighboring users is allowed. Hence, spatial-domain resource management is important to manage user behavior and to optimize overall network energy efficiency instead of that of individual ones. As discussed in Section 19.6, multiple-input multipl-output (MIMO) can provide significant energy efficiency improvement. A network with cooperation among users is a virtual MIMO system, in that users themselves provide the spatial degrees of freedom, and can be constructed to enhance network energy efficiency. On the other hand, cooperation requires signaling overhead and consumes additional energy. Cooperation based on inaccurate channel state information may also be harmful. Cooperation can also cause transmission delay that may impact throughput adversely and thus hurt energy efficiency. However, delay can be exploited for energy-efficient link adaptation, as extending transmission duration may improve energy efficiency.

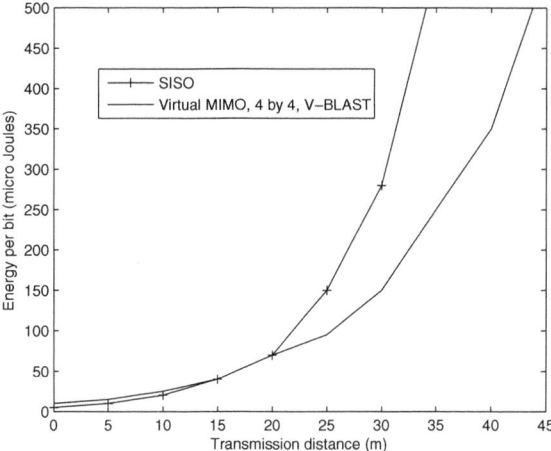

Figure 20.1 Energy consumption transmitting one bit in 4×4 V-BLAST-based virtual MIMO and SISO with both optimized M-QAM modulation

There has been some research in user cooperation for energy efficiency. It has been observed that significant energy savings can be achieved and the savings grow almost linearly with distance when either transmitter or receiver cooperation is allowed [172]. Furthermore, it is also observed that cooperation can even reduce delay within certain transmission ranges. This is because cooperation enables a higher order of modulation to increase the data transmission rate and reduces packet transmission time and delay. In [174], an energy-efficient virtual MIMO communication architecture based on V-BLAST receiver processing is proposed by assuming receiver cooperation. As shown in Figure 20.1, the proposed virtual MIMO architecture can offer significant energy savings over traditional single-input single-output (SISO)-based wireless sensor networks.

The energy consumption for reliable data transmission grows exponentially with distance, which is much faster than the linear relationship between energy and distance. Thus, it is more energy-efficient to send data using several shorter intermediate hops than using a long hop, if the energy to compute the route is negligible [104]. Hence, relays are effective in saving energy. However, relay incurs transmission delay and energy consumption of relay nodes. Therefore, in some scenarios, it is advantageous to use long hops [134]. The optimal selection of relay nodes should be a tradeoff between source-node performance and relay cost to enhance overall network energy efficiency.

20.2   Energy-efficient OFDMA in flat-fading channels

Multiple access is achieved in OFDMA by assigning subchannels to individual users based on QoS requirements and channel conditions. This allows simultaneous data transmission of several users. In this section, we introduce energy-efficient OFDMA communication with flat-fading channels.

Consider uplink transmission in an OFDMA network with one base station (BS) and multiple users. Let N and K denote the numbers of users and subchannels, respectively. Let c_i denote the number of subchannels assigned to user i. Each subchannel will be assigned to one user exclusively at each frame slot. Hence,

$$\sum_{i=1}^{N} c_i \leq K. \tag{20.1}$$

Define r_i as the achievable data rate on each subchannel by user i, then the data rate of user i is

$$R_i = r_i c_i. \tag{20.2}$$

Let $P_{T_i}(R_i)$ denote the power-rate relationship of user i and we assume $P_{T_i}(R_i)$ to be monotonically increasing and strictly convex and $P_{T_i}(R_i) = 0$. One example is

$$P_{T_i}(R_i) = (e^{\frac{R_i}{W}} - 1)N_o W/g. \tag{20.3}$$

The energy efficiency of user i is

$$U_i(R_i) = \frac{R_i}{P_i(R_i)} = \frac{R_i}{P_{Ci} + P_{T_i}(R_i)}. \tag{20.4}$$

The following proposition can be easily proved.

PROPOSITION 20.1   *For energy-efficient transmission, the data rate on each subchannel decreases with the increase in the number of subchannels assigned to a user, while the energy efficiency increases with it.*

Proof   We drop the user index as only one user is considered in the proof. $R = cr$ and $P_T(R) = c\overline{P}_T(r) = c\overline{P}_T(\frac{R}{c})$, where $\overline{P}_T(r)$ is the transmit power on each subchannel, and is monotonically increasing and strictly convex in r. According to Theorem 19.1, we have $R^* \overline{P}_T'(\frac{R^*}{c}) = P_C + c\overline{P}_T(\frac{R^*}{c})$, which is equivalent to $r^* \overline{P}_T'(r^*) - P_T(r^*) = \frac{P_C}{c}$. The left-hand side is increasing in r^* while the right-hand side is decreasing in c. Hence, the modulation order on each subchannel should decrease with the increasing number of subchannels assigned. The proof that the energy efficiency increases with the number of subchannels assigned is straightforward and is omitted.   □

The BS allocates subchannels to improve the overall network energy efficiency, which is defined in Section 3.3.4. Based on the allocation results, each user further adjusts power and coding on each subchannel for energy-efficient transmission. As shown in Proposition 20.1, the energy efficiency of each user increases with the number of allocated subchannels. However, the entire system bandwidth cannot be allocated exclusively to one user as it affects adversely the energy efficiency of other users as well as that of the overall network. This is illustrated in Figure 20.2, where a two-user OFDMA network with ten subchannels is considered. We can see that there is an optimal subchannel allocation that maximizes network energy efficiency. Hence, subchannel allocation is critical in determining the overall network energy efficiency.

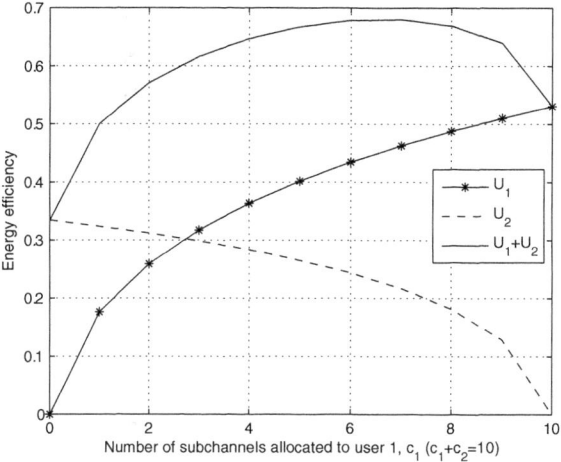

Figure 20.2 Energy efficiency of a two-user OFDMA network

20.2.1 Resource allocation without fairness

Define set $\mathfrak{c} = \{c_1, c_2, \ldots, c_N\}$ to be the set of numbers of subchannels assigned to each user. The subchannels should be assigned such that the network energy efficiency is maximized, i.e.

$$\mathfrak{c}^* = \arg\max_{\mathfrak{c}} \sum_i \frac{R_i}{P_i(R_i)} = \arg\max_{\mathfrak{c}} \sum_i U_i(R_i), \tag{20.5a}$$

subject to

$$R_i = r_i c_i, \tag{20.5b}$$

and

$$\sum_{i=1}^{N} c_i \leq K. \tag{20.5c}$$

Define $P_{T_i}(r_i)$ as the transmit power on each subchannel when the supported data rate is r_i for user i. $P_i(R_i) = P_{C_i} + c_i P_{T_i}(r_i)$. We have

$$\begin{aligned}
U_i(R_i) &= \frac{R_i}{P_i(R_i)} = \frac{c_i r_i}{P_{C_i} + c_i P_{T_i}(r_i)} \\
&= \frac{r_i}{\frac{P_{C_i}}{c_i} + P_{T_i}(r_i)}.
\end{aligned} \tag{20.6}$$

Let $V_i(c_i) = \frac{r_i}{\frac{P_{C_i}}{c_i} + P_{T_i}(r_i)}$, which is strictly concave in c_i. Problem (20.5) is equivalent to

$$\mathfrak{c}^* = \arg\max_{\mathfrak{c}} \sum_i V_i(c_i), \tag{20.7a}$$

subject to

$$\sum_{i=1}^{N} c_i \leq K. \tag{20.7b}$$

Since $V_i(c_i)$ is strictly concave and therefore a unique globally optimal subchannel assignment exists.

The Lagrangian method can be used to find the solution of the above optimization problem. The Lagrange function associated with problem (20.7) is

$$L(\mathfrak{c}, \lambda) = \sum_{i=1}^{N} V_i(c_i) - \lambda \left(\sum_{i=1}^{N} c_i - K \right). \tag{20.8}$$

Let $\frac{\partial L}{\partial c_i} = V_i'(c_i) - \lambda = 0$, the optimal assignment for user i is

$$c_i^* = V_i'^{-1}(\lambda^*), \tag{20.9}$$

where $V_i'^{-1}(\cdot)$ is the inverse function of $V_i'(\cdot)$, and λ^* is given by

$$\sum_i c_i^* = \sum_i V_i'^{-1}(\lambda^*) = K. \tag{20.10}$$

The optimal solution in (20.9) may produce a fractional subchannel assignment, which is not desired. If we search integers, c_i's, for problem (20.5), then it can be treated as a utility-based resource allocation, and was thoroughly investigated in Part II. The practical sorting-search algorithm in Part II, Chapter 8, can be used to assign subchannels for the purpose of energy efficiency.

20.2.2   Resource allocation with fairness

In this section, we consider energy-efficient scheduling with proportional fairness constraint. The BS assigns subchannels to maximize proportional fair (PF) network energy efficiency, i.e.

$$\mathfrak{c}^* = \arg\max_{\mathfrak{c}} \prod_i \frac{R_i}{P_i(R_i)} = \arg\max_{\mathfrak{c}} \sum_i \log \left(\frac{R_i}{P_i(R_i)} \right), \tag{20.11a}$$

subject to

$$R_i = c_i r_i, \tag{20.11b}$$

and

$$\sum_{i=1}^{N} c_i \leq K. \tag{20.11c}$$

Define $W_i(c_i) = \log(V_i(c_i))$. Problem (20.11) is equivalent to

$$\mathfrak{c}^* = \arg\max_{\mathfrak{c}} \sum_i \log(V_i(c_i)) = \arg\max_{\mathfrak{c}} \sum_i W_i(c_i), \tag{20.12a}$$

subject to

$$\sum_{i=1}^{N} c_i \le K. \tag{20.12b}$$

Since $V_i(c_i) > 0$ is strictly concave,

$$\frac{\partial^2 W_i(c_i)}{\partial c_i^2} = \frac{\partial^2 \log(V_i(c_i))}{\partial c_i^2}$$

$$= \frac{V_i^{''}(c_i))V_i(c_i)) - \left[V_i^{'}(c_i))\right]^2}{V_i^2(c_i))} < 0. \tag{20.13}$$

Hence, $W_i(c_i)$ is strictly concave in c_i. Problem (20.12) is strictly concave, and a unique globally optimal assignment exists. Similar to (20.9), the optimal assignment is given by

$$c_i^* = W_i^{'-1}(\lambda^*), \tag{20.14}$$

where λ^* satisfies $\sum_i W_i^{'-1}(\lambda^*) = K$.

20.2.3 Performance comparisons

In this section, we present performance results for energy-efficient resource allocation schemes. System parameters are listed in Table 20.1. All the schedulers and corresponding transmission schemes are listed in Table 20.2. In PropTrad, we set the transmit power to be 15 dBm, 25 dBm, and 33 dBm, respectively. Users are randomly dropped within the cell and the channels experience both log-normal shadowing with standard deviation of 10 dB, and Rayleigh fading with unit average power gain. There are 96 subchannels, each with 10 kHz.

Table 20.1 System parameters

Carrier frequency	1.5 GHz
Subchannel bandwidth	10 kHz
BER	10^{-6}
Symbol number of data interval, l	100
Time duration of data interval, T_s	0.01s
Time duration of signaling interval, τ	0.001s
Thermal noise power, N_o	-141 dBW/MHz
User antenna height	1.6 m
BS antenna height	40 m
Environment	Macro cell in urban area
Circuit power, P_C	100 mW
Maximum transmit power	33 dBm
Propagation model	Okumura–Hata model
Fading	Flat-fading
Modulation	Uncoded M-QAM

Table 20.2 Scheduling and transmission schemes

Legend	Scheduler	Modulation
OptEE	Energy-efficient scheduler without fairness	Energy-efficient transmission
RREE	Round-robin scheduler	Energy-efficient transmission
RRTrad	Round-robin scheduler	2, 4, 8-QAM selected with equal probability
PropTrad	Proportional scheduler	Adaptive modulation with fixed transmit power
PropEE	Energy-efficient scheduler with proportional fairness	Energy-efficient transmission

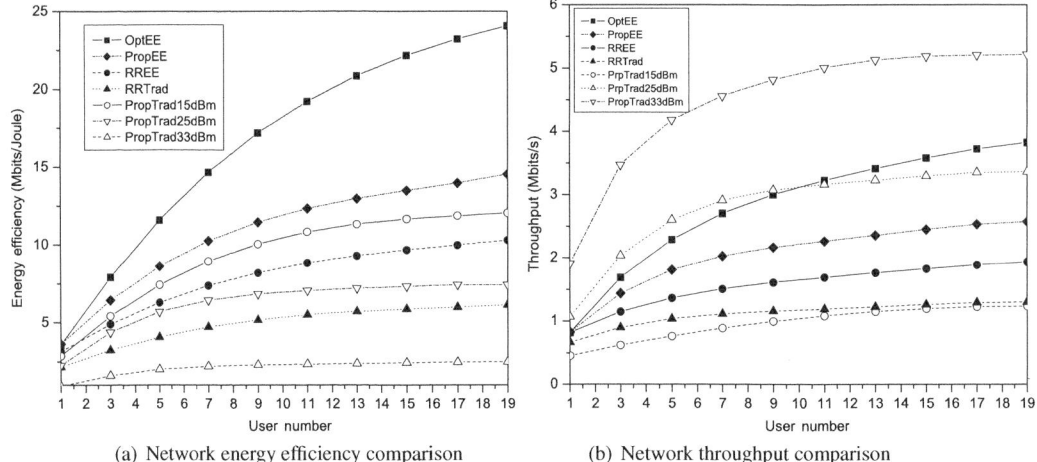

(a) Network energy efficiency comparison

(b) Network throughput comparison

Figure 20.3 Comparisons of different algorithms

Figure 20.3, compares energy efficiency and the corresponding throughput, respectively. We note that the energy-efficient scheduler without fairness performs with the highest energy efficiency and with similar throughput to the proportional scheduler with 25 dBm transmit power for adaptive modulation. Comparing the proportional scheduler with 25 dBm transmit power and the energy-efficient scheduler with proportional fairness, both of which guarantee fairness amongst users, we note that the energy-efficient scheduler has around 20% less instantaneous throughput than the proportional scheduler; however, it transmits about 100% more data than the proportional scheduler given a fixed amount of energy. Or equivalently, given a fixed amount of data, the energy-efficient scheduler saves 50% energy. Comparing the two round-robin schedulers, we note that the one with energy-efficient transmission always performs approximately 50% better than the one with fixed modulation for both energy efficiency and throughput and this comes from the adaptivity of both modulation and power by energy-efficient transmission to the channel status. We note that while energy-efficient scheduling can optimize the energy utilization, overall throughput is not optimized. Observing the performance of the proportional scheduler with different values of transmit power in

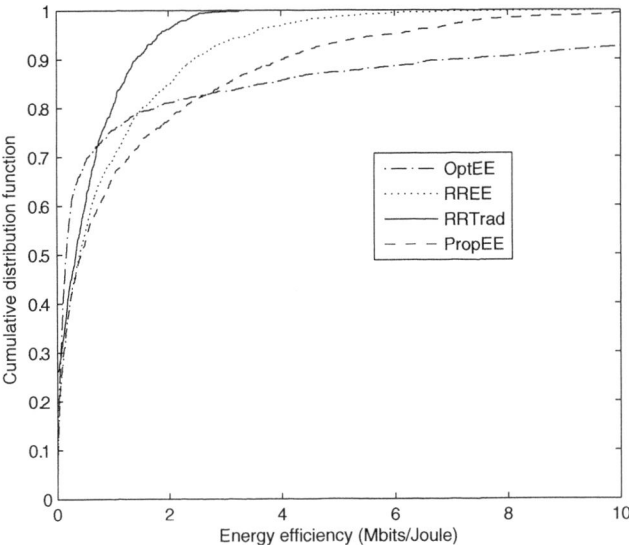

Figure 20.4 Fairness comparisons

20.3(a) and 20.3(b), we note that throughput increases with the transmit power, while the energy efficiency decreases. Energy efficiency and throughput efficiency do not necessarily agree.

Figure 20.4 compares the cumulative distribution functions of energy efficiency when 13 users are active. While the energy-efficient scheduler without fairness has the most percentage of users at high energy efficiency, the one with proportional fairness achieves the best fairness across different users with lowest percentage at low energy efficiency range. The energy-efficient scheduler with proportional fairness performs better than both round-robin schedulers, which means that, while assuring better fairness among all users, it also achieves higher energy efficiency.

20.3   Energy-efficient scheduling in frequency-selective channels

It is critical to consider overall network energy efficiency when performing frequency-domain resource management. The frequency selectivity of wideband wireless channels further accentuates this necessity. In this section, we study the scheduling in frequency-selective channels.

For frequency-selective fading channels, the problem is much more complex. The BS allocates subchannels to maximize the total energy efficiency across the whole network, i.e.

$$\mathcal{C}^* = \arg\max_{\mathcal{C}} \sum_i \frac{R_i}{P_{C_i} + P_{Ti}(\mathbf{R}_i)}, \tag{20.15a}$$

$$\text{subject to} \quad C_i \bigcap C_j = \varnothing, \forall i \neq j, \quad \text{and} \quad \bigcup_i C_i \subseteq \mathcal{K}, \tag{20.15b}$$

where $\mathcal{K}$ is the set of subchannels, $\mathcal{C} = \{C_1, C_2, \ldots, C_N\}$, and C_n is the subchannel assignment to the nth user. Subscripts i and j are used to distinguish users. P_{C_i} is the circuit power and $P_{T_i}(\mathbf{R}_i)$ is the transmission power. Problem (20.15) can be treated as a utility-based resource allocation. However, this is far from a trivial application of the utility scheduling schemes we introduced in Part II and is much more challenging in that the utility in (20.15) depends on a vector of rates instead of the sum rate. It can be treated as a multi-dimensional generalization of the traditional utility concept, whose solution is expected to be very complicated and still an open issue.

As complex approaches incur additional energy consumption, lower-complexity techniques for energy-efficient communications are highly desirable. In this section, we will introduce energy-efficient resource allocation schemes with the lowest complexity, i.e. closed-form scheduling algorithms, that perform close to the global optimum.

20.3.1   Time-averaged network energy efficiency

In the following, we focus on an uplink OFDMA system, as shown in Figure 20.5. Downlink can be derived in the same way. The methodology can be used in other systems as well to develop low-complexity energy-efficient solutions.

The BS assigns subchannels for each user to optimize overall network energy efficiency. Channels are assumed to be frequency-selective and with block-fading. Accurate channel state information is available to both BS and mobile users to optimize energy-efficient communications. The resource allocation settings are allowed to vary from one frame to another according to channel state information.

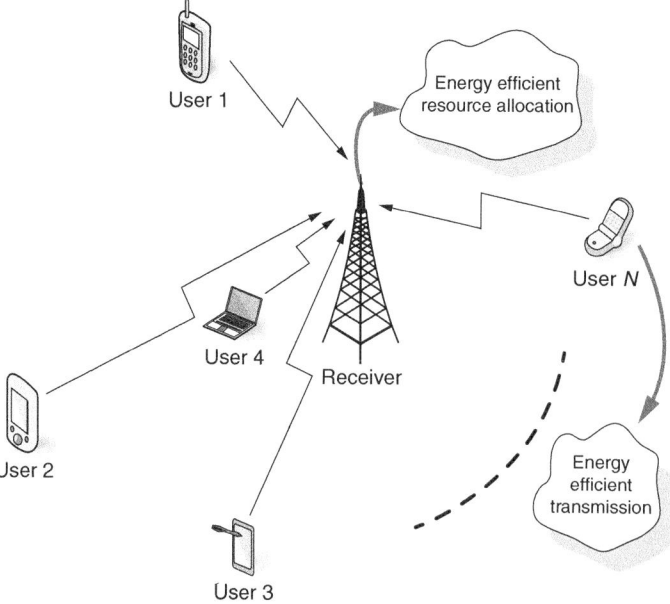

Figure 20.5 Network architecture

Consider a network with N users and K subchannels. Define the index set of all subchannels as $\mathcal{K} = \{1, 2, \ldots, K\}$ and the index set of subchannels assigned to user n at frame t to be $\mathcal{C}_n[t]$. Each subchannel is assigned to only one user in each frame. Consequently,

$$\mathcal{C}_i[t] \bigcap \mathcal{C}_j[t] = \emptyset, \forall i \neq j$$
$$\bigcup_i \mathcal{C}_i[t] \subseteq \mathcal{K}, \tag{20.16}$$

where $\emptyset$ defines an empty set. The data rate, power consumption, and energy efficiency of user n at frame t follows the discussion in Section 19.5.

To maximize network energy efficiency, the subchannels are allocated to maximize

$$U[t] = \sum_{n=1}^{N} u_n[t]. \tag{20.17}$$

In this case, while the sum energy efficiency is maximized, all subchannels may be assigned to the user with the best channel gain, which is completely unfair to other users. Hence, we also design a scheduler that maximizes the PF network energy efficiency, and the subchannels are allocated to maximize

$$V[t] = \sum_{n=1}^{N} \log(u_n[t]). \tag{20.18}$$

Sometimes, circuit power dominates power consumption in the above optimization, e.g. in short-range communications where low transmit power is needed to compensate for path loss. For example, consider a commercial 802.11 network adapter, Cisco Aironet 802.11a/b/g Wireless CardBus Adapter. As shown in [53], its operating voltage is 3.3 Volts and when it transmits at 54 Mbps, the current is 554 milliamps. Then the overall power consumption, including both transmit power and circuit power, is $3.3 \times 554 = 1828$ mW. However, as shown in [53], the transmit power for reliable data transmission can be only 20 mW when the data rate is 54 Mbps for a communications range of 13 meters. Assuming a power amplifier efficiency of 20%, the overall transmit power for reliable data transmission for this device is expected to be 100 mW. Hence, the circuit power consumption is $1828 - 100 = 1728$ mW. More examples can be easily found for other types of short-range communications, e.g. Bluetooth. In this case, maximizing energy efficiency is equivalent to maximizing throughput $T_n[t]$ as $P_n[t]$ is almost independent of transmit power allocation and rate adaptation. Correspondingly, (20.17) is equivalent to maximizing the sum of throughput weighted by the inverse of circuit power and (20.18) equals maximizing the product of throughput. The dependence of the optimization on circuit power will be further demonstrated later.

By considering time averaged bit-per-Joule metrics, we will introduce energy-efficient resource scheduling techniques in closed forms and the main results are summarized in Table 20.3. Please refer to the following sections for detailed definitions and their performances.

Table 20.3 Main results

Function	Formula
Link adaptation	$r_k^*[t] = \max\left(S'^{-1}\left(\frac{1}{u[t-1]}\frac{g_k[t]}{\sigma^2}\right), 0\right)$, and $p_k^*[t] = \frac{S(r_k^*[t])\sigma^2}{g_k[t]}, \forall k \in \mathcal{C}[t]$
Energy-efficient scheduler	$J(n,k) = \frac{r_{nk}[t]}{P_n[t-1]} - u_n[t-1]\frac{p_{nk}[t]}{P_n[t-1]}$, and $\mathcal{C}_n^* = \{k|J(n,k) > J(m,k), \forall m \neq n\}, \forall n$
PF energy-efficient scheduler	$J_f(n,k) = \frac{r_{nk}[t]}{T_n[t-1]} - \frac{p_{nk}[t]}{P_n[t-1]}$ and $\mathcal{C}_n^* = \{k|J_f(n,k) > J_f(m,k), \forall m \neq n\}, \forall n.$

20.3.2 Energy-efficient scheduler

In this section, the subchannels are assigned such that the sum energy efficiency $U[t]$ is maximized. Since $U[t-1]$ is fixed, it is equivalent to maximize

$$
\begin{aligned}
\triangle U &= U[t] - U[t-1] \\
&= \sum_{n=1}^{N} u_n[t] - \sum_{n=1}^{N} u_n[t-1] \\
&= \sum_{n=1}^{N} (u_n[t] - u_n[t-1]).
\end{aligned}
\tag{20.19}
$$

We can see that

$$
\begin{aligned}
u_n[t] - u_n[t-1] &= \frac{T_n[t]}{P_n[t]} - \frac{T_n[t-1]}{P_n[t-1]} \\
&= \frac{T_n[t]P_n[t-1] - P_n[t]T_n[t-1]}{P_n[t]P_n[t-1]}.
\end{aligned}
\tag{20.20}
$$

Substituting Equations (19.52) and (19.58) into (20.20), we have

$$
\begin{aligned}
&u_n[t] - u_n[t-1] \\
&= \left(P_n[t-1]\sum_{k\in\mathcal{C}_n[t]} r_{nk}[t] - \right. \\
&\quad \left. T_n[t-1]\left(\sum_{k\in\mathcal{C}_n[t]} p_{nk}[t] + p_{cn}[t]\right)\right) / (wP_n[t]P_n[t-1])
\end{aligned}
$$

$$= \sum_{k \in \mathcal{C}_n[t]} \frac{P_n[t-1]r_{nk}[t] - T_n[t-1]p_{nk}[t]}{wP_n[t]P_n[t-1]} - \frac{T_n[t-1]p_{cn}[t]}{wP_n[t]P_n[t-1]}$$

$$= \sum_{k=1}^{K} I_k(\mathcal{C}_n[t]) \frac{P_n[t]r_{nk}[t] - T_n[t-1]p_{nk}[t]}{wP_n[t]P_n[t-1]} - \frac{T_n[t-1]p_{cn}[t]}{wP_n[t]P_n[t-1]},$$

where the indicator function $I_k(\mathcal{C}_n)$ is defined as

$$I_k(\mathcal{C}_n) = \begin{cases} 1 & k \in \mathcal{C}_n, \\ 0 & \text{otherwise.} \end{cases} \tag{20.21}$$

Hence, the subchannel assignment is to maximize

$$\triangle U = \sum_{n=1}^{N} (u_n[t] - u_n[t-1])$$

$$= \sum_{n=1}^{N} \sum_{k=1}^{K} I_k(\mathcal{C}_n[t]) \frac{P_n[t-1]r_{nk}[t] - T_n[t-1]p_{nk}[t]}{wP_n[t]P_n[t-1]}$$

$$- \sum_{n=1}^{N} \frac{T_n[t-1]p_{cn}[t]}{wP_n[t]P_n[t-1]}$$

$$= \sum_{k=1}^{K} \sum_{n=1}^{N} I_k(\mathcal{C}_n[t]) \frac{P_n[t-1]r_{nk}[t] - T_n[t-1]p_{nk}[t]}{wP_n[t]P_n[t-1]}$$

$$- \sum_{n=1}^{N} \frac{T_n[t-1]p_{cn}[t]}{wP_n[t]P_n[t-1]}.$$

Define the allocation metric to be

$$J(n, k) = \frac{P_n[t-1]r_{nk}[t] - T_n[t-1]p_{nk}[t]}{P_n[t]P_n[t-1]}$$

$$\approx \frac{P_n[t-1]r_{nk}[t] - T_n[t-1]p_{nk}[t]}{P_n^2[t-1]} \tag{20.22}$$

$$= \frac{r_{nk}[t]}{P_n[t-1]} - u_n[t-1] \frac{p_{nk}[t]}{P_n[t-1]},$$

where $r_{nk}[t]$ is given by (19.65) and $p_{nk}[t]$ (19.66).

It is easy to see that $\triangle U$ is maximized by assigning subchannel k to the user with the highest allocation metric $J(n, k)$ on that subchannel, i.e. the optimal subchannel assignment is

$$\mathcal{C}_n^* = \{k | J(n, k) > J(m, k), \forall m \neq n\}, \forall n. \tag{20.23}$$

Note that in the above derivation, approximation is used in (20.22). If w is sufficiently large, the approximation error is close to zero and the scheduler is almost globally optimal.

When circuit power dominates power consumption, the allocation metric is

$$J_t(n, k) \approx \frac{r_{nk}[t]}{P_n[t-1]}. \tag{20.24}$$

Assume all users consume the same circuit power and $P_n[t-1]$ is the same for all users. Since the user with the maximum $r_{nk}[t]$ is the same as the one with the maximum signal to interference plus noise ratio (SINR) on that subchannel, the energy-efficient scheduler is equivalent to applying the traditional max-SINR scheduler on each subchannel to achieve the highest spectral efficiency [8], which is

$$C_n^* = \{k | r_{n,k} > r_{m,k}, \forall m \neq n\}, \forall n. \tag{20.25}$$

PF energy-efficient scheduler

In order to maximize the geometric mean of the energy efficiency of all users, the subchannels are assigned to maximize

$$V[t] = \sum_{n=1}^{N} \log(u_n[t]), \tag{20.26}$$

which is equivalent to maximizing

$$
\begin{aligned}
\triangle V &= V[t] - V[t-1] \\
&= \sum_{n=1}^{N} \log(u_n[t]) - \sum_{n=1}^{N} \log(u_n[t-1]) \\
&= \sum_{n=1}^{N} \left(\log\left(\frac{T_n[t]}{T_n[t-1]} \right) - \log\left(\frac{P_n[t]}{P_n[t-1]} \right) \right).
\end{aligned} \tag{20.27}
$$

Using the Taylor series expansion and the fact that $w \gg 1$, we have

$$
\begin{aligned}
\log\left(\frac{T_n[t]}{T_n[t-1]} \right) &= \log\left(1 - \frac{1}{w} + \frac{\frac{1}{w}\sum_{k \in C_n} r_{nk}[t]}{T_n[t-1]} \right) \\
&\approx \log\left(1 - \frac{1}{w} \right) + \frac{\sum_{k \in C_n} r_{nk}[t]}{T_n[t-1](w-1)}.
\end{aligned} \tag{20.28}
$$

Similarly, we have

$$
\begin{aligned}
&\log\left(\frac{P_n[t]}{P_n[t-1]} \right) \\
&\approx \log\left(1 - \frac{1}{w} \right) + \frac{\sum_{k \in C_n} p_{nk}[t] + p_{cn}[t]}{P_n[t-1](w-1)}.
\end{aligned} \tag{20.29}
$$

Hence, $\triangle V$ can be expressed as

$$
\begin{aligned}
\triangle V &= \sum_{n=1}^{N} \left(\frac{\sum_{k \in C_n[t]} r_{nk}[t]}{T_n[t-1](w-1)} - \frac{\sum_{k \in C_n[t]} p_{nk}[t] + p_{cn}[t]}{P_n[t-1](w-1)} \right) \\
&= \sum_{n=1}^{N} \sum_{k=1}^{K} \left(I_k(C_n[t]) \left(\frac{r_{nk}[t]}{T_n[t-1](w-1)} \right. \right. \\
&\quad \left. \left. - \frac{p_{nk}[t]}{P_n[t-1](w-1)} \right) \right) - \sum_{n=1}^{N} \frac{p_{cn}[t]}{P_n[t-1](w-1)}
\end{aligned}
$$

$$= \sum_{k=1}^{K} \sum_{n=1}^{N} \left(I_k(\mathcal{C}_n[t]) \left(\frac{r_{nk}[t]}{T_n[t-1]} \right. \right.$$

$$\left. \left. - \frac{p_{nk}[t]}{P_n[t-1]} \right) / (w-1) \right) - \sum_{n=1}^{N} \frac{p_{cn}[t]}{P_n[t-1](w-1)}.$$

Define the allocation metric to be

$$J_f(n,k) = \frac{r_{nk}[t]}{T_n[t-1]} - \frac{p_{nk}[t]}{P_n[t-1]}, \tag{20.30}$$

where $r_{nk}[t]$ is given by (19.65) and $p_{nk}[t]$ (19.66). Therefore, $\triangle V$ is maximized by assigning subchannel k to the user with the highest allocation metric $J_f(n,k)$ on that subchannel, i.e. the optimal subchannel assignment achieving proportional fairness is

$$\mathcal{C}_n^* = \{k | J_f(n,k) > J_f(m,k), \forall m \neq n\}, \forall n. \tag{20.31}$$

In the above derivation, approximations are used in (20.28) and (20.29). If w is sufficiently large, the approximation error is zero and the scheduler is almost globally optimal.

When the circuit power dominates the power consumption, the allocation metric is

$$J_{tf}(n,k) \approx \frac{r_{nk}[t]}{T_n[t-1]}, \tag{20.32}$$

and the energy-efficient scheduler is equivalent to applying the traditional proportional fair scheduler [165, 152] on each subchannel, i.e.

$$\mathcal{C}_n^* = \{k | J_{tf}(n,k) > J_{tf}(m,k), \forall m \neq n\}, \forall n. \tag{20.33}$$

20.3.3 Network performance

In this section, we compare these schemes with the global optima to evaluate the sub-optimality gap. For link adaptation, the global optimum is the solution that globally maximizes (19.60). Since $u[t]$ in (19.60) is strictly quasiconcave in the data rate vector $\mathbf{r}[t]$, a local optimal solution is also globally optimal [66] and iterative algorithms can be used to obtain the global optimal link adaptation. For energy-efficient schedulers, the global optimum is the solution that globally maximizes $U[t]$ in (20.17) or $V[t]$ in (20.18). We exhaustively search all possible subchannel assignments as well as the corresponding optimal link adaptation. The solution that achieves the highest $U[t]$ or $V[t]$ is globally optimal. The weight of the exponentially weighted low-pass filter, the number of subchannels, and the number of users in the system determine approximation accuracy in deriving the closed-form approaches in this chapter. Hence, we will focus on their impact on system energy efficiency performance.

International telecommunication union (ITU) multi-path pedestrian channel A [95] is used to model frequency-selective fading. Capacity approaching coding is assumed. Figure 20.6 compares the low-complexity suboptimal approaches with the global optimal approaches for energy-efficient link adaptation when there are ten subchannels in the system. The energy efficiency of the link adaptation is normalized by the energy efficiency of the global optimal solution. We show the normalized energy efficiency with different weights, w. ϵ is the transmit power to circuit power ratio, assuming the transmit power is allocated such that the average achieved spectral efficiency is one b/s/Hz. The circuit power is varied to have different ϵ values. From the figure, the link adaptation performs closely to the global optimum, with a performance loss of less than 2% when $w > 10$. Figure 20.7 further demonstrates normalized energy efficiency when the system has different numbers of subchannels. We can see that while more subchannels result in less approximation accuracy of the low-complexity energy-efficient link adaptation scheme, larger w can be used to ensure a small suboptimality gap.

Figure 20.8 shows the normalized energy efficiency of different schedulers in a three-user network. Since we need to exhaustively search all possible subchannel assignments as well as the corresponding optimal link adaptation to find the global optimal solution, the complexity grows exponentially. To reduce the complexity, the network is configured to have eight subchannels. As shown in Figure 20.8, the performance loss of the low-complexity closed-form schedulers and link adaptation is within 5% when $w > 10$. Table 20.4 further demonstrates the impact of the number of users on system performance when the network has eight subchannels.

As shown in Figures 20.6 and 20.8, both weight, w, and the transmit to circuit power ratio, ϵ, impact the system performance. The selection of the weight determines the approximation accuracy of (19.63), (20.22), (20.28), and (20.29). For example, in (20.22), $P_n[t]$ is approximated by $P_n[t-1]$. It is easy to see that when $w = \infty$, $P_n[t] = P_n[t-1]$ according to (19.58) and there is no approximation error. However,

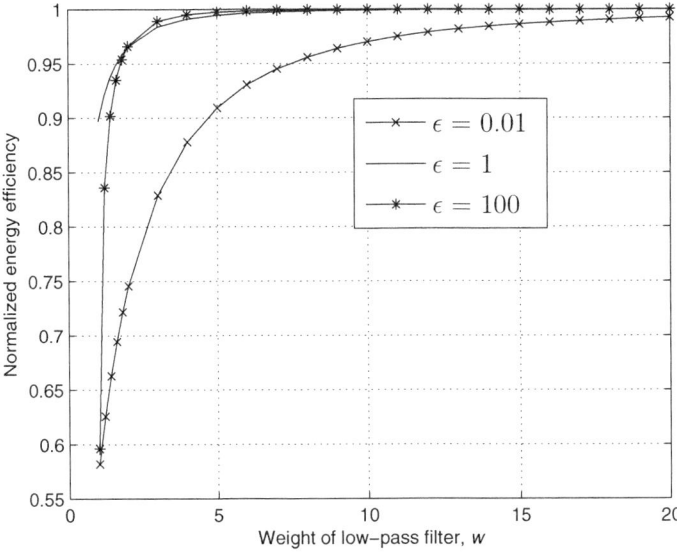

Figure 20.6 Normalized energy efficiency of a single link varying the weight of low-pass filer, w (ϵ: transmit to circuit power ratio)

Table 20.4 Impact of the number of users

Number of users	Normalized energy efficiency of Max-AM scheduler	Normalized energy efficiency of Max-GM scheduler
2	0.958	0.9952
3	0.9394	0.9931
4	0.9337	0.9926

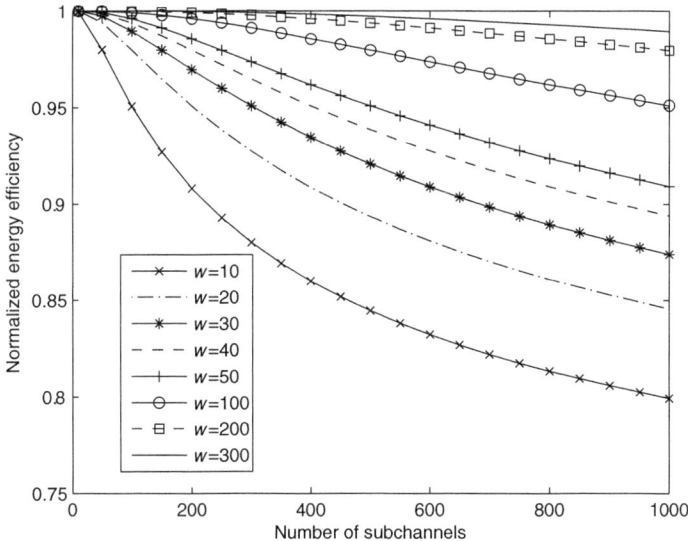

Figure 20.7 Normalized energy efficiency of a single link varying the number of subchannels ($\epsilon = 1$)

whenever w is finite, there is an approximation error and this error decreases with w. This can be seen if we define the error to be

$$err = |P_n[t] - P_n[t-1]|$$

in which $P_n[t]$ is the true value and $P_n[t-1]$ is the approximate value of $P_n[t]$ in (20.22). Referring to (19.58),

$$err = \frac{|-P_n[t-1] + p_n[t] + p_{cn}[t]|}{w}$$

and *err* decreases with w. Furthermore, the approximation error determines the accuracy of the allocation metric (third line of (20.22)). Hence, the selection of w impacts system performance. Similar analysis can be applied to (19.63), (20.28), and (20.29). The impact of the transmit power to circuit power ratio, ϵ, is the same as the impact of circuit power. When circuit power is larger, energy-efficient link adaptation will choose a higher data rate. Intuitively, this is because when circuit power is larger, energy-efficient link adaptation tends to transmit at a higher data rate to reduce the transmission time such that the circuit energy consumption can be reduced. Hence, when circuit power

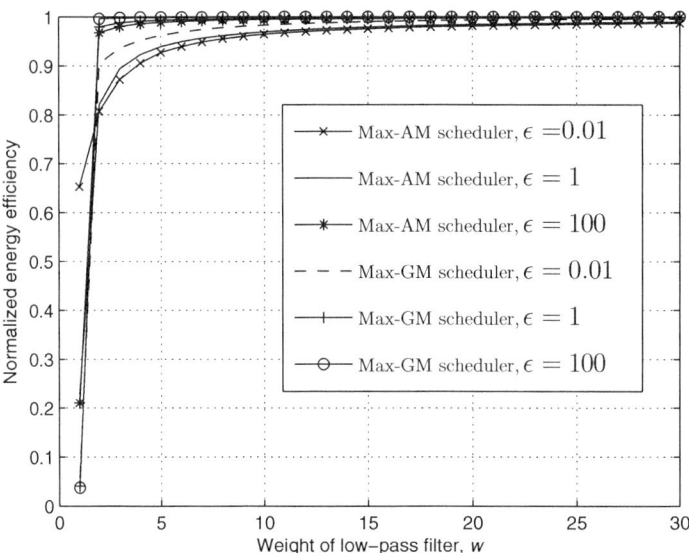

Figure 20.8 Normalized average energy efficiency of a three-user network (w: weight of low-pass filer; ϵ: the average transmit power to circuit power ratio; Max-AM: max arithmetic-mean energy-efficient scheduler; and Max-GM: max geometric-mean energy-efficient scheduler)

is larger, the actual optimal energy-efficient data rate is also larger. Furthermore, the optimal transmit power is also larger according to (19.66). However, to obtain closed-form solutions, we have approximated that $T[t] \approx T[t-1]$ and $P[t] \approx P[t-1]$. Given w, the approximation error depends on the actual optimal data rate and transmit power in $T[t]$ and $P[t]$. Hence, circuit power impacts approximation accuracy and thus system performance. Besides w and ϵ, the number of users and subchannels in the system also impact system performance. Note that approximation is used on each subchannel for each user to achieve closed-form link adaptation on all subchannels. When there are multiple users and subchannels in the system, the approximation errors on all sub-channels of all users will be accumulated to impact the behavior of both scheduling and link adaptation. Hence, the suboptimality gap also depends on the number of users and subchannels in the system, as shown in Table 20.4 and Figure 20.7. However, the approximation accuracy can always be improved by using higher values of w.

21 Distributed energy-efficient wireless resource management

In the previous two chapters, we discussed energy-efficient transmission and centralized resource scheduling techniques. As multiple users need to reuse the same spectrum at different locations, interference among different links determines overall network performance. Interference is one major factor that limits system performance, especially as more aggressive frequency reuse is used in future wireless networks. This motivates the design of advanced distributed medium access control (MAC) schemes to manage orthogonal network resources and the use of power control for interference management among users reusing the same frequency. In this chapter, we will first introduce distributed energy-efficient MAC designs. We will discuss the key design principle and the impact of traffic load on the network energy efficiency when distributed MAC protocols are used. Then we study the fundamental impact of interference on network energy efficiency. We first assume ideal collaboration for a two-user network and disclose several basic laws of energy-efficient designs in interference-limited scenarios. Then we study in detail the distributed power optimization for energy-efficient communications.

21.1 Distributed energy-efficient MAC design

Each wireless device may have five operating modes: transmit, receive, idle, sleep, and off. The main functions of each mode are listed below.

(i)   Transmit mode: sends data.
(ii) Receive mode: receives data.
(iii) Idle: all transceiver components are on and ready to send or receive data.
(iv) Sleep: the major transceiver circuit components are turned off with a very limited portion remaining on to listen to demands outside.
(v)   Off: power completely off.

In the five modes, transmit, receive, and idle modes can be called active modes because all circuit components are running and these components will consume circuit power. The difference in the amount of energy consumed in these modes is significant. An example is given in Table 21.1, which lists the power consumption of several commercial 802.11 transceivers in all operation modes [164]. We can see that while in the sleep mode power consumption stays the same, the power consumption in other

Table 21.1 Power consumption of a wireless transceiver

	802.11b	802.11a	802.11g
Sleep mode	132 mW	132 mW	132 mW
Idle mode	544 mW	990 mW	990 mW
Receive mode	726 mW	1320 mW	1320 mW
Transmit mode	1089 mW	1815 mW	1980 mW
Data rate	11 Mbit/s	54 Mbit/s	54 Mbit/s

modes has been increasing with each new standard that supports higher data rate. The power consumption in the transmit mode will be even much higher for longer-distance communications, such as in cellular networks, as more power is needed to compensate path losses, which grow exponentially with communication distances.

There are many sources of energy consumption from a MAC perspective. Below we list some examples.

- Traffic usually arrives sporadically but the transceiver circuit components need to be kept on even when there are no data so that the transceiver can respond to traffic quickly. Traffic examples include: email, webpage browsing, VoIP, etc. The device usually stays in the idle mode for a very long time, running all circuits without any transceiving activities. The energy consumption in the idle mode can be much larger than the one in other modes.
- In a wireless cellular network, the receiver has to have power on all the time so as to receive data. In the downlink of cellular networks, mobile devices have to stay on so that they can receive system messages from the base station. In addition, a device may have to decode packets destined for other users and therefore waste energy. In the uplink, the base station also needs to be turned on even if there are no users being served so that future connection requests from mobile users can always be heard and the states of mobile users can be monitored.
- In wireless networks using random access protocols, collisions may occur and the data sent out will no longer be useful. The energy in transmission and reception will be wasted. Therefore, it is necessary to avoid collisions or failures in packet receptions as much as possible.
- When mobile devices switch between different modes, it takes some time and significant amounts of transition energy are. In addition consumed to the turnaround time within the devices, the network also needs to be notified of the changes, which will take a much longer time.

21.1.1   General rules of distributed MAC design

Observing the above analysis, the main purpose of the energy-efficient distributed MAC design is to enable mobile devices to stay in the least-power-consuming mode, e.g. the sleep mode, as long as possible. With this design, the devices can deactivate as many functional circuit units as possible. However, there are issues related to switching off and on circuit components. For example, restarting will cause delay in responding to a traffic request. It may incur even higher energy consumption because of higher restarting current. So there is a tradeoff between performance and energy saving.

If there is no activity for a certain amount of time, a device can enter sleep mode. A timer can therefore be used to determine when a component can be turned off. This timer can be either statically or dynamically configured, depending on the scheduling algorithm, and will affect system energy consumption. On the other hand a device needs to wake up to respond to user or network activity from time to time. If the device is completely off, it will not be able to hear any activity request from outside and never wake up again. A small portion of circuits therefore needs to remain on and monitor incoming requests. On the other hand, it takes non-negligible time to wake up and the latency can be too long. Therefore, it is necessary to predict when responses are needed and wake up in advance.

Communications with large packet sizes and stable traffic flows can enter the sleep mode when there no traffic for a while to save energy. Since less mode switching is needed, little performance loss is expected in mode switching. For communications with sporadic and frequent traffic arrivals, transitions between modes can be too frequent, and sleeping whenever there are no data in the buffer will hurt performance while not saving any energy. One improvement is to buffer the traffic and schedule transmission such that it can continuously transmit or receive data and sleep longer, thereby staying in one mode as long as possible. For example, the data for different applications can be buffered, bundled, and delivered together to reduce the number of transitions.

To minimize the active time of all devices while ensuring the success of packet transmission, a distributed energy-efficient MAC protocol should command in advance when each device should send or receive data so that the device can choose to sleep in the remaining time. There are many ways of designing energy-efficient MACs. For example, a base station can buffer the data and periodically broadcast a message indicating which mobile devices need to receive the buffered data. Each mobile device has to wake up to receive the broadcast message. If there are data for a device, it will wake up and receive the data; otherwise it will sleep again. The synchronization between the base station and mobile devices enables mobile devices to sleep and wake up just in time for communications.

Distributed energy-efficient MAC design is widely used in wireless sensor networks. In wireless sensor networks, each sensor node may consist of one or more sensors, an embedded processing unit, and a low-power radio. Applications of sensor networks include earthquake monitoring in desert areas, traffic control, industrial automation, etc. Most sensor nodes are usually battery-powered and energy efficiency is thus of paramount importance because replacing batteries for all nodes frequently is difficult or even impossible. To save energy, sensor nodes will keep silent for most of the time and become active for data transmission after detecting something of interest. A typical example is sensor-MAC (S-MAC) [202]. To save power, S-MAC implements a periodic listen and sleep protocol, which is illustrated in Figure 21.1. Each node communicates with other nodes in the listen period and sleeps in the sleep period. When node sleeps, it will turn off its radio to save power. With periodic design, a large amount of energy consumption can be saved by avoiding unnecessary idle listening, especially when

Figure 21.1 Periodic listen and sleep

traffic load is low. To make the protocol more flexible, a parameter, duty cycle, is defined as the ratio of the listen-period to a complete sleep and listen cycle. The duty cycle can be adjusted from 1% to 100%. The listen period is divided into two parts, SYNC and DATA periods. The SYNC period is used to solve synchronization problems between neighboring nodes and the DATA period is used for data transmission. To ensure the success of packet transmission, S-MAC uses a carrier sense multiple access with collision avoidance (CSMA/CA) protocol with request to send/clear to send (RTS/CTS) to avoid collisions.

21.1.2   Impact of traffic load on energy consumption

As discussed in Section 5.2.2, slotted Aloha is one of the most basic distributed MAC protocols. In this section, we analyze the energy efficiency performance of slotted Aloha and demonstrate the impact of multi-user resource sharing on network energy efficiency. To simplify the analysis, we will make some rough approximations where necessary.

Consider the model in Figure 21.2, which describes the packet flows in one of the N users in the system. For each user, packets arrive according to a Poisson process with rate $\lambda_i = \lambda/N$, which is the average number of packet arrivals in one time slot. As soon as a packet arrives, it is transmitted in the coming time slot and the transmission probability p is determined by the arrival process. At most, one packet may be transmitted in one time slot and $\lambda_i \leq 1$. The transmission succeeds if there are no transmission attempts by other users. If it succeeds, the packet will arrive at the receiver and leave the system. A failed packet will be stored in the delay buffer. Packet k will remain there D_k time slots. D_k are assumed to be independent and identically distributed. Upon leaving the delay buffers, the packets are transmitted immediately. Here we make the approximation that messages leave the delay buffer following another Poisson process with rate σ_i. Besides this process is independent of the arrival process. With this approximation, the packet arrivals will be memoryless, independent of whether the packet is new or has been waiting in the buffer for transmission. According to the law of large numbers, this approximation is accurate if

$$E[D_k] = E[D] \gg 1, \tag{21.1}$$

where D denotes the stationary process D_k. The compound process of retransmitted and new messages will form a new Poisson process with intensity γ_i given by

$$\gamma_i = \lambda_i + \sigma_i. \tag{21.2}$$

Let q denote the probability that user i makes a successful transmission in a slot. If the system is in a stable equilibrium, the average number of packets arriving at the system

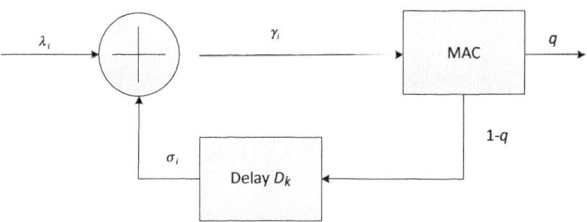

Figure 21.2 Simplified transmitter model in a slotted Aloha system

has to be equal to the average number of departing packets, otherwise the system will either contain an ever-increasing number of packets or become empty. This condition can be expressed as

$$qy_i = \lambda_i. \tag{21.3}$$

The condition for a successful transmission is that no other user is transmitting in the same time slot. Since all users are identical, the probability that a user will transmit will be the same for all terminals. Let p denote the transmission probability. Thus,

$$q = (1 - p)^{N-1}. \tag{21.4}$$

The probability p can be expressed as

$$
\begin{aligned}
p &= Pr\{\text{at least one arrival in the slot}\} \\
&= 1 - Pr\{\text{no arrival in the slot}\} \\
&= 1 - e^{-\gamma_i}.
\end{aligned} \tag{21.5}
$$

Inserting this result into (21.4) yields

$$q = (1 - p)^{N-1} = e^{-(N-1)\gamma_i} = e^{-\gamma} e^{\gamma_i}, \tag{21.6}$$

where $\gamma = N\gamma_i$. Now combine (21.3) and (21.6) to get

$$\lambda_i = q\gamma_i = \gamma_i e^{-\gamma} e^{\gamma_i}. \tag{21.7}$$

Summing (21.7) over all N identical users yields

$$\lambda = \sum_i^N \lambda_i = N\gamma_i e^{-\gamma} e^{\gamma_i} = \gamma e^{-\gamma(1-\frac{1}{N})}. \tag{21.8}$$

For a system with many users and sufficiently large N,

$$\lambda = \gamma e^{-\gamma}. \tag{21.9}$$

Equation (21.8) describes the relationship between the total packet arrival rate, λ, and the total rate of transmission attempts, γ. Figure 21.3 illustrates an example where N is sufficiently large. The right-hand side (RHS) of (21.9) cannot be arbitrarily large and λ has a maximal value given by

$$\lambda \leq \max \gamma e^{\gamma} = e^{-1}. \tag{21.10}$$

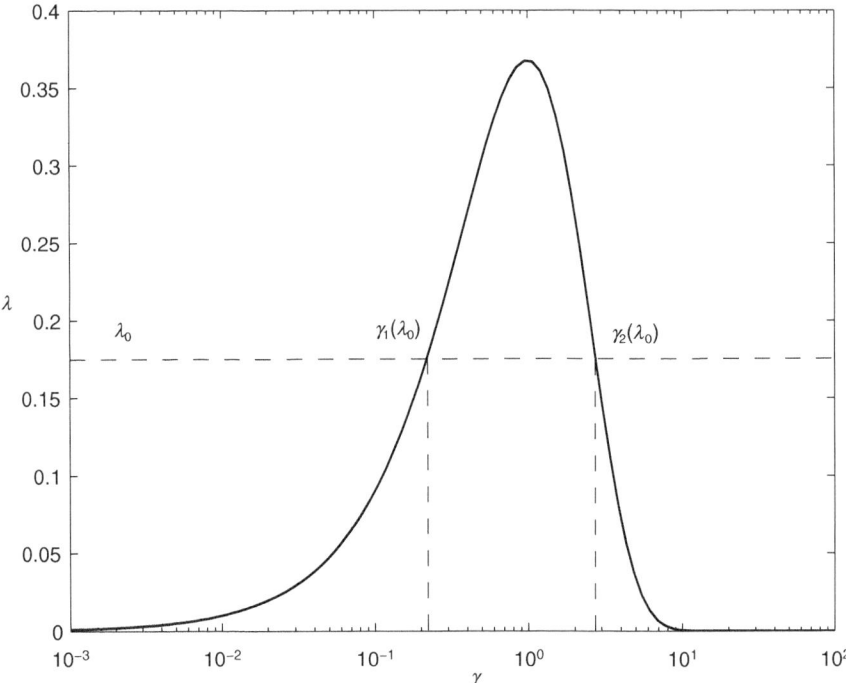

Figure 21.3 Slotted Aloha

The maximal throughput is

$$\lambda^* = e^{-1} \approx 36.8\%. \tag{21.11}$$

Note that the above value is for the case with sufficiently many users in the network. When there is a small number of users, the maximal throughput may be higher, which can be determined by (21.8).

Now we analyze the energy efficiency of the system. Assume all the users are uniformly distributed over a circular cell area accessing a base station (BS) in the center. The cell area has a radius R. The users use a power control scheme such that a constant signal power, P_r, is received at the BS. The signal level decays as the αth power of the distance, i.e. $P_r = \frac{P_T}{d^\alpha}$. If a user sends a packet in a time slot, the full time slot will be used for data transmission. The power amplifier efficiency of each user is ξ. When sending data in the active mode, each user consumes P_C circuit power. Each user goes into sleep mode immediately when there is no data transmission and consumes only P_S circuit power. Similarly, a sleeping user can switch to the active mode and send data immediately when it needs to send a packet.

The cumulative distribution function (CDF) of the transmission power of all users is

$$Pr(P_T \leq P) = Pr\left(d^\alpha \leq \frac{P}{P_r}\right) = \frac{\sqrt{P}}{R^2 \sqrt{P_r}}. \tag{21.12}$$

Each device needs to be active only in the slots when there are packets to be sent. If there is only one user in the network, the average power consumption is

$$\overline{P} = \lambda \left(\frac{P_T}{\xi} + P_C \right) + (1 - \lambda)P_S, \tag{21.13}$$

and the CDF of the average power consumption is

$$Pr(\overline{P} \leq P) = Pr \left(\lambda \left(\frac{P_T}{\xi} + P_C \right) + (1 - \lambda)P_S \leq P \right)$$

$$= \frac{\sqrt{\left(\frac{P-(1-\lambda)P_S}{\lambda} - P_C \right) \xi}}{R^2 \sqrt{P_r}}. \tag{21.14}$$

When there are N users in the network, the active time slots of each user consist of those for both new transmissions and retransmissions. Therefore, γ determines the probability that a user will be in the active mode or not in each time slot. γ is determined by (21.8). As illustrated in Figure 21.3, for a certain traffic arrival rate $\lambda = \lambda_0$, there might be two γ values: one is comparatively low and smaller than 1, whereas the other is considerably higher. These two values correspond to states of equilibrium in the network. The state corresponding to the smaller γ value has less transmission activities and almost all transmissions are successful. In the other state with much larger γ, there are a large number of backlogged packets, causing frequent collisions. The attempt rate γ is very high with a very low success probability q. If the system is in the equilibrium state with lower γ value, the CDF of the average power consumption is

$$Pr(\overline{P} \leq P) = Pr \left(\gamma \left(\frac{P_T}{\xi} + P_C \right) + (1 - \gamma)P_S \leq P \right)$$

$$= \frac{\sqrt{\left(\frac{P-(1-\gamma)P_S}{\gamma} - P_C \right) \xi}}{R^2 \sqrt{P_r}}, \tag{21.15}$$

Otherwise, $\gamma > 1$, indicating all users are always in the transmission state and the CDF of the average power consumption is

$$Pr(\overline{P} \leq P) = Pr \left(\left(\frac{P_T}{\xi} + P_C \right) \leq P \right)$$

$$= \frac{\sqrt{(P - P_C) \xi}}{R^2 \sqrt{P_r}}. \tag{21.16}$$

Because of the stochastic variations in traffic arrival, the system may move between the two equilibrium states. The long-term average power consumption will thus depend on how often the two equilibrium states exist.

Figure 21.4 illustrates the CDFs of average power consumption when there is either one or the maximum number of users in the network. In the second case, $\gamma = 1$. As shown in the figure, almost ten times more power is consumed when there is the maximum number of users in the network.

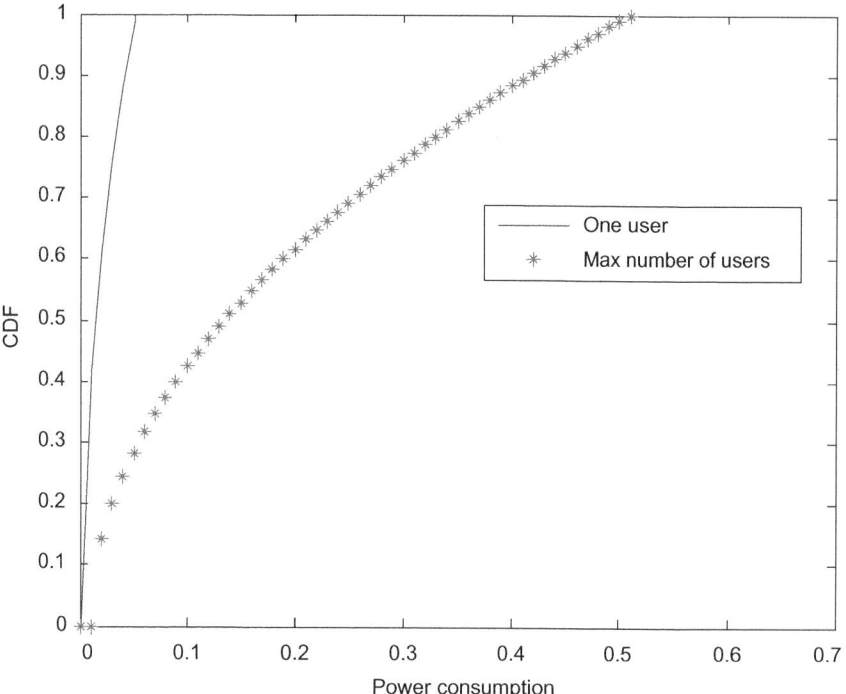

Figure 21.4 CDF of average power consumption ($\alpha = 4, P_C = 4mW, P_S = 1mW, \xi = 0.5,$ $R = 100m, \lambda_i = 0.025$)

21.2   Energy-efficient communications in special regimes

Distributed energy-efficient MAC protocols are used to allow users that heavily interfere with each other to send data using orthogonal resources to avoid interference. In general, the level of interference may vary significantly depending on channel conditions, and power control is essential in controlling network-level interference and thus overall network performance. In addition, power control also determines network energy utilization. In the following two sections, we study the fundamental impact of interference on network energy efficiency. In this section, we assume ideal collaboration for a two-user network and introduce several basic laws of energy-efficient designs in several special regimes.

When interference is considered, the solution to maximizing network energy efficiency is difficult to obtain, as the objective function, in general, is non-concave in transmission power. To gain some insight, we investigate the case where two users transmit simultaneously on the same frequency channel. We assume both users have complete network knowledge and cooperate to maximize network energy efficiency,

$$u(p_1, p_2) = \frac{r_1}{p_1 + p_c} + \frac{r_2}{p_2 + p_c}, \tag{21.17}$$

where

$$r_1 = w \log \left(1 + \frac{p_1 g_1}{p_2 g_{21} + \sigma^2} \right) \quad \text{and} \quad r_2 = w \log \left(1 + \frac{p_2 g_2}{p_1 g_{12} + \sigma^2} \right), \quad (21.18)$$

in which w is the signal bandwidth, p_i the transmission power of link i, g_i the channel gain of link i, g_{ij} the interference channel gain from link i to link j, σ^2 the noise power, and p_c the circuit power. As u is non-concave in p_1 and p_2, finding the global maximum is intractable. However, we can get some effective approaches by restricting our attention to some special regimes. The observations below can be easily extended to general wireless networks.

21.2.1 Circuit power dominated regime

In this regime, circuit power dominates power consumption, i.e. $p_c \gg p_n$ for $n = 1, 2$. This is usually true for short-range communications as small transmit power is needed to compensate for path loss. In this case, we have

$$u(p_1, p_2) \approx \frac{w}{p_c} \left(\log \left(1 + \frac{p_1 g_1}{p_2 g_{21} + \sigma^2} \right) + \log \left(1 + \frac{p_2 g_2}{p_1 g_{12} + \sigma^2} \right) \right). \quad (21.19)$$

Hence, maximizing energy efficiency is equivalent to maximizing sum network capacity, which has been discussed in the literature [23, 57]. The optimal solution takes on the form of binary power control where each user either shuts down or transmits with full power [23]. Whether two users transmit simultaneously or exclusively depends on interference strength.

21.2.2 Transmit power dominated regime

When the circuit power is negligible, e.g. in extremely long-distance communications where transmit power should be strong enough to compensate for large path loss,

$$u(p_1, p_2) \approx \frac{w \log \left(1 + \frac{p_1 g_1}{p_2 g_{21} + \sigma^2} \right)}{p_1} + \frac{w \log \left(1 + \frac{p_2 g_2}{p_1 g_{12} + \sigma^2} \right)}{p_2}. \quad (21.20)$$

It is obvious that the second term in $u(p_1, p_2)$ is strictly decreasing in p_1. To determine the first term, we need to verify that $F(p) = \frac{\log(1 + ap)}{p}$, $\forall a > 0$, is strictly decreasing in p, i.e.

$$\frac{\partial F(p)}{\partial p} = \frac{ap - \log(1 + ap) - ap \log(1 + ap)}{p^2(1 + ap)} < 0, \quad (21.21)$$

which is equivalent to $G(p) = ap - \log(1 + ap) - ap \log(1 + ap) < 0$. $G(0) = 0$. Besides $\frac{\partial G(p)}{\partial p} = -a \log(1 + ap) < 0$. Hence, $G(p) < 0, \forall p > 0$. Thus, $F(p)$ is strictly decreasing in p.

We can see that $u(p_1, p_2)$ is strictly decreasing with both p_1 and p_2. Hence, the optimal solution is to allocate as little power as possible. However, the above conclusion holds only when the circuit power is negligible. When the transmit power is comparable to the circuit power, other approaches are needed to determine the optimal power.

21.2.3   Noise dominated regime

Now we look at the problem from a different perspective. When noise is much stronger than interference, we have

$$u(p_1, p_2) \approx \frac{w \log \left(1 + \frac{p_1 g_1}{\sigma^2}\right)}{p_1 + p_c} + \frac{w \log \left(1 + \frac{p_2 g_2}{\sigma^2}\right)}{p_2 + p_c}. \tag{21.22}$$

Hence, the problem is decoupled and the sum network energy efficiency is maximized when each user adapts its transmission to maximize its own energy efficiency, which has been discussed thoroughly in Chapter 19.

21.2.4   Interference dominated regime

In the interference dominated regime, interference is much stronger than noise, i.e. $p_1 g_{12} \gg \sigma^2$ and $p_2 g_{21} \gg \sigma^2$ for any feasible p_1 and p_2 that support reliable transmission. To be specific, we require that $p_1 g_{12} \gg \sigma^2$ and $p_2 g_{21} \gg \sigma^2$ are significant enough that the *interference-to-noise ratio* (INR) and signal-to-interference-plus-noise ratio (SINR) of each user satisfies

$$INR > 1 + SINR. \tag{21.23}$$

Note that Equation (21.23) does hold when the interference is strong enough since INR increases with interference power while SINR decreases with it. The interference dominated regime exists when different transmissions are close to each other, e.g. closely coupled. Hence,

$$u(p_1, p_2) \approx \frac{w \log \left(1 + \frac{p_1 g_1}{p_2 g_{21}}\right)}{p_1 + p_c} + \frac{w \log \left(1 + \frac{p_2 g_2}{p_1 g_{12}}\right)}{p_2 + p_c}. \tag{21.24}$$

Since we are considering the interference dominated regime, whenever users 1 and 2 are sending data, $p_1 g_{12} \gg \sigma^2$ and $p_2 g_{21} \gg \sigma^2$ and $INR > 1 + SINR$ due to close coupling between these transmissions. In wireless communications, radio links exhibit a threshold effect where link quality is acceptable where the signal-to-noise ratio must exceed certain thresholds [111]. This indicates that the power allocation should not be too small. We assume that feasible p_1 and p_2 satisfies $p_1 \geq \widehat{p}_1$ and $p_2 \geq \widehat{p}_2$; otherwise, the user is shut down. Besides, in the interference dominated regime, $\widehat{p}_1 g_{12} \gg \sigma^2$ and $\widehat{p}_2 g_{21} \gg \sigma^2$. We compare two schemes. The first is to let both users send data simultaneously and the other is to shut down one user. First, we will show that when both users transmit, for user 1,

$$\frac{2w \log \left(1 + \frac{p_1 g_1}{p_2 g_{21} + \sigma^2}\right)}{p_1 + p_c} < \frac{w \log \left(1 + \frac{p_1 g_1}{\sigma^2}\right)}{p_1 + p_c}, \tag{21.25}$$

which is equivalent to showing that $\left(1 + \frac{p_1 g_1}{p_2 g_{21} + \sigma^2}\right)^2 < 1 + \frac{p_1 g_1}{\sigma^2}$. This inequality equals

$$1 + \frac{p_1 g_1}{p_2 g_{21} + \sigma^2} < \frac{p_2 g_{21}}{\sigma^2}. \tag{21.26}$$

Since $INR > SINR + 1$, (21.26) holds and so is (21.25). Similarly, we have the same result for user 2. When both users are sending, the maximum energy efficiency is

$$\max_{\substack{p_1 \geq \hat{p}_1 \gg \frac{\sigma^2}{g_{12}} \\ p_2 \geq \hat{p}_2 \gg \frac{\sigma^2}{g_{21}}}} \left(\frac{w \log\left(1 + \frac{p_1 g_1}{p_2 g_{21} + \sigma^2}\right)}{p_1 + p_c} + \frac{w \log\left(1 + \frac{p_2 g_2}{p_1 g_{12} + \sigma^2}\right)}{p_2 + p_c} \right). \tag{21.27}$$

Assume the above maximum energy efficiency is obtained by p_1° and p_2°. According to (21.25),

$$\frac{w \log\left(1 + \frac{p_1^\circ g_1}{p_2^\circ g_{21} + \sigma^2}\right)}{p_1^\circ + p_c} < \frac{w \log\left(1 + \frac{p_1^\circ g_1}{\sigma^2}\right)}{2(p_1^\circ + p_c)} \leq \frac{1}{2} \max_{p_1} \left(\frac{w \log(1 + \frac{p_1 g_1}{\sigma^2})}{p_1 + p_c} \right) \tag{21.28}$$

Similarly,

$$\frac{w \log\left(1 + \frac{p_2^\circ g_2}{p_1^\circ g_{12} + \sigma^2}\right)}{p_2^\circ + p_c} < \frac{1}{2} \max_{p_2} \left(\frac{w \log(1 + \frac{p_2 g_2}{\sigma^2})}{p_2 + p_c} \right).$$

Suppose $g_1 \geq g_2$. It is easy to see that

$$\max_{p_1} \left(\frac{w \log(1 + \frac{p_1 g_1}{\sigma^2})}{p_1 + p_c} \right) \geq \max_{p_2} \left(\frac{w \log(1 + \frac{p_2 g_2}{\sigma^2})}{p_2 + p_c} \right). \tag{21.29}$$

Comparing the above inequalities, we can see that

$$\max_{p_1} \left(\frac{w \log(1 + \frac{p_1 g_1}{\sigma^2})}{p_1 + p_c} \right) > \frac{w \log(1 + \frac{p_2^\circ g_2}{p_1^\circ g_{12} + \sigma^2})}{p_2^\circ + p_c} + \frac{w \log(1 + \frac{p_2^\circ g_2}{p_1^\circ g_{12} + \sigma^2})}{p_2^\circ + p_c}. \tag{21.30}$$

Therefore, $u(p_1, p_2)$ is maximized using an on–off power control scheme, i.e. time sharing, which lets the user with higher channel gain transmit data and shut down the other user. The extension of the conclusion to the generic multi-user case is straightforward.

We can easily generalize the conclusion. In the interference-dominated regime, orthogonal resource allocation that completely avoids interference is necessary to achieve the highest network energy efficiency. This indicates that how to allocate orthogonal resources, e.g. time slots, which is determined by access protocol designs, will determine how energy efficient the network would be. Conventional MAC protocols, which were discussed thoroughly in Part III, that improve spectrum efficiency can therefore be used to improve network energy efficiency as well. However, the main difference lies in the link adaptation and energy-efficient transmission techniques, introduced in Chapter 19, which should be applied after the MAC has granted access resources. An example is illustrated in Figure 21.5, where the average INR ratio is 20 dB. In this figure, network energy efficiency is maximized by shutting down user 2 and choosing the right power for user 1 to maximize its energy efficiency.

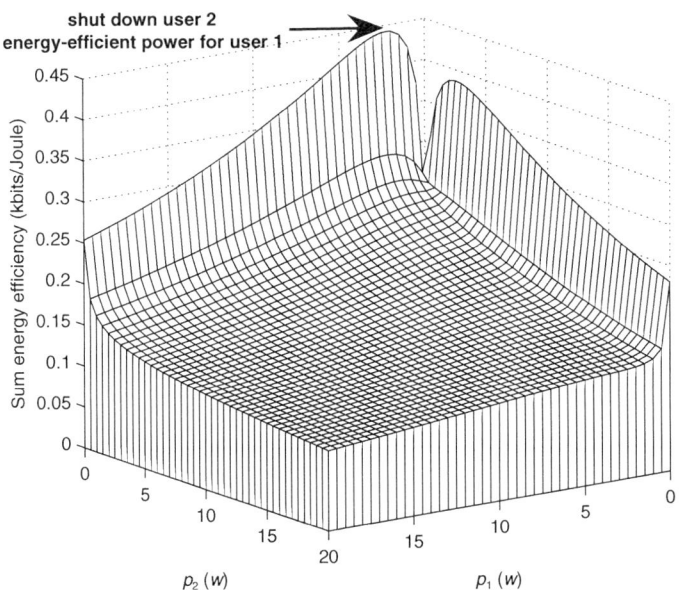

Figure 21.5 Sum energy efficiency and transmit powers in interference dominated regime

21.3   Distributed energy-efficient power control in frequency-selective channels

In this section, we introduce interference-aware energy-efficient power optimization in frequency-selective channels. Consider a system with K subchannels. Each of them experiences independent and flat-fading and *additive white Gaussian noise* (AWGN). There are N users, each consisting of a pair, the transmitter and the receiver, and operating on these subchannels. All users interfere with each other. Accurate channel state information is available to any pair of transmitter and receiver. Define the signal power attenuation of user i at subchannel k to be $g_{ii}^{(k)}$ and the interference power gain from the transmitter of user i to the receiver of user j at subchannel k to be $g_{ij}^{(k)}$. The noise power on each subchannel is σ^2. The power allocation of user n on all subchannels is defined by vector $\mathbf{p}_n = [p_n^{(1)} p_n^{(2)} \cdots p_n^{(K)}]$. The interference on all subchannels of user n is denoted by vector $\mathbf{I}_n = [I_n^{(1)} I_n^{(2)} \cdots I_n^{(K)}]$, where

$$I_n^{(k)} = \sum_{i=1, i \neq n}^{N} p_i^{(k)} g_{in}^{(k)}. \tag{21.31}$$

Consequently, the SINR, $\eta_n^{(k)}$, of user n at subchannel k can be expressed as

$$\eta_n^{(k)} = \frac{p_n^{(k)} g_{nn}^{(k)}}{\sum_{i=1, i \neq n}^{N} p_i^{(k)} g_{in}^{(k)} + \sigma^2}. \tag{21.32}$$

The data rate at subchannel k of user n, $r_n^{(k)}$, is assumed to be a function of $\eta_n^{(k)}$ and can be expressed as

$$r_n^{(k)} = R(\eta_n^{(k)}),\qquad(21.33)$$

where $R()$ is assumed to be strictly concave and increasing in SINR with $R(0) = 0$. For capacity approaching coding, $r_n^{(k)} = w\log(1 + \eta_n^{(k)})$, where w is the bandwidth of each subchannel.

Let the data rate vector of user n across the K subchannels be

$$\mathbf{r}_n = [r_n^{(1)}, r_n^{(2)}, \ldots, r_n^{(K)}],$$

then the overall data rate is

$$r_n = \sum_{k=1}^{K} r_n^{(k)}.\qquad(21.34)$$

The total transmission power is

$$p_n = \sum_{k=1}^{K} p_n^{(k)}.\qquad(21.35)$$

The energy efficiency of user n is

$$u_n = \frac{r_n}{p_n + p_c}.\qquad(21.36)$$

and the network energy efficiency is

$$u = \sum_{n=1}^{N} u_n,\qquad(21.37)$$

which is a function of $p_n^{(k)}$ for all n and k.

We need to determine power allocation of all users to optimize the overall network energy efficiency subject to the interference scenario. As introduced in the previous section, the solution maximizing sum network energy efficiency is difficult to obtain as the objective function is in general non-concave in $p_n^{(k)}$. More users and subchannels in the system will result in more local maximums, and searching for the globally optimal power allocation would be a daunting task. Even if the globally optimal solution can be found, it is still impractical since the central controller requires complete network knowledge, including interference channel gains. Hence, we consider distributed power optimization in this chapter.

21.3.1 Non-cooperative energy-efficient power optimization game

In this section, we assume no cooperation among users and all users apply the same policy using their own local information. We model the non-cooperative energy-efficient power optimization from a game-theory perspective and then investigate the existence and uniqueness of its equilibrium.

The power control is modeled by a non-cooperative game in game theory [55]. Rooted in economics, game theory has been broadly applied in wireless communications for random access and power control optimization [58, 57].

Consider the power allocation of user n and define the power vectors of other users to be vector

$$\mathbf{p}_{-n} = (\mathbf{p}_1, \mathbf{p}_2, \ldots, \mathbf{p}_{n-1}, \mathbf{p}_{n+1}, \ldots, \mathbf{p}_N). \tag{21.38}$$

Given the power allocation of all other users, $\mathbf{p}_{-n}$, the best response of the power allocation of user n is given by

$$\mathbf{p}_n^o = f_n(\mathbf{p}_{-n}) = \arg\max_{\mathbf{p}_n} u_n(\mathbf{p}_n, \mathbf{p}_{-n}), \tag{21.39}$$

where u_n is given by (21.36) and is a function of both $\mathbf{p}_n$ and $\mathbf{p}_{-n}$. $f_n(\mathbf{p}_{-n})$ is called the best response function of user n. The existence and uniqueness of $\mathbf{p}_n^o$, i.e. the best response, is assured by Equation (19.6) in Section 19.4.2 in the previous chapter, which is also summarized in Appendix A.14 when we prove Theorem 21.3 of this chapter.

Note that non-cooperative power control is not efficient in terms of SE optimization since users tend to act selfishly by increasing their transmission power beyond what is reasonable [58]. Hence, pricing mechanisms are introduced to regulate aggressive power transmissions by individuals so as to produce a more socially beneficial outcome to improve the sum throughput of all users [57]. Different from spectral efficiency optimal power control, energy-efficient power optimization desires a power setting that is greedy in energy efficiency but thrifty of power. Furthermore, Problem (21.39) is equivalent to

$$\begin{aligned} \mathbf{p}_n^o &= \arg\max_{\mathbf{p}_n} \log(u_n(\mathbf{p}_n, \mathbf{p}_{-n})) \\ &= \arg\max_{\mathbf{p}_n} \left(\log(r_n) - \log(p_n + p_c)\right), \end{aligned} \tag{21.40}$$

which implies that energy-efficient power control can be regarded as a variation of the traditional spectral-efficient power control with power pricing [57]. Since this power-conservative expression is socially favorable in interference-limited scenarios, energy-efficient power control is desirable to reduce interference and improve throughput in a non-cooperative setting.

Each user optimizes their power independently. The variation in power allocation of one user impacts that of all others. Equilibrium is the condition of a network in which competing influences are balanced assuming invariant channel conditions. Its properties are important to network performance. Hence, we characterize the equilibrium of non-cooperative energy-efficient power optimization in the following three sections.

21.3.2   Existence of equilibrium

In a non-cooperative game, a set of strategies is said to be at Nash equilibrium if no user can gain individually by unilaterally altering its own strategy. Define the equilibrium as

$$\mathbf{p}^* = (\mathbf{p}_1^*, \mathbf{p}_2^*, \ldots, \mathbf{p}_N^*). \tag{21.41}$$

Nash equilibrium can be strictly described by the following definition.

DEFINITION 21.1 *In an energy-efficient non-cooperative game, an equilibrium is a set of power allocations where no user can unilaterally improve its energy efficiency by choosing a different set of power allocations, i.e.*

$$\mathbf{p}^* = f(\mathbf{p}^*) = (f_1(\mathbf{p}^*_{-1}), f_2(\mathbf{p}^*_{-2}), \dots, f_N(\mathbf{p}^*_{-N})), \tag{21.42}$$

where $f(\mathbf{p})$ is the network response function.

The network response relies on the energy efficiency of all users. The following lemma can be easily proven.

LEMMA 21.2 *$u_n(\mathbf{p}_n, \mathbf{p}_{-n})$ is strictly quasiconcave in $\mathbf{p}_n$.*

Based on Lemma 21.2, the existence of the equilibrium $\mathbf{p}^*$ is given in Theorem 21.3. A necessary and sufficient condition for a set of power allocation to be an equilibrium is summarized in Theorem 21.3 and is proved in Appendix A.14.

THEOREM 21.3 (Existence) *There exists at least one equilibrium $\mathbf{p}^*$ in the non-cooperative energy-efficient power optimization game defined by (21.39). A set of power allocations of all users, $\mathbf{p}^* = (\mathbf{p}_1^*, \mathbf{p}_2^*, \dots, \mathbf{p}_N^*)$, is an equilibrium if and only if it satisfies that for any subchannel i of any user n:*

(i) $\quad R'(0)\gamma_n^{(i)*} \geq \dfrac{\sum_{j \neq i} r_n^{(j)*}}{p_c + \sum_{j \neq i} p_n^{(j)*}}, \quad \dfrac{\partial u_n(\mathbf{p}_n, \mathbf{p}_{-n})}{\partial p_n^{(i)}}\bigg|_{\mathbf{p}_n = \mathbf{p}_n^*} = 0, \text{ i.e. } R'(\gamma_n^{(i)*} p_n^{(i)*})\gamma_n^{(i)*} = u(\mathbf{p}_n^*, \mathbf{p}_{-n}^*);$

(ii) $\quad$ *otherwise,* $p_n^{(i)*} = 0,$

$\quad\quad$ *where* $\gamma_n^{(i)*} = \dfrac{g_{nn}^{(i)}}{\sum_{j=1, j \neq n}^{N} p_j^{(i)*} g_{jn}^{(i)} + \sigma^2}.$

In *(i)* of Theorem 21.3, $\dfrac{\sum_{j \neq i} r_n^{(j)*}}{p_c + \sum_{j \neq i} p_n^{(j)*}}$ is the overall energy efficiency of user n when subchannel i is idle. Subchannel i should be used only if using it can improve the overall energy efficiency of user n and this is determined by how good the state of subchannel i is. If $R'(0)\gamma_n^{(i)*} \geq \dfrac{\sum_{j \neq i} r_n^{(j)*}}{p_c + \sum_{j \neq i} p_n^{(j)*}}$, which is equivalent to $g_{nn}^{(i)} \geq \dfrac{\sum_{j \neq i} r_n^{(j)*}}{p_c + \sum_{j \neq i} p_n^{(j)*}} \dfrac{\sum_{j=1, j \neq n}^{N} p_j^{(i)*} g_{jn}^{(i)} + \sigma^2}{R'(0)}$, the state of subchannel i is good enough for data transmission and the power allocated to it satisfies $R'(\gamma_n^{(i)*} p_n^{(i)*})\gamma_n^{(i)*} = u(\mathbf{p}_n^*, \mathbf{p}_{-n}^*)$. When the power allocations of all users satisfy *(i)* or *(ii)* in Theorem 21.3, the network is in an equilibrium state.

21.3.3 Uniqueness of equilibrium in flat-fading channels

In this section, we discuss the uniqueness of the equilibrium. First, we consider a special case where there is a single subchannel in a network and

$$p_n^o = f_n(\mathbf{p}_{-n}) = \arg\max_{p_n} u_n(p_n, \mathbf{p}_{-n}). \tag{21.43}$$

In this case, the properties of the response functions are stated in Lemma 21.4 and are proved in Appendix A.15.

LEMMA 21.4   *When there is only one subchannel, the power allocation, i.e. the response functions, of all users satisfy:*

- *Concavity: $f_n(\mathbf{p}_{-n})$ is strictly concave in $\mathbf{p}_{-n}$.*
- *Positivity: $f_n(\mathbf{p}_{-n}) > 0$.*
- *Monotonicity: If $\mathbf{p}_{-n} \succ \mathbf{q}_{-n}, f_n(\mathbf{p}_{-n}) > f_n(\mathbf{q}_{-n})$.*
- *Scalability: For all $\alpha > 1$, $\alpha f_n(\mathbf{p}_{-n}) > f_n(\alpha \mathbf{p}_{-n})$.*

where $\succ$ defines vector inequality and each element of the vector satisfies the inequality.

Note that monotonicity indicates that increasing interference results in increasing transmission power while scalability indicates that variation in transmission power is always less than that of interference power. This assures convergence to a unique equilibrium.

The properties in Lemma 21.4 can be extended to networks with multiple subchannels where all subchannels experience the same channel gain, i.e. flat-fading channels. This can be done by defining $f_n(\mathbf{p}_{-n})$ to be the optimal total transmission power on all subchannels.

THEOREM 21.5 (Uniqueness)   *When the channel experiences flat-fading, there exists one and only one equilibrium $\mathbf{p}^*$ in the non-cooperative energy-efficient power optimization game defined by (21.39).*

Proof   It has been shown in [159] that non-cooperative power control with positivity, monotonicity, and scalability has a unique fixed point $\mathbf{p} = f(\mathbf{p})$. Hence, we have the above theorem.   □

21.3.4   Uniqueness of equilibrium in frequency-selective channels

When there are multiple subchannels that experience frequency selective fading, whether there is a unique equilibrium depends on channel conditions.

As an example, consider a network with two users. Let $p_c = 1, w = 1, \sigma^2 = 1, g_{11}^{(1)} = g_{11}^{(2)} = g_{22}^{(1)} = g_{22}^{(2)} = 1, g_{12}^{(1)} = g_{21}^{(1)} = 1e^{-10}, g_{12}^{(2)} = g_{21}^{(2)} = 1e^{10}$. In the following, we show that there are at least two equilibria for this network. We first show that one of the equilibria has the form $\mathbf{p}_1^* = [p_\alpha p_\beta]$ and $\mathbf{p}_2^* = [p_\gamma 0]$, where p_α, p_β, and p_γ are positive. We only need to verify that there exist p_α, p_β, and p_γ that satisfy Theorem 21.3. Suppose $\mathbf{p}_2^* = [p_\gamma 0]$. After some calculation, it is easy to see that $\sigma^2 \gg p_\gamma g_{21}^{(1)}$ and $\eta_1^{(1)} \approx \frac{p_\alpha g_{11}^{(1)}}{\sigma^2}$. Hence, both subchannels of user 1 have approximately the same SINR condition. Thus, in equilibrium, transmission powers on the two subchannels of user 1 are almost the same. Besides, they cannot be zero. Hence, both are positive and satisfy the first condition of Theorem 21.3. Assume $\mathbf{p}_1^* = [p_\alpha p_\beta]$. Now we verify $\mathbf{p}_2^*$. Since user 2 does not transmit on the second subchannel, $\frac{\sum_{j \neq 1} r_n^{(j)*}}{p_c + \sum_{j \neq 1} p_n^{(j)*}} = 0$ and the first condition of Theorem 21.3 should be satisfied. Hence, positive power is allocated on the first subchannel in the equilibrium of user 2. Regarding the second subchannel,

$\gamma_n^{(2)*} = \dfrac{g_{nn}^{(2)}}{p_\beta g_{12}^{(2)} + \sigma^2} - \dfrac{1}{p_\beta 1 e^{10} + 1} \to 0$. Hence, $\dfrac{\sum_{j \neq 2} r_2^{(j)*}}{p_c + \sum_{j \neq 2} p_n^{(j)*}} > R'(0) \gamma_n^{(2)*} \to 0$ and condition 2 of Theorem 21.3 is satisfied. Hence, $\mathbf{p}_2^* = [p_\gamma 0]$. Numerical methods can be used to determine the exact values of p_α, p_β, and p_γ.

Due to the symmetry of network conditions, $\mathbf{p}_1 = [p_\gamma \; 0]$ and $\mathbf{p}_2 = [p_\alpha \; p_\beta]$ must be another equilibrium. Hence, the network has at least two equilibria. When there are more users and subchannels, more equilibria may exist. However, when the interfering channels satisfy a certain condition, there will be a unique equilibrium.

Based on Theorems 17.8 and 17.9, a sufficient condition to assure a unique equilibrium of the non-cooperative energy-efficient power optimization is given below.

THEOREM 21.6 (Uniqueness) *In frequency-selective channels, the non-cooperative energy-efficient power optimization game defined by (21.39) has a unique equilibrium if for any user n, $\|f_n(\mathbf{p}_{-n}) - f_n(\check{\mathbf{p}}_{-n})\| < \|\mathbf{p}_{-n} - \check{\mathbf{p}}_{-n}\|$ for any different $\mathbf{p}_{-n}$ and $\check{\mathbf{p}}_{-n}$ or $\left\| \dfrac{\partial \mathbf{I}_n}{\partial \mathbf{p}_{-n}} \right\| < \dfrac{1}{\sup_{I_n} \left\| \dfrac{\partial f_n}{\partial I_n} \right\|}$.*

Note that the above theorem only gives sufficient conditions of uniqueness that may not be necessary ones. For example, for a single-channel network, due to the strict concavity of $f_n(\mathbf{p}_{-n})$, $\sup_{I_n} \left\| \dfrac{\partial f_n}{\partial I_n} \right\| = \dfrac{\partial f_n}{\partial I_n} \Big|_{I_n = 0}$. However, for all interference channel gains, there is always a unique equilibrium, as shown in Theorem 21.5.

21.3.5 Conservative nature of power control

In this section, we investigate the behavior of energy-efficient power optimization. To facilitate analysis and gave insight, consider a symmetric single-channel network. There are N users, all experiencing the same channel power gain g. All interference channels have the same power gain $\tilde{g}$. To characterize the interference level, we need to use a metric that is independent of transmission powers. Define the network coupling factor

$$\alpha = \frac{\tilde{g}}{g}, \tag{21.44}$$

which characterizes at what level different links interfere with each other. Higher α represents a heavier interfering scenario. According to Theorem 21.5, the equilibrium state of non-cooperative energy-efficient power optimization is unique. Due to the assumption of network symmetry, all users transmit with the same power in the equilibrium. Define the transmission power of all users to be p. Figure 21.6 shows the transmission power in the equilibrium when the network has different couplings and numbers of users. We can see that the equilibrium power decreases with either user number or α, and automatically alleviates network interference.

Note that this is dramatically different from conventional distributed spectral efficiency-maximization power control schemes. With conventional schemes, the transmitting power will be increased to maintain QoS requirements after detecting higher interference power. This increases interference to other users, which will act the same way and increase their transmitting powers, leading to disastrous results if no

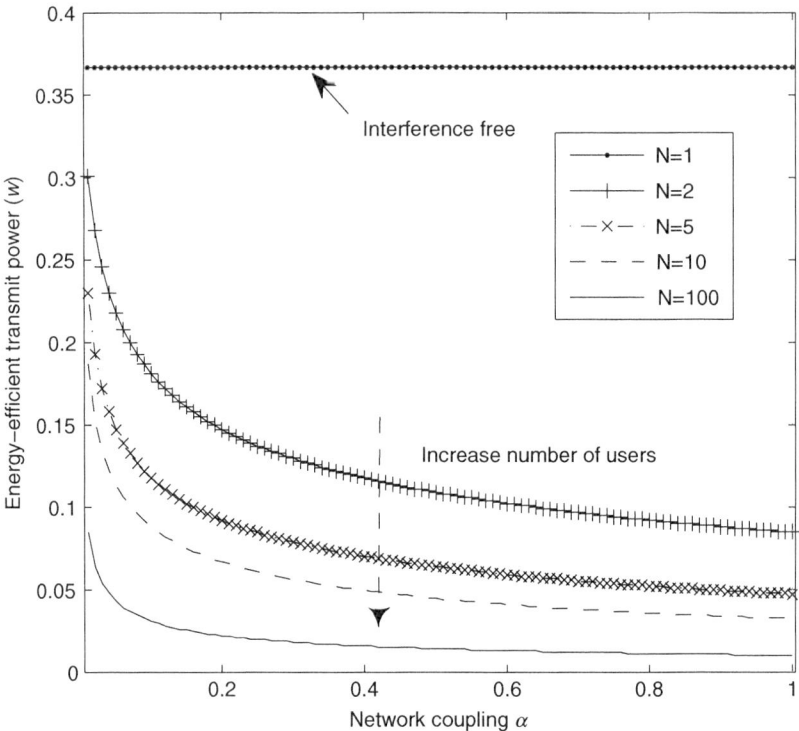

Figure 21.6 Non-cooperative energy-efficient power optimization in the equilibrium($P_c = 1, g = 1, \sigma^2 = 0.01$)

power regulation is applied. So conventional schemes behave in an aggressive way in competing for wireless resources. On the other hand, energy-efficient power control acts conservatively. The increase in interference power will cause the transmitter to choose a lower transmitting power, which reduces the interference to other users. From another perspective, when the transmitter detects lower interference, it will increase its power. Therefore, distributed energy-efficient power control, though selfish, is a harmonious approach from a holistic network standpoint and has the potential to improve network spectral efficiency as well.

21.3.6   Spectral efficiency and energy efficiency improvement

In this section, we present simulation results for an interference-limited uplink OFDMA cellular network. The network consists of seven hexagonal cells and the center cell is surrounded by the other six. The frequency-reuse factor is one. Users are uniformly dropped into each cell at each simulation trial. The system parameters are listed in Table 21.2. The base station schedules subchannels to maximize different network performance metrics. All schedulers and corresponding power control schemes are listed in Table 21.3. The energy-efficient scheduling algorithms in Section 20.2 are used. We

Table 21.2 System parameters

Carrier frequency	1.5 GHz
Number of subchannels	96
Subchannel bandwidth	10 kHz
Target BER	10^{-3}
Thermal noise power, N_o	−141 dBW/MHz
Circuit power, P_C	100 mW
Maximum transmission power	33 dBm
Propagation model	Okumura–Hata model
Shadowing	Log-normal
Fading	Rayleigh flat-fading
Modulation	Uncoded M-QAM

Table 21.3 Scheduling and power control

Legend	Scheduler	Power control
OptEE	Energy-efficient scheduler w/o fairness	Energy-efficient power optimization
PropEE	Energy-efficient scheduler w/ proportional fairness	Energy-efficient power optimization
Trad-Prop	Traditional proportional fair	Fixed power
S-Pwr	Traditional proportional fair	Adaptive power control

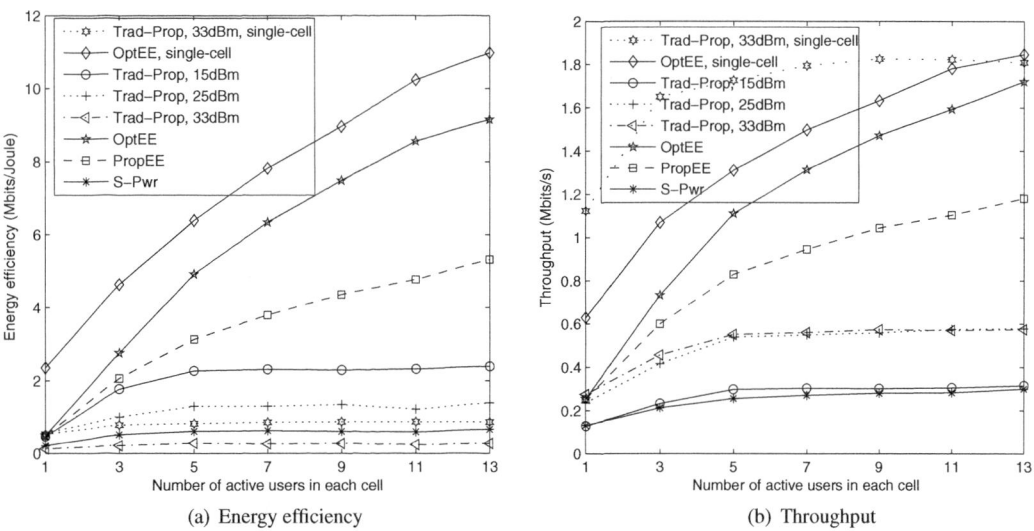

(a) Energy efficiency　　　　　　　(b) Throughput

Figure 21.7 Performance comparison of different schemes

also implement a soft power control scheme proposed in 802.16m [31]. In this scheme, transmission power is controlled based on interference strength and path loss. The parameters in the soft power control scheme are selected to maximize the throughput of cell-edge users, while not hurting the throughput of other users too much.

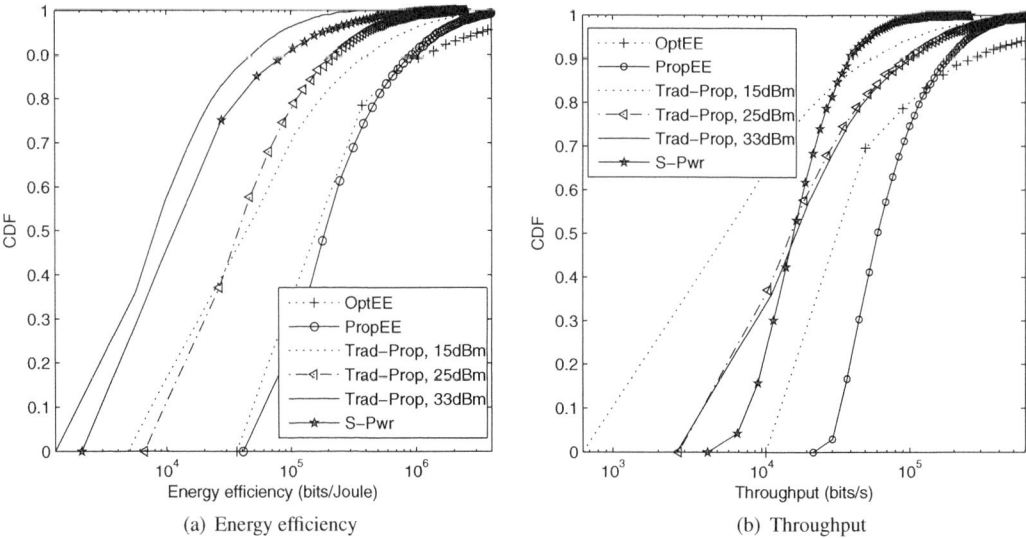

Figure 21.8 Comparison of different schemes

Figure 21.7 compares the average sum network energy efficiency and the corresponding throughput performance respectively. For fixed-power transmission, the transmission powers are shown in the legends. We can see that transmitting with the highest power brings the highest interference and causes significant throughput loss for the traditional scheduler. In contrast, energy-efficient power control effectively reduces network interference and has much less throughput loss. While the results in Section 19.4 show that energy efficiency and throughput efficiency do not necessarily agree for an interference-free single-cell scenario, the situation is different for a multi-cell interference-limited network. Here energy-efficient schemes optimize both throughput and energy utilization and exhibit an improved spectral efficiency tradeoff. Figure 21.8 further shows the *cumulative distribution functions* (CDFs) of energy efficiency and throughput when there are nine users in the network. We can see that the energy efficiency schemes not only improve the sum energy efficiency and throughput, but also uniformly improve the performance of all users in the cell.

22 Energy-efficient cellular network design

In this chapter, we discuss energy-efficient designs for wireless cellular networks, which is a major concern for operators to reduce operational expenditure and to reduce environmental impact. In cellular networks, base stations dominate energy consumption and drain approximately 60–80% of overall network energy consumption. We will discuss the energy-efficient design of cellular base stations (BSs). There are several current technologies for environmentally friendly and energy-efficient cellular networks. A few examples are:

(i)   Link level: energy-efficient transmission, dynamic mode switching, low-power circuit design, etc.
(ii) Multi-user level: energy-efficient scheduling and resource management.
(iii) Network level: energy-efficient topological approaches from deployment to operation and energy-efficient mobility management.

In previous chapters, we have discussed both link and multi-user level energy-efficient designs. In this chapter, the focus is on network-level solutions. We will first introduce some fundamental tradeoffs in using wireless resources. Then we investigate energy-efficient network deployment and operation technologies.

22.1   Fundamental tradeoffs in network resource utilization

Both spectrum and energy efficiency emphasize communication quality in the sense that successful data transmission will help improve both metrics. From this perspective, energy efficiency also requires spectral efficiency improvement and vice versa. On the other hand, energy and spectrum are two independent fundamental resources and both are needed to meet quality of service (QoS) requirements. Tradeoff is always there and depends on network choice. Any network design can always choose to optimize the utilization of one resource over the other. If one resource is redundant and the other is not, the network will need to optimize network behavior towards the non-redundant resource. If both are redundant, then the system can be operated with the best QoS experience. If both are stringent, the network has to choose between them. Energy and spectrum efficiency are equally important and there is no clear advantage of one metric over the other. Which metric is more desired will depend on network needs. To facilitate network design, in this section we will give a thorough investigation of the fundamental tradeoffs in wireless resource utilization from different aspects, especially the relationship between energy and spectrum efficiency in different network environments.

22.1.1   Spectral and energy efficiency in single-user systems

In this section, we investigate the tradeoff between spectral and energy efficiency in a single-user system.

When there is only one user in the network, the spectral efficiency (*SE*) of the system is

$$\eta_{SE} = \log_2\left(1 + \frac{pg}{WN_0}\right). \tag{22.1}$$

Assuming first an ideal transmitter with zero circuit power, the energy efficiency (*EE*) of the system is

$$\eta_{EE} = \frac{W\log_2\left(1 + \frac{pg}{WN_0}\right)}{p}. \tag{22.2}$$

Therefore, spectrum and energy efficiency can be characterized by the following equation:

$$\eta_{EE} = \frac{\eta_{SE}g}{(2^{\eta_{SE}} - 1)N_0}. \tag{22.3}$$

This relationship is illustrated in Figure 22.1 and η_{EE} is strictly decreasing in η_{SE}.

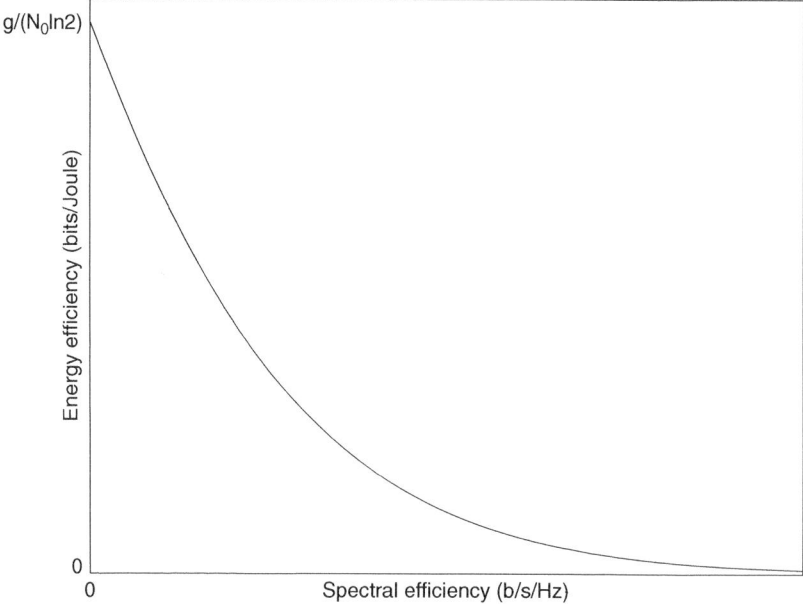

Figure 22.1 Tradeoff between energy efficiency and spectral efficiency with zero circuit power

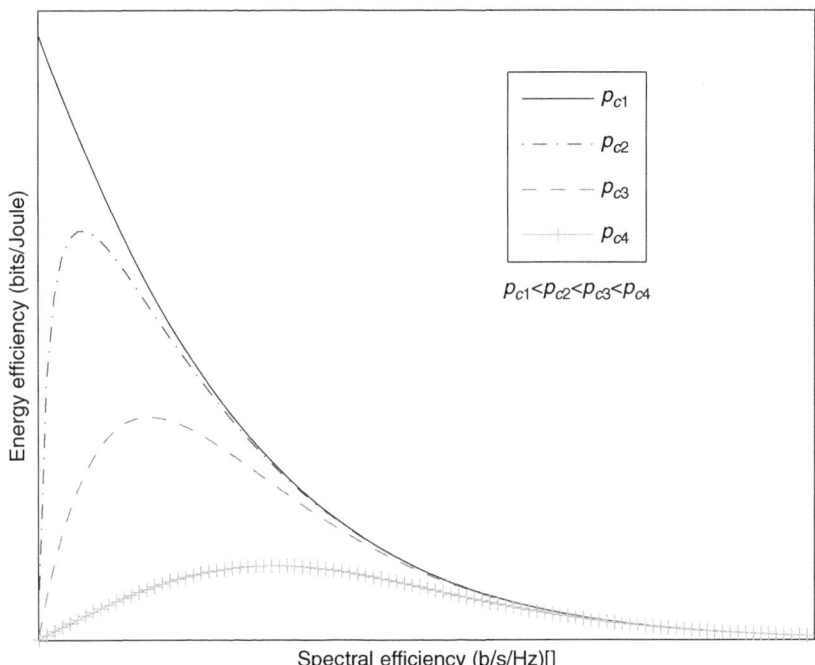

Figure 22.2 Tradeoff between energy efficiency and spectral inefficiency in practice

In practice, circuit power is not zero and the energy efficiency is

$$\eta_{EE} = \frac{W \log_2 \left(1 + \frac{pg}{WN_0}\right)}{p + p_c}. \tag{22.4}$$

Therefore, we have the following tradeoff relationship:

$$\eta_{EE} = \frac{\eta_{SE}}{(2^{\eta_{SE}} - 1)N_0/g + p_c/W}, \tag{22.5}$$

which is illustrated in Figure 22.2, where $P_{c1} = 0$.

22.1.2 Spectral and energy efficiency in multi-user systems with orthogonal selective channels

Assume there are K orthogonal subchannels, in the network each with bandwidth W_i, $i = 1, ..., K$. All the K subchannels are used and the power allocated on each subchannel is positive. We do not consider subchannels not used or allocated zero power.

The K active subchannels can be used by one transmitter, e.g. an orthogonal frequency division multiplexing (OFDM) or multiple-input multiple-output (MIMO) transmitter, or multiple transmitters, e.g. orthogonal frequency division multiple access (OFDMA), multiple-user MIMO (MU-MIMO), coordinated multiple-point transmission (CoMP), where each transmitter is assigned a subset of the K subchannels. For example, in a cellular network with two cells, one using a 10-subcarrier OFDMA to serve three downlink users and one 2×2 MU-MIMO serving two uplink users. They operate on different frequency bands and do not interfere with each other. So there are $10 + 2 = 12$ subchannels. Another example is a two-user time division multiple access (TDMA) network, where one user sends packets on even time slots and the other odd slots. In total, they use 1000 time slots, in which no packets are transmitted in 20 slots because of deep fading. In this case, there are 980 subchannels.

In the following, we investigate the tradeoff between energy and spectral efficiency. Network spectral efficiency is

$$\eta_{SE} = \sum_{i=1}^{K} \log_2 \left(1 + \frac{p_i g_i}{W_i N_0} \right), \tag{22.6}$$

where i indicates the corresponding subchannel properties. The overall network transmission power consumption is

$$p_T = \sum_{i=1}^{K} p_i. \tag{22.7}$$

Let $\psi(p_T) = [\psi_1(p_T), \psi_2(p_T), ..., \psi_K(p_T)$ denote the mapping from the total network transmission power to the transmission power on each subchannel. For example, for a single OFDM transmitter with the total transmission power limit p_T, the optimal power allocation on each subchannel that achieves the channel capacity is the water-filling power allocation given by

$$p_i = \psi_i(p_T) = \left[\gamma - \frac{N_o W_i}{g_i} \right], \tag{22.8}$$

where γ is given by (22.7). Note that the subchannels with zero power allocation have been excluded. It can be easily proven that $\psi(p_T)$ is concave in p_T. In general, we assume $\psi(p_T)$ is concave in p_T.

Further, network spectral efficiency, η_{SE}, is strictly concave in $\psi(p_T)$, which is concave in P_T. η_{SE} is strictly increasing in each argument. According to the vector composition rule that preserves concavity, we obtain the following theorem readily summarizing the overall network power–rate relationship:

THEOREM 22.1   *If the power allocation policy $\psi(p_T)$ is concave, network spectral efficiency $\eta_{SE}(P_T)$ is a strictly concave function in total network transmission power, P_T. Besides, $\eta_{SE}(0) = 0$ and $\eta_{SE}(P_T)$ is monotonically increasing in P_T.*

Reversing the above theorem, we have the following rate–power relationship:

THEOREM 22.2   *If the power allocation policy $\psi(p_T)$ is concave, overall network power consumption $P_T(\eta_{SE})$ is a strictly convex function of network spectral efficiency, η_{SE}. Besides, $P_T(0) = 0$ and $P_T(\eta_{SE})$ is monotonically increasing in η_{SE}.*

Network energy and spectral efficiency can be characterized by the following equation:

$$\eta_{EE} = \frac{W \cdot \eta_{SE}}{P_T(\eta_{SE})}, \tag{22.9}$$

which is decreasing and strictly convex in η_{SE}. An example is already given in Figure 22.1. The tightest energy efficiency upper bound is

$$\widehat{\eta}_{EE} = \frac{W}{P'_T(0)}. \tag{22.10}$$

We have the following theorem:

THEOREM 22.3   *If the power allocation policy $\psi(p_T)$ is concave, network energy efficiency, without considering circuit power, is a strictly convex function of spectral efficiency. Besides, energy efficiency decreases with spectral efficiency and is tightly upper bounded by $\frac{W}{P'_T(0)}$.*

When circuit power is considered, network energy and spectral efficiency is related through

$$\eta_{EE} = \frac{W \cdot \eta_{SE}}{P_T(\eta_{SE}) + P_c}, \tag{22.11}$$

where P_c is the cumulative circuit power of all transmitters. It can be easily proven that η_{EE} is strictly quasiconcave in η_{SE} and Theorem 19.1 can be used to find optimal network spectral efficiency for highest network energy efficiency. Figure 22.2 illustrates several examples.

22.1.3   Spectral and energy efficiency in multi-user systems with interference channels

In this section, we consider a multi-user system where different users interfere with each other, as discussed in Section 21.3. Non-cooperative power control is used. To facilitate analysis and obtain insights, consider a symmetric single-channel network. There are N users, all experiencing the same channel power gain g. All interference channels have the same power gain $\widetilde{g}$. To characterize the interference level, we need to use a metric that is independent of transmission powers. Define the network coupling factor

$$\alpha = \frac{\widetilde{g}}{g}, \tag{22.12}$$

which characterizes at what level different links interfere with each other. Higher α represents a heavier interfering scenario. According to Theorem 21.5, the equilibrium state of non-cooperative energy-efficient power optimization is unique. Due to the assumption of network symmetry, all users transmit with the same power in the equilibrium. Define the transmission power of all users to be p.

The overall network energy efficiency will be

$$u(p) = \sum_{n=1}^{N} \frac{w \log\left(1 + \frac{pg}{\sum_{i,i\neq n} p\widetilde{g} + \sigma^2}\right)}{p + p_c}$$

$$= \frac{Nw \log\left(1 + \frac{p}{(N-1)\alpha p + \frac{\sigma^2}{g}}\right)}{p + p_c}, \tag{22.13}$$

and the network spectral efficiency will be

$$r(p) = N \log \left(1 + \frac{p}{(N-1)\alpha p + \frac{\sigma^2}{g}} \right). \tag{22.14}$$

With non-cooperative spectral-efficient power control, every user allocates power to selfishly maximize its spectral efficiency. Without power limit, the transmission power tends to infinity in the equilibrium. Besides, we can see that $r(p)$ is strictly increasing in p. Hence, the maximum network spectral efficiency is obtained in the equilibrium and the upperbound is

$$r_{SE} = \lim_{p \to \infty} r(p) = N \log \left(1 + \frac{1}{(N-1)\alpha} \right) \tag{22.15}$$

with the corresponding energy efficiency $u_{SE} = \lim_{p \to \infty} u(p) = 0$, which is completely energy inefficient, and non-cooperative spectral efficiency optimal power control is not desired for energy efficiency.

With non-cooperative energy-efficient power optimization, the network energy efficiency at the equilibrium is $u_{EE} = u(p^*)$ with the corresponding spectral efficiency $r_{EE} = r(p^*)$. Hence, the spectral efficiency penalty of energy-efficient power optimization is

$$r_{tr} = r_{SE} - r_{EE} = N \log \left(1 + \frac{1}{(N-1)\alpha} \right) - r(p^*). \tag{22.16}$$

In an interference-free scenario, i.e. $N = 1$ or $\alpha = 0$, the penalty is infinite. Otherwise, whenever interference exists, it is bounded.

To further understand the tradeoff, Figure 22.3 illustrates a case when two users transmit with the same power and interfere with each other. Curves with markers indicate the relationship between transmission power and spectral efficiency when the network has different couplings, while those without markers indicate the corresponding energy efficiency. When $\alpha = 0$, arbitrary spectral efficiency can be obtained by choosing enough transmission power. When $\alpha > 0$, regions beyond the spectral efficiency upperbound are not achievable. Furthermore, energy efficiency is much more sensitive to power selection than spectral efficiency. For example, when $\alpha = 0.1$, the transmission power is chosen to be -3 dBw for energy-efficient power optimization. The spectral efficiency achieved is 4.2 b/s/Hz, while the energy efficiency is 2.8 bits/Joule. If we further increase the transmission power, the energy efficiency decreases very fast while the spectral efficiency only improves slightly. Hence, in interference-limited scenarios, increasing transmission power beyond the optimal power for energy efficiency has little spectral efficiency improvement but significantly hurts energy efficiency. Furthermore, power optimization to achieve the highest energy efficiency will also have reduced the spectral efficiency penalty with the increase in α. These indicate that energy-efficient communications have significant advantages, especially in heavy interference environments, e.g. cell-edge communications.

In this section, we have made some simple assumptions to gain intuitive understanding about the tradeoff between energy and spectral efficiency. In practice, different users will experience different levels of interference from each other. In addition, some users may experience interference because they share the same channels, and

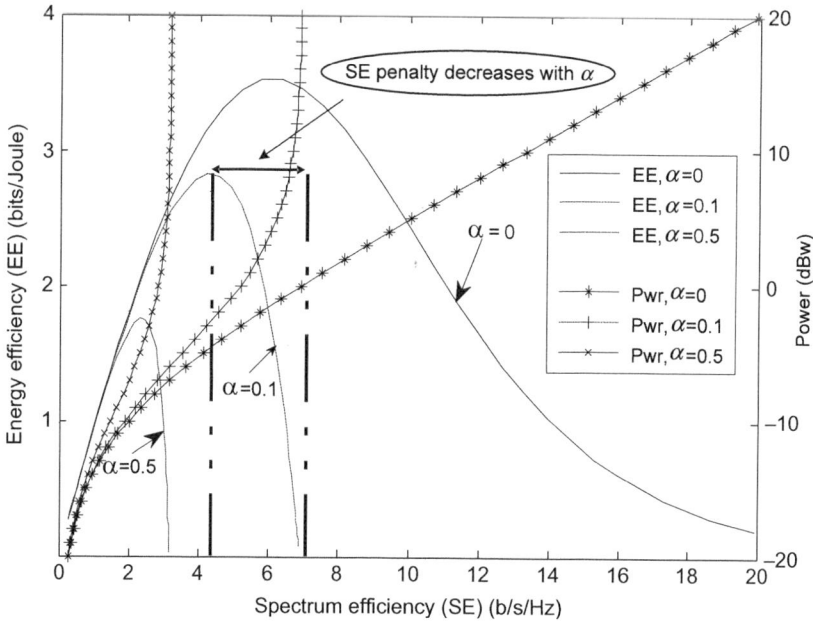

Figure 22.3 Tradeoff of (EE) and (SE) with different interfering scenarios: left Y axis is for curves without markers and indicates the achieved energy efficiency while the spectral efficiency is as in X axis; right Y axis is for curves with markers and indicates the required transmission power to achieve the spectral efficiency given in X axis. ($p_c = 1, g = 1, \sigma^2 = 0.01, N = 2$).

others experience no interference as they use orthogonal channels. The tradeoff between energy and spectral efficiency in general will be the statistical average of all effects that have been discussed in these three subsections.

22.2   Energy-efficient homogeneous network deployment

Deployment cost consists of both capital and operational expenditure. Capital expenditure refers to cost in building infrastructure, e.g. network equipment, site construction, and so on, while operational expenditure is about energy consumption, maintenance, site lease, etc. Both types of costs need to be estimated in network planning. Energy efficiency is the dominant factor in network operation and in this section we discuss how to deploy cellular networks in an energy-efficient way.

In the downlink of a coverage-limited cellular network, the received power at the distance, d, from the BS is

$$p_r = \frac{p}{d^\alpha},$$ (22.17)

where p is the transmission power and α the path loss exponent. The interference is negligible and the signal-to-noise ration (SNR) is

$$\Gamma = \frac{p}{d^\alpha N_0 W},$$ (22.18)

where N_0 is the noise spectral density. Suppose there is always a cell-edge user, with $d = R_{cell}$, that desires a minimum data rate r_0, which is given by

$$r_0 = W \log_2 \left(1 + \frac{\Gamma}{\theta} \right). \tag{22.19}$$

Solving the above equation for transmission power, we have

$$p = \left(2^{\frac{r_0}{W}} - 1 \right) \theta R_{cell}^\alpha N_0 W. \tag{22.20}$$

From (22.20), the required transmission power for a certain coverage increases exponentially with the cell coverage R_{cell}. Now let us take a look at the total network energy consumption in a certain area. The number of BS in a service area is given by

$$N_{BS} = \frac{A}{\pi R_{cell}^2}, \tag{22.21}$$

where A is the area size.

The total required transmission power in a cellular network to meet the coverage requirement can be written as a function of N_{BS} as follows

$$P_{total} = \left(2^{\frac{r_0}{W}} - 1 \right) \theta N_0 W \left(\frac{A}{\pi} \right)^{\frac{\alpha}{2}} N_{BS}^{1 - \frac{\alpha}{2}}. \tag{22.22}$$

We can see that the total transmission power of a mobile radio network with a fixed service area decreases with the number of BS in the network for any $\alpha > 2$. In practice, α is always bigger than 2. Therefore, if we consider only transmission power, the cell sizes should be as small as possible. Hence, a high-density deployment strategy with many micro base stations is the most energy efficient. The estimated network energy efficiency can be defined as

$$\eta_{EE} = \frac{r_0 N_{BS}}{\left(2^{\frac{r_0}{W}} - 1 \right) \theta N_0 W \left(\frac{A}{\pi} \right)^{\frac{\alpha}{2}} N_{BS}^{1 - \frac{\alpha}{2}}} = \frac{r_0 N_{BS}^{\frac{\alpha}{2}}}{\left(2^{\frac{r_0}{W}} - 1 \right) \theta N_0 W \left(\frac{A}{\pi} \right)^{\frac{\alpha}{2}}}, \tag{22.23}$$

which grows exponentially with the number of BSs deployed.

The above conclusion is true only for transmission power minimization. In practice, in addition to communications energy consumption, each BS also consumes computation energy for signal processing, site cooling, backhaul, and so on [176]. Then total network power consumption would be modeled by

$$P_{net} = \frac{P_{total}}{\zeta} + N_{BS} p_c + p_o = \left(2^{\frac{r_0}{W}} - 1 \right) \theta N_0 W \left(\frac{A}{\pi} \right)^{\frac{\alpha}{2}} N_{BS}^{1 - \frac{\alpha}{2}} / \zeta + N_{BS} p_c + p_o. \tag{22.24}$$

Here P_{net} is no longer always decreasing in N_{BS}. A high-density deployment strategy may not be the most energy efficient.

It can be easily proved that (22.24) is strictly convex and there exists a unique finite optimal N_{BS} that minimizes (22.24), as shown in Figure 22.4. The optimal N_{BS} and the corresponding minimum network power consumption can be found by setting the first-order derivative to be zero and the optimal N_{BS} is

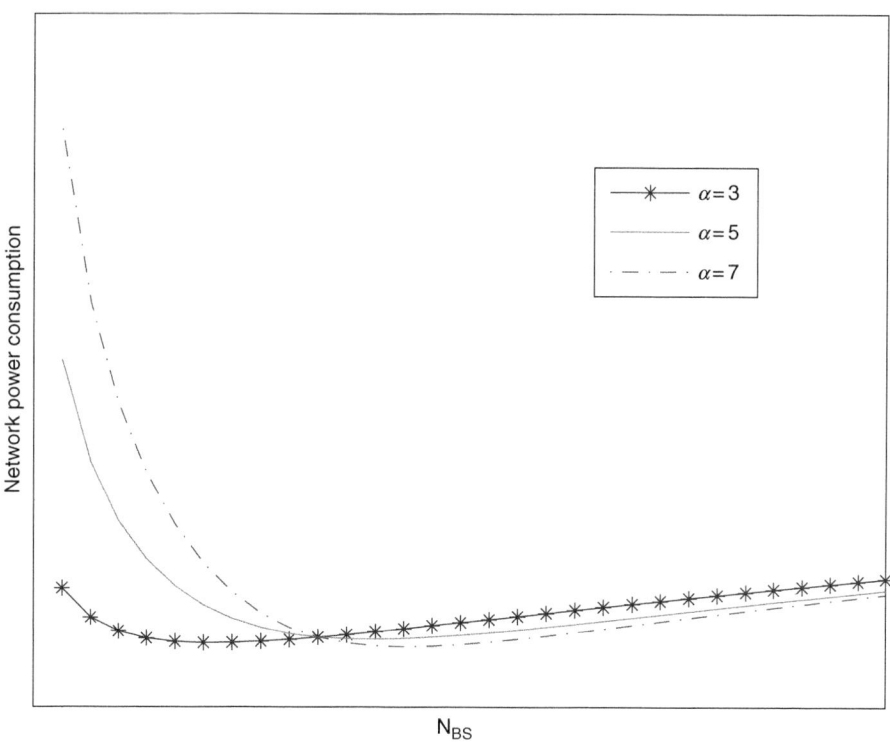

Figure 22.4 Network power consumption and BS density relationship with A cell-edge rate requirement

$$\frac{N_{BS}^*}{A} = \frac{1}{\pi} \left(\frac{p_\beta}{p_c} \right)^{\frac{2}{\alpha}} \left(\frac{\alpha}{2} - 1 \right)^{\frac{2}{\alpha}},$$ (22.25)

where $p_\beta = \left(2^{\frac{r_0}{W}} - 1 \right) \theta N_0 W / \zeta$, the desired receiving power of a cell-edge user if the power amplifier efficiency is one. Correspondingly, the minimum power consumption to serve the area is

$$\frac{p_{net}^*}{A} = \frac{1}{\pi} p_\beta^{\frac{2}{\alpha}} p_c^{1 - \frac{2}{\alpha}} \left[\left(\frac{\alpha}{2} - 1 \right)^{\frac{2}{\alpha} - 1} \frac{\alpha}{2} \right].$$ (22.26)

And, the network energy efficiency is

$$\eta_{EE} = \frac{r_0 N_{BS}^*}{p_{net}^*} = \frac{r_0 (1 - \frac{2}{\alpha})}{p_c}.$$ (22.27)

We can see the network energy efficiency is only affected by the BS circuit power, rate requirement, and the wireless channel property.

The impact of the path loss exponent on the optimal BS density (22.25) and minimum power consumption (22.26) is not straightforward and depends on p_β and p_c. Figure 22.5 illustrates the optimal BS density with respect to different α values when $\frac{p_\beta}{p_c}$ has different values. When p_β equals p_c, N_{BS}^* first increases and then decreases. However, p_β is usually much smaller than p_c in real cellular networks and we can

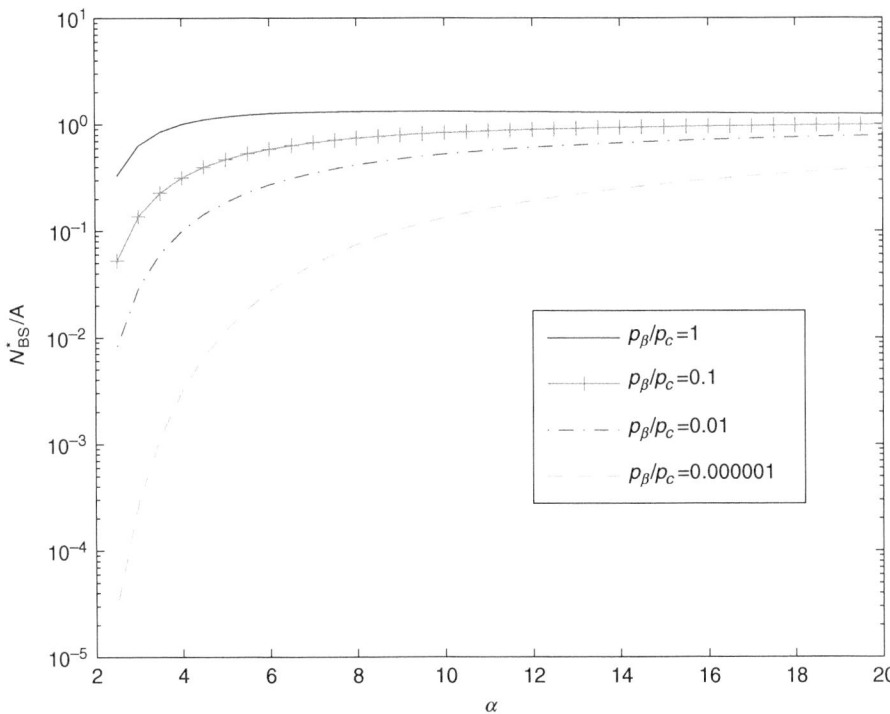

Figure 22.5 Optimal BS density and path loss exponent

conclude that, in practice, the optimal BS density increases with the path loss exponent. The proof can be easily done. Similarly, we can also conclude that, in practice, the minimum power consumption to serve a unit area increases with the path loss exponent.

Given (22.25), (22.27), and the above discussions, the following conclusions can be derived immediately.

PROPOSITION 22.4   *The optimal BS density increases with the data rate requirement r_0, coding gap θ, and noise density N_0, and decreases with system bandwidth W, power amplifier efficiency ζ, and circuit power p_c. In practice, it also increases with the path loss exponent.*

PROPOSITION 22.5   *The minimum power consumption to serve a unit area increases with the data rate requirement r_0, coding gap θ, noise density N_0, or circuit power p_c, and decreases with system bandwidth W or power amplifier efficiency ζ. In practice, it also increases with the path loss exponent.*

22.3   Energy-efficient heterogeneous network deployment

A macro cell usually covers a large area and therefore is not efficient in providing broadband services. According to the discussion in Section 22.2, one way to improve energy efficiency is to decrease the coverage of cells and thus reduce the signal propagation loss

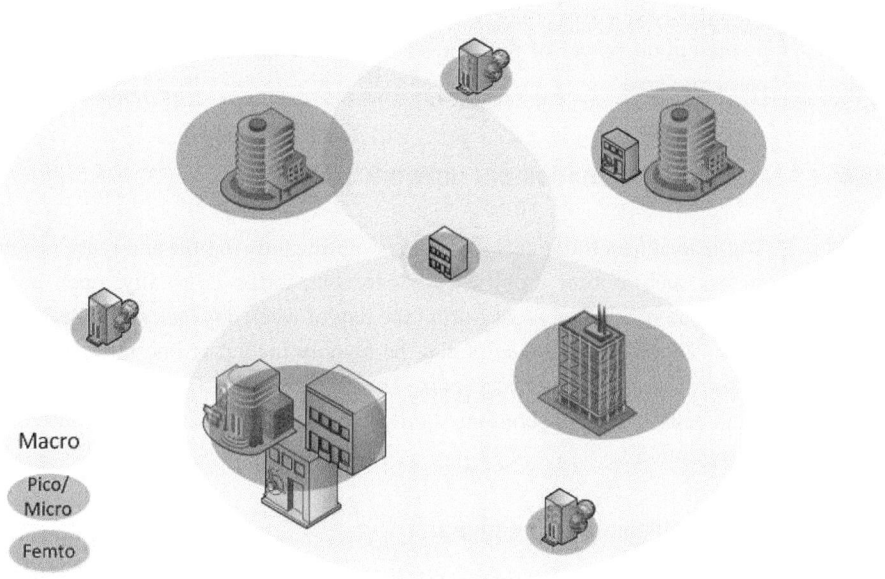

Macro

Pico/
Micro

Femto

Figure 22.6 An example of heterogeneous networks

to reduce transmission power. Heterogeneous cellular deployment with small cells, e.g. micro, pico, or femto cells, under umbrella macro cells can therefore be used to improve network energy efficiency. An example of an heterogeneous network is illustrated in Figure 22.6. The small cells are cells served with lower-power BSs. A micro or pico cell usually covers a range of one or several hundred meters and can be used to cover smaller hotspot areas with dense traffic, such as shopping malls, hotels, airports, train stations, etc. On the other hand, a femto cell can be used to cover a much smaller area like an individual house and the coverage can be only a few meters or ten meters. Small cells are much more power efficient than macro cells. For example, a typical femto cell may consume only 5W in total compared to several thousand watts that would be needed to support a macro cell. Deploying too many small cells may be expensive because of the growing expenditure on site rental, base station equipment, and maintenance. As discussed in Section 22.2, the computation energy for signal processing, site cooling, backhaul, and so on increases linearly with the number of sites. Therefore, there is a tradeoff in determining the deployment density of heterogeneous networks, and a similar strategy as in Section 22.2 can be used to find the optimal solution. However, the BS density in heterogeneous networks should vary in different areas depending on the channel condition, system bandwidth, and hetnet types. To be more specific, the optimal BS density is still given by

$$\frac{N^*_{BS}}{A} = \frac{1}{\pi} \left(\frac{p_\beta}{p_c} \right)^{\frac{2}{\alpha}} \left(\frac{\alpha}{2} - 1 \right)^{\frac{2}{\alpha}}, \tag{22.28}$$

where the area A should be the local area where the same type of BSs are deployed, W is the system bandwidth for the HetNet BS, N_0 should consist of both thermal noise and interference from other types of adjacent HotNet BSs and the umbrella macro cells, p_c the circuit power of the HetNet BS, and α the channel path loss exponent for this particular area.

22.4   Energy-efficient cellular network operation

Traffic load in cellular networks varies significantly in time and space because of various factors such as user mobility. For example, traffic is usually much heavier in office areas than residential areas during the day of week days and the other way around after work. Therefore, many cells may be heavily loaded at one time and carry almost no traffic at another, and vice versa. The traffic fluctuations would be much higher for small cells. The BS operations should be adapted accordingly to reduce network energy consumption.

22.4.1   Energy-efficient cell breathing

Traditional access networks are planned based on the peak hour traffic, and static cell sizes are usually used. With static cell size deployment, the network does not adapt to the fluctuating traffic loads and always works with peak power consumption to achieve the highest performance. To save energy, cell size can therefore be adjusted depending on traffic in the network. This technique has been deployed in code division multiple access (CDMA) networks, in which cells with heavy loads will reduce their cell sizes by decreasing the transmission power through power control. Some users at cell edges will be handed over to adjacent cells with lower traffic and the traffic load is therefore shared across adjacent cells. Besides, handing over cell-edge users from heavy-traffic cells to light-traffic ones also saves transmission power. This is because these users will be allocated more spectrum resources in the light-traffic cells and therefore energy efficiency can be improved, according to Proposition 20.1. The technique is frequently called "cell-breathing." To achieve the most energy saving, network-level power management will be needed so that multiple cells can coordinate to decide on traffic handovers.

22.4.2   Energy-efficient BS sleeping

Running a BS consumes a considerable amount of energy and BS sleeping will save a significant amount of this energy. Since the traffic load varies according to a certain variation pattern over time and space, BSs can switch into the sleeping mode when its traffic load is low.

For example, a simple BS sleeping algorithm can be based on a threshold N, the number of active users in the cell. The algorithm is illustrated in Figure 22.7. Each BS remains in the sleeping mode until the number of new user requests accumulates to N. When it is in the active state, it continues serving the users even when there are less than N users and will not go to sleep unless there is no user in its coverage. Clearly, there is a tradeoff between BS energy saving and user performance. The larger N is, the more energy can be saved, and the worse the network performs in terms of

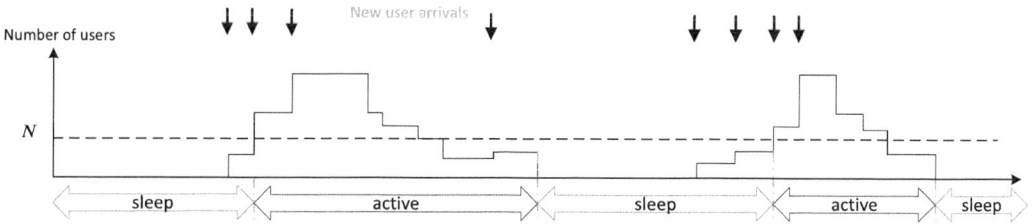

Figure 22.7 BS sleeping control based on the number of users in the area

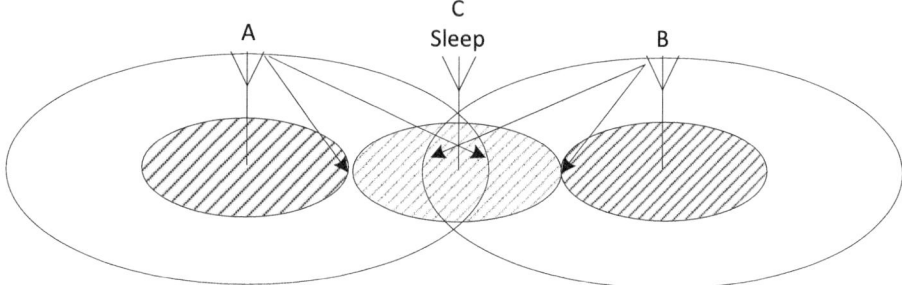

Figure 22.8 Coverage preserved by adjacent cells

network throughput, user outage, and so on. N should be properly chosen such that the maximum energy saving can be achieved without sacrificing network performance too much.

22.4.3   Cell size adaptation techniques

When some BSs are in sleeping mode or reducing cell size, their original coverage can be preserved by the remaining adjacent active cells. There are several techniques to enlarge the coverage of the adjacent active cells such that they cover the cells in sleep mode. For example, increasing the transmission power or adjusting the antenna tilt can expand their coverage, as shown in Figure 22.8. Furthermore, several adjacent BSs can cooperate to cover the area of cells in sleep mode using the so-called CoMP transmit/receive technology. With CoMP, the reception power can be increased because signals from several cells can be combined together for signal detection, therefore resulting in the expansion of cell coverage.

Furthermore, BS sleeping should be coupled with energy-efficient handover. When a BS enters the sleeping mode, the users originally in its serving area need to be handed over to a new BS. To select the serving BS, the traffic loads on different cells should be jointly considered so that the loads can be balanced. Besides the received signal quality also needs to be evaluated. In some cases, it might be beneficial to hand over as many users as possible to only one cell as more BSs would have better opportunities to sleep.

22.4.4   Other energy-efficient designs

There are many other techniques that should be redesigned to improve cellular network energy efficiency.

All device components in the network should be made more energy efficient. For example, in the BS, radio frequency (RF)-transmitters and antennas are located at different places and are connected using long coaxial cables, which adds several dBs loss in power transmission. Therefore, low attenuation RF cables should be used and the RF amplifier should be placed as close to the antenna as possible. In the BS, the power amplifier consumes almost 50% of the energy consumption, out of which 80–90% is wasted as heat [34]. Air conditioners are needed to cool down the heat, consuming even more energy. Most power amplifiers have efficiency between 5 and 20%. Existing BSs are very inefficient because of the need for amplification linearity and high peak-to-average powered ratios. The modulation and coding schemes in 3/4G standards are characterized by signal envelopes that vary significantly with PAPR larger than 10 dB. To achieve high linearity, PAs have to operate far below their saturation points, resulting in very low amplification efficiency. New PA architectures need to be developed to increase the efficiency to over 50% or higher.

While numerous efforts have been focused on energy-efficient communication technologies, new energy sources should also be developed to push forward the development of energy-efficient mobile networks. These renewable energy resources can be used as power sources of BSs, especially in remote rural areas. In these areas, the availability of electrical grids is poor and the electricity for BS operations is supplied by diesel powered generators, which not only are expensive to maintain, but also generate a significant amount of CO_2 emissions. In these cases, we can use renewable energy resources like solar, wind, and sustainable biofuels energy systems, which are much more viable at reducing overall network expenditure. For example, a one square meter solar panel can produce 10% of the energy needed for the operation of a 3G macro cell BS [17]. Many telecom companies worldwide have started developing solar-driven BSs.

23 Implementation in practice

In the past several years, many research groups have actively implemented energy-efficient communication techniques.

With energy-efficient communications, the system should assemble components that present a controllable tradeoff between performance and power consumption. Based on this flexibility, the system can adapt to dynamic environments and traffic conditions to avoid traditional worst-case communications and globally reduce power consumption. We have identified a set of parameters that influence system-level energy efficiency and performance, i.e. modulation and coding scheme, transmit power, access policy, and so on. Using these parameters as control knobs, energy management policies are systematically derived at design-time and calibrated at run-time in [164] to adapt the system configuration to the actual user quality of service (QoS) requirements and environment parameters. By exploiting these knobs of actual radio frequency (RF) components over a modified IEEE 802.11 medium access control (MAC), system lifetime is shown to be increased by a factor of 2 to 5 over conventional techniques. Duty-cycling techniques and careful choice of various design parameters has led to the development of a very energy-efficient sensor network radio [166].

Besides MAC layer implementations, there are also several efforts for physical (PHY) layer realizations. In [39], an orthogonal frequency division multiplexing (OFDM) transmitter design, shown in Figure 23.1, that effectively presents these characteristics is presented as well as its control strategy. The system consists of the following three stages: an *inphase and quadrature* (I/Q) direct-upconversion mixer, a driver amplifier, and an external power amplifier. Both the driver amplifier and power amplifier are made flexible in terms of controlling output power, linearity, and DC power consumption. To optimally calibrate system parameters, a controller is designed to translate the high-level transmit power and linearity requirement in optimal

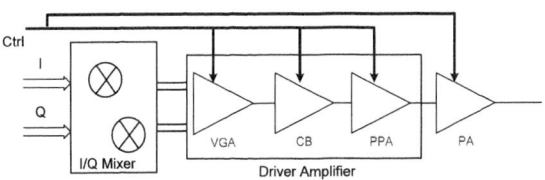

Figure 23.1 Block diagram of the analog transmitter

circuit settings. The system-level energy management technique postulated in [207] is applied on the transmitter architecture and its control subsystem in [39]. Based on measurements carried out on the physical realization of the transmitter, the benefit of the aforementioned system-level energy management technique has been re-evaluated. It is shown that the transmitter presents an energy-scalability up to 30%, which translates on average to a system-level energy efficiency improvement of up to 40%.

Rather than improvement to existing systems, work is also being undertaken to design new system architectures to comprehensively enhance network energy efficiency, such as the PicoRadio project at Berkeley [104, 70] and the μAMPs project at MIT [41]. The PicoRadio project at Berkeley [104] aims to design an architecture that will provide flexibility for low-energy multi-hop communications and the architecture is implemented in ASIC, FPGA, and ARM platforms. We have known that it is more energy-efficient to send a bit using several short intermediate hops than using one longer hop and the most energy-efficient routing policy is using an infinite number of hops, each over the smallest possible distance. Besides, appropriate selection of intermediate hops can also improve link quality and thus increase the probability of transmission success to save retransmission energy. Figure 23.2, from [70], shows the advantages of using intermediate nodes for packet forwarding in PicoRadio. The percent of packet success is the ratio of times a packet reached its destination. The deep fades in the dashed line indicates nulls in the radio signal. The fade goes away, as shown by the solid line, when a third node is added in a more advantageous location and can forward the packet to the destination. Obviously, the number of intermediate hops is limited by how many nodes lie between nodes, but there are more factors to take into account, e.g. energy dissipation in transceiver processing and retransmission. In [104], the optimal number of hops is determined by finding the best energy tradeoff between transmission, retransmission,

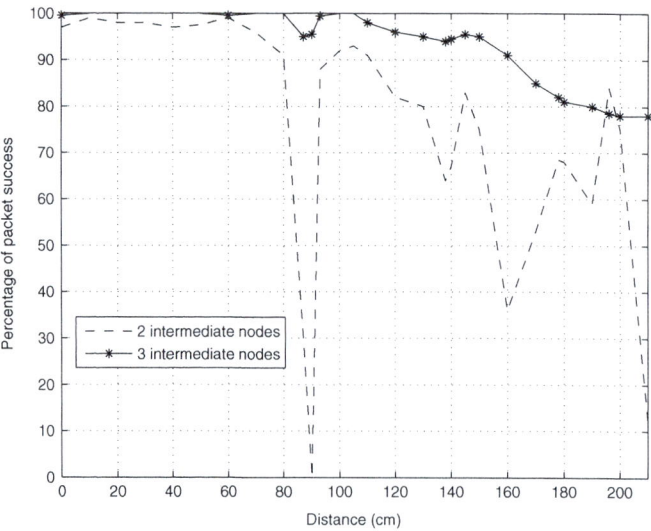

Figure 23.2 Effect of forwarding

Table 23.1 Energy consumption per FFT operation

	Non-scalable		Scalable	
FFT length	8-bit	16-bit	8-bit	16-bit
1024-point	1320 nJ	1448 nJ	575 nJ	1491 nJ
512-point	607 nJ	750 nJ	240 nJ	629 nJ
256-point	269 nJ	334 nJ	103 nJ	269 nJ
128-point	118 nJ	147 nJ	44 nJ	116 nJ

and overhead. The architecture consists of a parameterized and configurable physical layer to determine power control modes, modulation scheme, and bit rate for energy efficiency. This configurable architecture enables energy minimization opportunities in wireless networks to be efficiently realized in silicon. The MIT μAMPs project [41] focuses on architecture and circuit design techniques for energy-efficient communications of wireless microsensor systems with lower transmission distances ($<$ 10m) and lower bit rates (typically $<$ kbs). The μAMPs-1 sensor node processor uses dynamic voltage scaling to minimize energy consumption for a given performance requirement. The radio transmit power adjusts to one of six levels, depending on the physical location of the target nodes. Power consumption of the node varies from 3.5mW in the deepest sleep state up to almost 2W. To enable energy-awareness of the *fast Fourier transform* (FFT) algorithm, its implementation includes tunable structures, such as memory size and variable bit precision, to handle a variety of scenarios effectively. The energy scalable FFT architecture was simulated in a 0.18 μm CMOS process at 1.5-V operation and the simulated energy dissipated is summarized in Table 23.1 from [41], which shows a definite advantage for a scalable architecture over a non-scalable architecture. The scalable architecture is more energy-efficient for all but the high-quality point (1,024 point, 16-bit). At the high-quality point, the scalable design has a disadvantage due to the overhead logic. The scalable FFT processor was also fabricated in a standard 0.18 μm CMOS process and standard ASIC flow to demonstrate these energy-scalable architectural techniques. At 1.5-V operation, when compared to a StrongARM SA-1100 implementation, the FFT processor shows over a 350X measured energy reduction. Since the μAMPs project focuses on short-range communications with short packets, circuit components (frequency synthesizers, mixers, etc.), rather than the power amplifier, dominate power consumption. In order to reduce transmission start-up time, which is crucial in determining circuit power consumption, the energy-efficient transmitter uses a variable loop bandwidth method [75] for the phase-locked loop. Furthermore, similar to the PicoRadio project, energy efficiency in the μAMPs project is also enhanced through multi-hop routings and the energy-efficient number of hops is determined by both distance independent and dependent components.

To summarize, energy-efficient communication design not only saves energy, but also achieves long-term environmental sustainability and profitability. Besides technical developments that have been discussed in this book, non-technical improvements and implementations, such as pricing, marketing, government regulation, and law establishment, would also be important in the success of energy-efficient wireless revolution.

Appendix A Proofs of Theorems and Lemmas

A.1 Proof of Theorem 7.1

Proof If the $\bar{D}_1^*$s are optimal, then any change of allocation will not increase the average utility. Let $(f - \frac{1}{2}\triangle f, f + \frac{1}{2}\triangle f) \in D_1^*$. If $(f - \frac{1}{2}\triangle f, f + \frac{1}{2}\triangle f)$ be assigned to the other user, then the data rate of user 1 will be decreased by $\triangle r_1 = c_1(f)\triangle f$ while the data rate of user 2 will be increased by $\triangle r_2 = c_2(f)\triangle f$. But, the new average utility will be equal to or less than the optimal one, i.e.,

$$U_1(r_1^* - \triangle r_1) + U_2(r_2^* + \triangle r_2) \leq U_1(r_1^*) + U_2(r_2^*),$$

which is equivalent to

$$U_2(r_2^* + \triangle r_2) - U_2(r_2^*) \leq U_1(r_1^*) - U_1(r_1^* - \triangle r_1).$$

Dividing both sides by $\triangle f$, we have

$$\frac{U_2(r_2^* + \triangle r_2) - U_2(r_2^*)}{\triangle f} \leq \frac{U_1(r_1^*) - U_1(r_1^* - \triangle r_1)}{\triangle f}.$$

Since $\triangle r_1 = c_1(f)\triangle f$ and $\triangle r_2 = c_2(f)\triangle f$, we have

$$\frac{U_2(r_2^* + \triangle r_2) - U_j(r_2^*)}{\triangle r_2}c_2(f) \leq \frac{U_1(r_1^*) - U_1(r_1^* - \triangle r_1)}{\triangle r_1}c_1(f).$$

When $\triangle f \to 0$, $\triangle r_1 \to 0$ and $\triangle r_2 \to 0$. Consequently,

$$\lim_{\triangle r_2 \to 0} \frac{U_2(r_2^* + \triangle r_2) - U_j(r_2^*)}{\triangle r_2}c_2(f) \leq \lim_{\triangle r_1 \to 0} \frac{U_1(r_1^*) - U_1(r_1^* - \triangle r_1)}{\triangle r_1}c_1(f),$$

or

$$U_2'(r_2^*)c_2(f) \leq U_1'(r_1^*)c_1(f)f \in \bar{D}_1^*. \tag{A.1}$$

which implies, for any $f \in \bar{D}_1^*$,

$$\frac{c_2(f)}{c_1(f)} \leq \frac{U_1'(r_1^*)}{U_2'(r_2^*)}(= \alpha^*),$$

i.e. $f \in \bar{D}_1(\alpha^*)$ and $D_1^* \subseteq \bar{D}_1(\alpha^*)$.

Similarly, we can prove that

$$D_2^* \subseteq \bar{D}_2(\alpha^*).$$

Therefore,

$$D_1(\alpha^*) = [0,\ B] - \bar{D}_2(\alpha^*)$$
$$\subseteq [0,\ B] - D_2^*$$
$$= D_1^*$$

□

A.2 Proof of Theorem 7.3

Proof For a fixed subcarrier assignment D_i for all i, we define $p_i(f)$ for $i = 1, 2, \ldots, M$ as,

$$p_i(f) = \begin{cases} p(f) & f \in D_i \\ 0 & \text{otherwise} \end{cases}.$$

Using the Lagrangian method, the above optimization problem with the power constraint maximizes

$$\frac{1}{M} \sum_{i=1}^{M} U_i \left(\int_{D_i} \log_2 [1 + \beta p(f) \rho_i(f)\, df] \right) - \lambda' \left[\frac{1}{B} \int_0^B p(f)\, df - 1 \right],$$

or

$$\frac{1}{M} \sum_{i=1}^{M} \left\{ U_i \left(\int_{D_i} \log_2 [1 + \beta p_i(f) \rho_i(f)\, df] \right) - \lambda' \left[\frac{1}{B} \int_{D_i} p_i(f)\, df - 1 \right] \right\},$$

where $\lambda' \geq 0$.

With the Karush–Kuhn–Tucker (KKT) conditions [169], we have

$$\frac{1}{M} U_i'(r_i^*) \frac{\partial}{\partial p_i(f)} \log_2 \{1 + \beta p_i(f) \rho_i(f)\} - \frac{\lambda'}{B} \frac{\partial}{\partial p_i(f)} p_i(f) \Big|_{p_i(f)=p_i^*(f)} \quad \text{(A.1)}$$

$$= 0, \quad \text{for all } i, \lambda' \geq 0, \quad \text{(A.2)}$$

$$\lambda' \left[\sum_{i=1}^{M} \frac{1}{B} \int_{D_i} p_i(f)\, df - 1 \right] = 0. \quad \text{(A.3)}$$

(A.1) is equivalent to

$$U_i'(r_i^*) \frac{\beta \rho_i(f)}{1 + \beta \rho_i(f) p_i^*(f)} - \lambda' \frac{M}{\log_2(e)B} = 0, \quad \text{for all } i.$$

Let $\lambda = \lambda' \frac{M}{\log_2(e)B}$. Then the optimal power allocation for a fixed subcarrier assignment satisfies:

$$\begin{cases} p_i^*(f) = \left[\dfrac{U_i'(r_i^*)}{\lambda} - \dfrac{1}{\beta \rho_i(f)} \right]^+ & f \in D_i \\[2mm] \displaystyle\sum_{i=1}^{M} \dfrac{1}{B} \int_{D_i} p_i^*(f)\, df = 1. \end{cases}$$

or

$$
\begin{cases}
p^*(f) = \left[\dfrac{U_i'(r_i^*)}{\lambda} - \dfrac{1}{\beta \rho_i(f)} \right]^+ & f \in D_i \\[2ex]
\dfrac{1}{B} \displaystyle\int_0^B p^*(f)\,df = 1.
\end{cases}
$$

□

A.3   Proof of Theorem 7.5

Proof   Assume that the system has joint DSA and APA. Then $\forall\, \mathbf{r}^{(1)},\, \mathbf{r}^{(2)} \in \mathcal{C}_{DSA+APA}$, $\alpha \in [0, 1]$, we need to show that $\alpha \mathbf{r}^{(1)} + (1-\alpha)\mathbf{r}^{(2)} \in \mathcal{C}_{DSA+APA}$. $\mathbf{r}^{(1)} = [r_1^{(1)}, r_2^{(1)}, \ldots, r_M^{(1)}]^T$ is achieved with $D_m^{(1)}$ and $p^{(1)}(f)$, $\mathbf{r}^{(2)} = [r_1^{(2)}, r_2^{(2)}, \ldots, r_M^{(2)}]^T$ is achieved with $D_m^{(2)}$ and $p^{(2)}(f)$, where for $m \in \{1, 2, \ldots, M\}$. Of course, $D_m^{(1)}$ and $D_m^{(2)}$ satisfy (7.3) and (7.4); $p^{(1)}(f)$ and $p^{(2)}(f)$ yield (7.5). We represent those two power allocations as $\mathbf{p}^{(1)}$ and $\mathbf{p}^{(2)}$, respectively.

We define the measure of a frequency set as follows. When the frequency set $D = \bigcup_i [a_i, b_i]$, $b_i \le a_{i+1}$, the measure μ is given by $\mu(D) = \sum_i (b_i - a_i)$. For user m, we have

$$
r_m^{(1)} = \int_{D_m^{(1)}} c_m^{\mathbf{p}^{(1)}}(f)\,d\mu,
$$

$$
r_m^{(2)} = \int_{D_m^{(2)}} c_m^{\mathbf{p}^{(2)}}(f)\,d\mu,
$$

where $c_m^{\mathbf{p}}(f)$ denotes the achievable throughput of user m at frequency f with power allocation $\mathbf{p}$.

We divide $[0, B]$ into a family of sets F_ns so that

$$
\bigcup_n F_n = [0, B], \quad F_i \bigcap F_j = \varnothing \ i \ne j, \tag{A.1}
$$

$$
\max_{f \in F_n} \left\{ c_m^{\mathbf{p}^{(1)}}(f) \right\} - \min_{f \in F_n} \{ c_m^{\mathbf{p}^{(1)}}(f) \} \to 0 \text{ for all } m, n, \tag{A.2}
$$

$$
\max_{f \in F_n} \{ c_m^{\mathbf{p}^{(2)}}(f) \} - \min_{f \in F_n} \{ c_m^{\mathbf{p}^{(2)}}(f) \} \to 0 \text{ for all } m, n. \tag{A.3}
$$

(A.2) and (A.3) imply

$$
\max_{f \in F_n} p^{(1)}(f) - \min_{f \in F_n} p^{(1)}(f) \to 0 \text{ for all } n,
$$

$$
\max_{f \in F_n} p^{(2)}(f) - \min_{f \in F_n} p^{(2)}(f) \to 0 \text{ for all } n.
$$

Each F_n is divided into two subsets F_n^{α} and $F_n^{(1-\alpha)}$ that satisfy

$$
F_n^{\alpha} \bigcup F_n^{(1-\alpha)} = F_n, \quad F_n^{\alpha} \bigcap F_n^{(1-\alpha)} = \varnothing \tag{A.4}
$$

and $\mu(F_n^{\alpha}) = \alpha \mu(F_n)$.

If $F_n \in D_m$, we use $D_{m,n}$ to denote F_n. Thus,

$$r_m^{(1)} = \sum_n c_m^{\mathbf{p}^{(1)}}(n)\mu(D_{m,n}^{(1)})$$

$$r_m^{(2)} = \sum_n c_m^{\mathbf{p}^{(2)}}(n)\mu(D_{m,n}^{(2)})$$

In the same way, using $D_{m,n}^\alpha$ to denote $F_n^\alpha \in D_m$, we have

$$\int_{D_m^{(1),\alpha}} c_m^{\mathbf{p}^{(1)}}(f)\,d\mu = \sum_n c_m^{\mathbf{p}^{(1)}}(n)\mu(D_{m,n}^{(1),\alpha})$$

$$= \alpha r_m^{(1)}$$

$$\int_{D_m^{(2),(1-\alpha)}} c_m^{\mathbf{p}^{(2)}}(f)\,d\mu = \sum_n c_m^{\mathbf{p}^{(2)}}(n)\mu(D_{m,n}^{(2),(1-\alpha)})$$

$$= (1-\alpha)r_m^{(2)}$$

where
$$D_m^{(1),\alpha} = \bigcup_n D_{m,n}^{(1),\alpha}$$

$$D_m^{(2),(1-\alpha)} = \bigcup_n D_{m,n}^{(2),(1-\alpha)}.$$

Therefore, with the new frequency assignment $D_m = D_m^{(1),\alpha} \cup D_m^{(2),(1-\alpha)}$ and the new power allocation

$$p(f) = \begin{cases} p^{(1)}(f) & f \in D_m^{(1),\alpha} \\ p^{(2)}(f) & f \in D_m^{(2),(1-\alpha)} \end{cases},$$

the new data rate for user m is

$$r_m = \int_{D_m^{(1),\alpha}} c_m^{\mathbf{p}^{(1)}}(f)\,d\mu + \int_{D_m^{(2),(1-\alpha)}} c_m^{\mathbf{p}^{(2)}}(f)\,d\mu$$

$$= \alpha r_m^{(1)} + (1-\alpha)r_m^{(2)}$$

Furthermore, due to (A.1) and (A.4), the D_ms satisfy (7.3) and (7.4). In addition,

$$\frac{1}{B}\int_0^B p(f)\,d\mu = \frac{1}{B}\sum_m \int_{D_m} p(f)\,d\mu$$

$$= \frac{1}{B}\sum_m \int_{D_m^{(1),\alpha}} p^{(1)}(f)\,d\mu + \frac{1}{B}\sum_m \int_{D_m^{(2),(1-\alpha)}} p^{(2)}(f)\,d\mu$$

$$= \frac{\alpha}{B}\sum_m \int_{D_m^{(1)}} p^{(1)}(f)\,d\mu + \frac{1-\alpha}{B}\sum_m \int_{D_m^{(2)}} p^{(2)}(f)\,d\mu$$

$$= \alpha\frac{1}{B}\int_0^B p^{(1)}(f)\,d\mu + (1-\alpha)\frac{1}{B}\int_0^B p^{(2)}(f)\,d\mu$$

$$\leq 1$$

Therefore, there are feasible frequency assignment and power allocation schemes such that $\alpha\mathbf{r}^{(1)} + (1-\alpha)\mathbf{r}^{(2)} \in \mathcal{C}$.

Let $p^{(1)}(f) = p^{(2)}(f)$ in the above proof. Then we have that the achievable data rate region is convex when only DSA is used. Let $D_m^{(1)} = D_m^{(2)}$ for all m in the above proof. Similarly, we have that the achievable data rate region is also convex when only APA is used. □

A.4   Proof of Lemma 9.1

We only prove the case with finite channel states here. For the continuous channel state distributions, the major idea of the proof is still straightforward and very similar to that for finite channel states, but the technicality seems intricate due to measure-theoretic complications.

Proof   Let $\mathcal{J}$ represent the finite channel state set, and π_j be the stationary probability of state j, $j \in \mathcal{J}$. $T_j(t)$ denotes the subintervals of $[0, t]$ during which the channel state is j. $|T_j(t)|$ is the total length of these subintervals. Due to the ergodicity of the channel states, there exists a time t' such that for any small value $\delta > 0$,

$$\frac{|T_j(t')|}{t'} \leq \pi_j + \delta, \tag{A.1}$$

$$\text{and} \quad \liminf_{t \to \infty} \frac{\int_{\tau=0}^{t} \mathbf{r}(\tau)\, d\tau}{t} \leq \frac{\int_{\tau=0}^{t'} \mathbf{r}(\tau)\, d\tau}{t'} + \delta.$$

Thus,

$$\liminf_{t \to \infty} \frac{\int_{\tau=0}^{t} \mathbf{r}(\tau)\, d\tau}{t} \leq \sum_{j \in \mathcal{J}} \frac{|T_j(t')|}{t'} \frac{1}{|T_j(t')|} \int_{\tau \in T_j(t')} \mathbf{r}(\tau)\, d\tau + \delta.$$

According to (9.20), there exists a stationary policy $\mathcal{R}(j)$ such that

$$\frac{1}{|T_j(t')|} \int_{\tau \in T_j(t')} \mathbf{r}(\tau)\, d\tau \leq \mathcal{R}(j). \tag{A.2}$$

It follows from (A.1) and (A.2) that

$$\liminf_{t \to \infty} \frac{\int_{\tau=0}^{t} \mathbf{r}(\tau)\, d\tau}{t} \leq \sum_{j \in \mathcal{J}} (\pi_j + \delta)\mathcal{R}(j) + \delta$$

$$= \sum_{j \in \mathcal{J}} \pi_j \mathcal{R}(j) + \delta(|R(j)| + 1).$$

Since $\sum_{j \in \mathcal{J}} \pi_j \mathcal{R}(j) \in \tilde{\mathcal{C}}$, let $\delta \to 0$, then

$$\liminf_{t \to \infty} \frac{\int_{\tau=0}^{t} \mathbf{r}(\tau)\, d\tau}{t} \in \tilde{\mathcal{C}}.$$

A.5   Proof of Lemma 9.3   □

Proof Let

$$\mathbf{r}^*(\mathbf{H}) = \arg\max_{\mathbf{r}\in\mathcal{C}(\mathbf{H})} \mathbf{w}^T \mathbf{r}.$$

In other words,

$$\mathbf{w}^T(\mathbf{r}_1 - \mathbf{r}^*(\mathbf{H})) \le 0, \quad \mathbf{r}_1 \in \mathcal{C}(\mathbf{H}), \tag{A.1}$$

$$\mathbf{w}^T(\mathbf{r}_2 - \mathbf{r}^*(\mathbf{H})) \le 0, \quad \mathbf{r}_2 \in \mathcal{C}(\mathbf{H}). \tag{A.2}$$

Let $\mathbf{r}' = \alpha\mathbf{r}_1 + (1 - \alpha)\mathbf{r}_2$, where $\alpha \in (0, 1)$. Then $\mathbf{r}'$ is in the convex hull of $\mathcal{C}(\mathbf{H})$. Because of (A.1) and (A.2), it follows that

$$\mathbf{w}^T(\mathbf{r}' - \mathbf{r}^*(\mathbf{H})) \le 0, \quad \mathbf{r}' \in cov(\mathcal{C}(\mathbf{H})).$$

Taking expectation on both sides, we have

$$\mathbf{w}^T(\tilde{\mathbf{r}}' - \tilde{\mathbf{r}}^*) \le 0, \quad \tilde{\mathbf{r}}' \in \tilde{\mathcal{C}},$$

which is equivalent to

$$\tilde{\mathbf{r}}^* = \arg\max_{\tilde{\mathbf{r}}\in\tilde{\mathcal{C}}} \mathbf{w}^T\tilde{\mathbf{r}}.$$

□

A.6   Proof of Lemma 9.5

Proof

$$Q_i[n] - \bar{Q}_i[n] = Q_i[n] - \{(1 - \rho_w)\bar{Q}_i[n - 1] + \rho_w Q_i[n]\}$$
$$= (1 - \rho_w)\{Q_i[n] - \bar{Q}_i[n - 1]\}. \tag{A.1}$$

Define

$$\xi_i'[n] = Q_i[n] - Q_i[n - 1]$$
$$= -\min(Q_i[n - 1], r_i[n]T_s) + a_i[n]. \tag{A.2}$$

Form (A.1), we have the following recurrence formula for $Q_i[n] - \bar{Q}_i[n]$,

$$Q_i[n] - \bar{Q}_i[n] = (1 - \rho_w)\{Q_i[n - 1] - \bar{Q}_i[n - 1]\} + (1 - \rho_w)\xi_i'[n]. \tag{A.3}$$

Since $\bar{Q}_i[0] = Q_i[0]$, it follows from the recursive relationship in (A.3) that

$$Q_i[n] - \bar{Q}_i[n] = \sum_{j=0}^{n-1}(1 - \rho_w)^{n-j}\xi_i'[j + 1].$$

We have $|Q_i[n] - \bar{Q}_i[n]| \le \sum_{j=0}^{n-1}(1 - \rho_w)^{n-j}|\xi_i'[j + 1]|$, and

$$\mathbb{E}\{|Q_i[n] - \bar{Q}_i[n]|\} \leq \sum_{j=0}^{n-1} (1 - \rho_w)^{n-j} \mathbb{E}\{|\xi_i'[j+1]|\}.$$

It follows from (A.2) that

$$\mathbb{E}\{|\xi_i'[j]|\} \leq \mathbb{E}\{r_i[j]\}T_s + \mathbb{E}\{a_i[j]\}$$

$$\leq (R_{\text{total}} + \lambda_i)T_s$$

$$< \infty,$$

where R_{total} is the maximum expected sum capacity of the system. Obviously, for $n < \infty$, $\mathbb{E}\{|Q_i[n] - \bar{Q}_i[n]|\}$ is bounded. Therefore, we need to consider the asymptotic case in which $n \to \infty$. In this case,

$$\mathbb{E}\{|Q_i[n] - \bar{Q}_i[n]|\} \leq (R_{\text{total}} + \lambda_i)T_s \lim_{n\to\infty} \sum_{j=0}^{n-1}(1 - \rho_w)^{n-j}$$

$$= \frac{1 - \rho_w}{\rho_w}(R_{\text{total}} + \lambda_i)T_s$$

$$< \infty.$$

□

A.7   Proof of Theorem 11.4

Proof   According to the results of extreme value theory in Section 11.1, we have to show that

$$\lim_{r\to\infty} \frac{d}{dr}\left[\frac{1 - F_R(r)}{f_R(r)}\right] = 0,$$

if

$$\lim_{\gamma\to\infty} \frac{d}{d\gamma}\left[\frac{1 - F_\Gamma(\gamma)}{f_\Gamma(\gamma)}\right] = 0. \tag{A.1}$$

Since

$$\frac{1 - F_R(r)}{f_R(r)} = \frac{1 - F_\Gamma(T^{-1}(r))}{f_\Gamma(T^{-1}(r))\left(T^{-1}\right)'(r)},$$

we have

$$\frac{d}{dr}\left[\frac{1 - F_R(r)}{f_R(r)}\right]$$

$$= -1 - \frac{\left[1 - F_\Gamma(T^{-1}(r))\right]\left[f_\Gamma'(T^{-1}(r))((T^{-1})'(r))^2 + f_\Gamma(T^{-1}(r))\left(T^{-1}\right)''(r)\right]}{\left[f_\Gamma(T^{-1}(r))\left(T^{-1}\right)'(r)\right]^2}$$

$$= -1 - \underbrace{\frac{[1 - F_\Gamma(T^{-1}(r))]f_\Gamma'(T^{-1}(r))}{f_\Gamma^2(T^{-1}(r))}}_{\text{Part I}} - \underbrace{\frac{[1 - F_\Gamma(T^{-1}(r))]\left(T^{-1}\right)''(r)}{f_\Gamma(T^{-1}(r))\left[\left(T^{-1}\right)'(r)\right]^2}}_{\text{Part II}} \quad \text{(A.2)}$$

Because $T^{-1}(r)$ is monotonically increasing with x and $T^{-1}(r) \to \infty$ as $r \to \infty$,

$$\lim_{r\to\infty} \frac{[1 - F_\Gamma(T^{-1}(r))]f_\Gamma'(T^{-1}(r))}{f_\Gamma^2(T^{-1}(r))} = \lim_{\gamma\to\infty} \frac{[1 - F_\Gamma(\gamma)]f_\Gamma'(\gamma)}{f_\Gamma^2(\gamma)}$$

It is easy to check that

$$\frac{d}{d\gamma}\left[\frac{1 - F_\Gamma(\gamma)}{f_\Gamma(\gamma)}\right] = -1 - \frac{[1 - F_\Gamma(\gamma)]f_\Gamma'(\gamma)}{f_\Gamma^2(\gamma)}.$$

Thus, we have

$$\lim_{r\to\infty} \text{Part I} = \lim_{r\to\infty} \frac{d}{d\gamma}\left[\frac{1 - F_\Gamma(\gamma)}{f_\Gamma(\gamma)}\right]. \quad \text{(A.3)}$$

Let $\tilde{T}^{-1}(r) = \frac{2^{\frac{x}{B}}}{\beta}$. Due to the fact that

$$\left(T^{-1}\right)''(r) = \frac{\ln 2}{B}\left(T^{-1}\right)'(r),$$

and $(\tilde{T}^{-1})'(r) = (T^{-1})'(r)$, it follows that

$$\lim_{r\to\infty} \text{Part II} = \lim_{r\to\infty} \frac{\ln 2[1 - F_\Gamma(T^{-1}(r))]}{Bf_\Gamma(T^{-1}(r))\left(T^{-1}\right)'(r)}, \quad \text{(A.4)}$$

$$= \lim_{r\to\infty} \frac{\ln 2[1 - F_\Gamma(T^{-1}(r))]}{Bf_\Gamma(T^{-1}(r))\left(\tilde{T}^{-1}\right)'(r)}. \quad \text{(A.5)}$$

Since $\tilde{T}^{-1}(r) = T^{-1}(r) + \frac{1}{\gamma}$ and $\tilde{T}^{-1}(r) \to \infty$ as $r \to \infty$,

$$\lim_{r\to\infty} \frac{1 - F_\Gamma(\tilde{T}^{-1}(r))}{f_\Gamma(\tilde{T}^{-1}(r))} = \lim_{r\to\infty} \frac{1 - F_\Gamma(T^{-1}(r))}{f_\Gamma(T^{-1}(r))},$$

if (A.1) holds. Thus, we have

$$\lim_{r\to\infty} \text{Part II} = \lim_{r\to\infty} \frac{\ln 2[1 - F_\Gamma(\tilde{T}^{-1}(r))]}{Bf_\Gamma(\tilde{T}^{-1}(r))\left(\tilde{T}^{-1}\right)'(r)}$$

$$= \lim_{r\to\infty} \frac{\ln 2[1 - F_\Gamma(\tilde{T}^{-1}(r))]}{Bf_\Gamma(\tilde{T}^{-1}(r))\tilde{T}^{-1}(r)\frac{\ln 2}{B}}$$

$$= \lim_{\gamma\to\infty} \frac{1 - F_\Gamma(\gamma)}{f_\Gamma(\gamma)\gamma} \quad \text{(A.6)}$$

Combining (A.3) and (A.6), we obtain

$$\lim_{r\to\infty} \frac{d}{dr}\left[\frac{1 - F_R(r)}{f_R(r)}\right] = \lim_{\gamma\to\infty} \frac{d}{d\gamma}\left[\frac{1 - F_\Gamma(\gamma)}{f_\Gamma(\gamma)}\right] + \lim_{\gamma\to\infty} \frac{1 - F_\Gamma(\gamma)}{f_\Gamma(\gamma)\gamma}. \quad \text{(A.7)}$$

According to L'Hospital's rule, for a function $g(x)$ such as $g(x) \to \infty$ as $x \to \infty$, if $\lim_{x \to \infty} g'(x) = 0$, then $\lim_{x \to \infty} \frac{g(x)}{x} = 0$. Equation (A.1) results in

$$\lim_{\gamma \to \infty} \frac{1 - F_\Gamma(\gamma)}{f_\Gamma(\gamma)\gamma} = 0,$$

Therefore, we obtain

$$\lim_{r \to \infty} \frac{d}{dr} \left[\frac{1 - F_R(r)}{f_R(r)} \right] = 0.$$

Since

$$F_R^{-1}(x) = T(F_\Gamma^{-1}(x)),$$

$$= B \log_2(1 + \beta F_\Gamma^{-1}(x)),$$

we can obtain the normalizing constants (11.18) and (11.19) according to the results of extreme value theory in Section 11.1. $\qquad\square$

A.8 Proof of Equation (11.23)

Proof Let $X \geq 0$ be a random variable with distribution function $F(x)$ and $\mathbb{E}\{X\}$ is finite. The expected residual life of X is given by

$$R(t) = \mathbb{E}\{X - t | X \geq t\}$$

$$= \frac{1}{1 - F(t)} \int_t^\infty 1 - F(x)dx.$$

Theorem 2.1.3 and Lemma 2.7.2 in [82] show that if $F(x)$ is in the domain of the Gumbel distribution,

$$b_M = R(a_M), \tag{A.1}$$

and

$$\lim_{t \to \infty} \frac{R(t)}{t} = 0. \tag{A.2}$$

Since a_M monotonically increases with M, (A.1) and (A.2) directly indicates that

$$\lim_{M \to \infty} \frac{b_M}{a_M} = \lim_{M \to \infty} \frac{R(a_M)}{a_M}$$

$$= \lim_{t \to \infty} \frac{R(t)}{t}$$

$$= 0$$

$\qquad\square$

A.9   Proof of Theorem 15.1

Proof   We prove that the two conditions of the definition hold for CAD-MAC.

(1)   Suppose two links, (i,j) and (k,l), that have won the contention have collision and the transmission of user i interferes with the reception of user l. First, (i,j) and (k,l) should not have won the contention at the same CRS since the REQUESTs of the two links collide at user l and user l will not acknowledge SUCCESS. If (i,j) receives SUCCESS first, the OCCUPIED signal of user i will prevent user l from acknowledging SUCCESS. If (k,l) wins first, the broadcasting of SUCCESS by user l will prevent user i from sending REQUEST. Hence, condition 1 always holds.

(2)   To verify condition 2, suppose that there exists a link (I, J) that has not won access and does not collide with any link that has won. Besides, within the interference range of link (I, J), no other link could win as otherwise, after that link wins, link (I, J) should not contend and the contention is completely resolved. There are two possibilities. (1) User I does not send any REQUEST all the time or (2) whenever user I sends a REQUEST, it collides with that of other links. We show in the following that both have zero possibility:

- User I does not send any REQUEST all the time. This indicates that $h_{IJ} < \widehat{H}_{IJ}[k]$ for all $k > K$, where $K > 0$. Obviously, nobody that interferes with user J should send anything. Hence, user J will keep on sending IDLE signals to user I and $\widehat{H}_{IJ}[k]$ will be lowered successively. It is easy to see that in this case $\lim_{k \to \infty} \widehat{H}_{IJ}[k] = \widehat{H}_{IJ}^m$. Hence, the probability that $h_{IJ} < \widehat{H}_{IJ}[k]$ for all $k > K$ is zero and sooner or later user I will send a REQUEST and win.

- Whenever (I, J) sends a REQUEST, it collides with others. Define the CRSs that (I, J) sends REQUESTs by $\mathcal{C} = \{c_1, c_2, \ldots\}$, where $c_1 < c_2 < \ldots$. Suppose there are N links that collide with (I, J). According to (15.20), the contention probability of any link using ADs I or II is $\frac{1}{2}$ after sending the first REQUEST. If using AD III, the contention probability is zero. We consider the CRSs after all the interfering links have sent the first REQUEST and define $N_k \leq N$ to be the number of interfering links that contends with probability $\frac{1}{2}$ in CRS k. The probability that (I, J) keeps on contending and never succeeds is given by

$$\Pr\{(I, J) \text{ never wins}\}$$

$$= \lim_{|\mathcal{C}| \to \infty} \prod_{k \in \mathcal{C}} \Pr\{\text{at least one interferer contends in CRS } k\}$$

$$= \lim_{|\mathcal{C}| \to \infty} \prod_{k \in \mathcal{C}} \left(1 - \left(1 - \frac{1}{2}\right)^{N_k}\right) \tag{A.1}$$

$$\leq \lim_{|\mathcal{C}| \to \infty} \prod_{k \in \mathcal{C}} \left(1 - \left(\frac{1}{2}\right)^{N}\right) = \lim_{|\mathcal{C}| \to \infty} \left(1 - \left(\frac{1}{2}\right)^{N}\right)^{|\mathcal{C}|} < \sigma \tag{A.2}$$

for any $\sigma > 0$. Hence, the probability that (I, J) never resolves its contention is zero; i.e., with probability one, (I, J) always wins the contention when none of its neighbors can win and the network contention within the interference range of (I, J) can always be resolved.

Theorem 15.1 follows immediately.   □

A.10   Proof of Theorem 15.2

Proof   Suppose there are N links and in CRS 1 each has the contention probability

$$p_{ij}[1] = \frac{1}{N}.$$

According to (15.20), the contention probability in CRS k is

$$p_{ij}[k] = \begin{cases} \frac{1}{N}, \text{IDLE in all the previous CRSs}, \\ \frac{1}{2}, \text{otherwise}. \end{cases} \tag{A.1}$$

Define by $\bar{k}_n$ the average number of CRSs necessary to resolve the collision involving n links. From (A.1), these links will contend with probability $\frac{1}{2}$ if they still contend in the following CRSs. Hence,

$$\bar{k}_n = \left(\frac{1}{2}\right)^n \left[\sum_{i=2}^{n} \binom{n}{i}(\bar{k}_i + 1) + \binom{n}{0}(\bar{k}_n + 1) + \binom{n}{1}1 \right], \tag{A.2}$$

where $\left(\frac{1}{2}\right)^n \binom{n}{i}$ is the probability that i users have their gains above the thresholds and on average $\bar{k}_i$ additional CRSs are needed if $i > 1$. If $i = 0$, all users have their gains below the thresholds and are involved in the following contention. If $i = 1$, the contention is resolved. It has been proved in [210] that $\bar{k}_n$ in (A.2) satisfies

$$\log_2(n) \leq \bar{k}_n \leq \log_2(n) + 1 \tag{A.3}$$

for all n. Before a collision happens, all users may have channel gains so low that several CRSs are necessary for them to lower their thresholds successively until some users are allowed to send REQUESTs. Hence, the average number of CRSs necessary for completely resolving the network contention is

$$\bar{K}_N = \sum_{i=0}^{\infty} \left(\left(1 - \frac{1}{N}\right)^{Ni} \left(\sum_{n=1}^{N} \binom{N}{n} \left(\frac{1}{N}\right)^n \left(1 - \frac{1}{N}\right)^{N-n} (\bar{k}_n + i + 1) \right) \right), \tag{A.4}$$

where $(1 - \frac{1}{N})^{Ni}$ is the probability that all users have their gains below their thresholds in all the first i CRSs and $\binom{N}{n}(\frac{1}{N})^n(1 - \frac{1}{N})^{N-n}$ is the probability that in the $i + 1$st CRS n users send REQUESTs and collide. Let

$$M_N = \sum_{n=1}^{N} \binom{N}{n} \left(\frac{1}{N}\right)^n \left(1 - \frac{1}{N}\right)^{N-n} (\bar{k}_n + 1).$$

Then,

$$M_N \leq \sum_{n=1}^{N} \binom{N}{n} \left(\frac{1}{N}\right)^n \left(1 - \frac{1}{N}\right)^{N-n} (\log_2(n) + 1)$$

$$< \sum_{n=1}^{N} \binom{N}{n} \left(\frac{1}{N}\right)^n \left(1 - \frac{1}{N}\right)^{N-n} (n + 1)$$

$$= 2 - \left(1 - \frac{1}{N}\right)^N . \tag{A.5}$$

Hence, $\overline{K}_N$ equals

$$\overline{K}_N = M_N \sum_{i=0}^{\infty} \left(1 - \frac{1}{N}\right)^{Ni} + \sum_{i=0}^{\infty} i \left(1 - \frac{1}{N}\right)^{Ni}$$

$$= \frac{M_N}{1 - (1 - \frac{1}{N})^N} + \frac{(1 - \frac{1}{N})^N}{(1 - (1 - \frac{1}{N})^N)^2}$$

$$< \frac{2 - (1 - \frac{1}{N})^N}{1 - (1 - \frac{1}{N})^N} + \frac{(1 - \frac{1}{N})^N}{(1 - (1 - \frac{1}{N})^N)^2}$$

$$= 1 + \frac{1}{[1 - (1 - \frac{1}{N})^N]^2}$$

$$< 1 + [1 - e^{-1}]^{-2} . \tag{A.6}$$

Hence, $\overline{K}_N$ is bounded for all N and the right-hand side of (A.4) converges. A tighter bound is

$$\overline{K}_N \leq \frac{\widehat{M}_N}{1 - (1 - \frac{1}{N})^N} + \frac{(1 - \frac{1}{N})^N}{(1 - (1 - \frac{1}{N})^N)^2},$$

where

$$\widehat{M}_N = \sum_{n=1}^{N} \binom{N}{n} \left(\frac{1}{N}\right)^n \left(1 - \frac{1}{N}\right)^{N-n} (\log_2(n) + 1).$$

As N goes to infinity, using computer calculation, we have

$$\overline{K}_N < \overline{K}_\infty \leq 2.43. \tag{A.7}$$

$\square$

A.11   Proof of Theorem 15.3

Proof Let K be the number of CRSs necessary to completely resolve network contention in a frame slot and $\mathcal{K} = \{1, 2, \ldots, K\}$ the corresponding set of CRSs. Let L be the number of links winning the contention and $\mathcal{K}_l, i = 1, \ldots, L$, the corresponding set of

CRSs that the lth winning link is involved in the contention. Assume that $\mathcal{K}_l, l = 1, \ldots, L$ are independently and identically distributed and independent of L. Obviously,

$$\mathcal{K} = \bigcup_i \mathcal{K}_i \text{ and } K = |\mathcal{K}| \leq \sum_i |\mathcal{K}_i|, \tag{A.1}$$

where $|\mathcal{X}|$ is the cardinality of set $\mathcal{X}$. Define the contention coexistence factor as

$$\beta = \frac{\mathbf{E}(\sum_i |\mathcal{K}_i|)}{\mathbf{E}(|\mathcal{K}|)}. \tag{A.2}$$

It is easy to see that β is the average number of simultaneous resolutions in each CRS. For example, if all users interfere with all others, then $L = 1$ and $\beta = 1$, meaning only one resolution in each CRS. If a network consists of L groups of users and the communication of any group does not interfere with that of any other group, then these L groups can resolve the contention within each group to produce L winners. If we further assume $\mathcal{K}_1, \mathcal{K}_2, \ldots, \mathcal{K}_L$ are independently and identically distributed, then $\beta = L$, indicating L simultaneous resolutions in each CRS on average. Then we have

$$\overline{K} = \mathbf{E}(K) = \mathbf{E}\left(|\mathcal{K}|\right) = \frac{\mathbf{E}(\sum_i^L |\mathcal{K}_i|)}{\beta}. \tag{A.3}$$

Furthermore, L is a stopping time for K_i and according to Wald's equation [175], we have

$$\overline{K} = \frac{\mathbf{E}(|\mathcal{K}_i|)\mathbf{E}(L)}{\beta}. \tag{A.4}$$

Obviously, from Theorem 15.2, $\mathbf{E}(|\mathcal{K}_i|) < \overline{K}_\infty$. Hence,

$$\overline{K} < \frac{2.425 \cdot \mathbf{E}(L)}{\beta}. \tag{A.5}$$

$\square$

A.12   Proof of Theorem 17.9

Proof   For any two power vectors $\mathbf{p}_{-n}$ and $\check{\mathbf{p}}_{-n}$, define the function

$$\mathfrak{F}_n(\theta) = F_n(\check{\mathbf{p}}_{-n} + \theta(\mathbf{p}_{-n} - \check{\mathbf{p}}_{-n})). \tag{A.1}$$

It is clear that

$$\mathfrak{F}_n(0) = F_n(\check{\mathbf{p}}_{-n}) \text{ and } \mathfrak{F}_n(1) = F_n(\mathbf{p}_{-n}); \tag{A.2}$$

By the chain rule, we know that

$$\frac{\partial \mathfrak{F}_n}{\partial \theta} = (\mathbf{p}_{-n} - \check{\mathbf{p}}_{-n}) \frac{\partial F_n}{\partial(\check{\mathbf{p}}_{-n} + \theta(\mathbf{p}_{-n} - \check{\mathbf{p}}_{-n}))}. \tag{A.3}$$

Hence, we have

$$F_n(\mathbf{p}_{-n}) - F_n(\check{\mathbf{p}}_{-n}) = \mathfrak{F}_n(1) - \mathfrak{F}_n(0)$$

$$= \int_0^1 \mathfrak{F}_n'(\theta)d\theta \qquad (A.4)$$

$$= (\mathbf{p}_{-n} - \check{\mathbf{p}}_{-n}) \int_0^1 \frac{\partial F_n}{\partial(\check{\mathbf{p}}_{-n} + \theta(\mathbf{p}_{-n} - \check{\mathbf{p}}_{-n}))}d\theta.$$

Thus, $\|F_n(\mathbf{p}_{-n}) - F_n(\check{\mathbf{p}}_{-n})\|$

$$= \left\| (\mathbf{p}_{-n} - \check{\mathbf{p}}_{-n}) \int_0^1 \frac{\partial F_n}{\partial(\check{\mathbf{p}}_{-n} + \theta(\mathbf{p}_{-n} - \check{\mathbf{p}}_{-n}))}d\theta \right\|$$

$$\leq \|(\mathbf{p}_{-n} - \check{\mathbf{p}}_{-n})\| \left\| \int_0^1 \frac{\partial F_n}{\partial(\check{\mathbf{p}}_{-n} + \theta(\mathbf{p}_{-n} - \check{\mathbf{p}}_{-n}))}d\theta \right\|$$

$$\leq \|(\mathbf{p}_{-n} - \check{\mathbf{p}}_{-n})\| \int_0^1 \left\| \frac{\partial F_n}{\partial(\check{\mathbf{p}}_{-n} + \theta(\mathbf{p}_{-n} - \check{\mathbf{p}}_{-n}))} \right\| d\theta$$

$$\leq \|(\mathbf{p}_{-n} - \check{\mathbf{p}}_{-n})\| \int_0^1 \left\| \sup_{\mathbf{p}_{-n}} \frac{\partial F_n}{\partial \mathbf{p}_{-n}} \right\| d\theta$$

$$= \|(\mathbf{p}_{-n} - \check{\mathbf{p}}_{-n})\| \left\| \sup_{\mathbf{p}_{-n}} \frac{\partial F_n}{\partial \mathbf{p}_{-n}} \right\|.$$

Besides, according to the chain rule,

$$\frac{\partial F_n}{\partial \mathbf{p}_{-n}} = \frac{\partial \mathbf{I}_n}{\partial \mathbf{p}_{-n}} \frac{\partial \tilde{F}_n}{\partial \mathbf{I}_{-n}}.$$

Hence, we have

$$\frac{\|F_n(\mathbf{p}_{-n}) - F_n(\check{\mathbf{p}}_{-n})\|}{\|\mathbf{p}_{-n} - \check{\mathbf{p}}_{-n}\|} \leq \sup_{\mathbf{p}_{-n}} \left\| \frac{\partial F_n}{\partial \mathbf{p}_{-n}} \right\|$$

$$= \sup_{\mathbf{p}_{-n}} \left\| \frac{\partial \mathbf{I}_n}{\partial \mathbf{p}_{-n}} \frac{\partial \tilde{F}_n}{\partial \mathbf{I}_{-n}} \right\|$$

$$\leq \left\| \frac{\partial \mathbf{I}_n}{\partial \mathbf{p}_{-n}} \right\| \sup_{\mathbf{I}_{-n}} \left\| \frac{\partial \tilde{F}_n}{\partial \mathbf{I}_{-n}} \right\|;$$

When

$$\left\| \frac{\partial \mathbf{I}_n}{\partial \mathbf{p}_{-n}} \right\| < \frac{1}{\sup_{\mathbf{I}_n} \left\| \frac{\partial \tilde{F}_n}{\partial \mathbf{I}_n} \right\|}, \qquad (A.5)$$

$$\frac{\|F_n(\mathbf{p}_{-n}) - F_n(\check{\mathbf{p}}_{-n})\|}{\|\mathbf{p}_{-n} - \check{\mathbf{p}}_{-n}\|} < 1. \qquad (A.6)$$

The uniqueness of equilibrium follows immediately from Theorem 17.8.   □

A.13 Proof of Lemma 19.5

Proof Define the upper contour sets of $U(\mathbf{R})$ as

$$S_\alpha = \{\mathbf{R} \succeq \mathbf{0} | U(\mathbf{R}) \geq \alpha\}, \tag{A.1}$$

where symbol $\succeq$ denotes vector inequality and $\mathbf{R} \succeq \mathbf{0}$ means each element of $\mathbf{R}$ is non-negative. According to Proposition C.9 of [66], $U(\mathbf{R})$ is strictly quasiconcave if and only if S_α is strictly convex for any real number α. When $\alpha < 0$, no points exist on the contour $U(\mathbf{R}) = \alpha$. When $\alpha = 0$, only $\mathbf{0}$ is on the contour $U(\mathbf{0}) = \alpha$. Hence, S_α is strictly convex when $\alpha \leq 0$. Now we investigate the case when $\alpha > 0$. S_α is equivalent to $S_\alpha = \{\mathbf{R} \succeq \mathbf{0} | \alpha P_C + \alpha P_T(\mathbf{R}) - R \leq 0\}$. Since $P_T(\mathbf{R})$ is strictly convex in $\mathbf{R}$, S_α is also strictly convex. Hence, we have the strict quasiconcavity of $U(\mathbf{R})$.

The partial derivative of $U(\mathbf{R})$ with r_i is

$$\frac{\partial U(\mathbf{R})}{\partial r_i} = \frac{P_C + P_T(\mathbf{R}) - RP'_T(\mathbf{R})}{(P_C + P_T(\mathbf{R}))^2} \triangleq \frac{\beta(r_i)}{(P_C + P_T(\mathbf{R}))^2}, \tag{A.2}$$

where $P'_T(\mathbf{R})$ is the first partial derivative of $P_T(\mathbf{R})$ with respect to r_i. According to Lemma 19.5, if r_i^* exists such that $\frac{\partial U(\mathbf{R})}{\partial r_i}\Big|_{r_i=r_i^*} = 0$, it is unique, i.e. if there is a r_i^* such that $\beta(r_i^*) = 0$, it is unique. In the following, we investigate the conditions when r_i^* exists.

The derivative of $\beta(r_i)$ is

$$\beta'(r_i) = -RP''_T(\mathbf{R}) < 0, \tag{A.3}$$

where $P''_T(\mathbf{R})$ is the second partial derivative of $P_T(\mathbf{R})$ with respect to r_i. Hence, $\beta(r_i)$ is strictly decreasing. According to the L'Hopital's rule, it is easy to show that

$$\begin{aligned}
\lim_{r_i \to \infty} \beta(r_i) &= \lim_{r_i \to \infty} (P_C + P_T(\mathbf{R}) - RP'_T(\mathbf{R})) \\
&= \lim_{r_i \to \infty} \left(\frac{P_C + P_T(\mathbf{R}) - RP'_T(\mathbf{R})}{r_i} r_i \right) \\
&= \lim_{r_i \to \infty} \left(\frac{P'_T(\mathbf{R}) - P'_T(\mathbf{R}) - RP''_T(\mathbf{R})}{1} r_i \right) \\
&= \lim_{r_i \to \infty} -P''_T(\mathbf{R})Rr_i < 0.
\end{aligned} \tag{A.4}$$

Besides,

$$\begin{aligned}
\lim_{r_i \to 0} \beta(r_i) &= \lim_{r_i \to 0} (P_C + P_T(\mathbf{R}) - RP'_T(\mathbf{R})) \\
&= P_C + P_T(\mathbf{R}_i^{(0)}) - R_i^{(0)} P'_T(\mathbf{R}_i^{(0)}),
\end{aligned} \tag{A.5}$$

where $\mathbf{R}_i^{(0)} = [r_1, r_2, \ldots, r_{i-1}, 0, r_{i+1}, \ldots, r_K]^T$ and $R_i^{(0)} = \sum_{j \neq i} r_j$.

(1°) When $P_C + P_T(\mathbf{R}_i^{(0)}) - R_i^{(0)} P'_T(\mathbf{R}_i^{(0)}) \geq 0$, $\lim_{r_i \to 0} \beta(r_i) \geq 0$. Together with (A.4), we see that r_i^* exists and $U(\mathbf{R})$ is first strictly increasing and then strictly decreasing in r_i.

(2°) When $P_C + P_T(\mathbf{R}_i^{(0)}) - R_i^{(0)} P_T'(\mathbf{R}_i^{(0)}) < 0$, $\lim_{r_i->0} \beta(r_i) < 0$. Together with (A.3) and (A.4), t_i^* does not exist. However, $U(\mathbf{R})$ is always strictly decreasing in r_i. Hence, $U(\mathbf{R})$ is maximized at $r_i = 0$.

Lemma 19.5 is readily obtained.   □

A.14   Proof of Theorem 21.3

Proof   In [105], it has been shown that a Nash equilibrium exists in a non-cooperative game if for any n (1) $\mathbf{p}_n$ is a non-empty, convex, and compact subset of some Euclidean space $\mathfrak{R}^L$ and (2) $u_n(\mathbf{p}_n, \mathbf{p}_{-n})$ is continuous and quasiconcave in $\mathbf{p}_n$, both of which are satisfied in our non-cooperative energy-efficient control game. Hence, the existence of the equilibrium immediately follows. In a point-to-point energy-efficient transmission, the necessary and sufficient condition for a data rate vector of user n, $\mathbf{r}_n^o = [r_n^{(1)o}, r_n^{(2)o}, \dots, r_n^{(K)o}]$, to be globally optimal is given by, for any subchannel i:

(i)   if $\frac{P_c + \sum_{j \neq i} p_n^{(j)}}{\sum_{j \neq i} r_n^{(j)}} \geq \left. \frac{\partial(\sum_j p_n^{(j)})}{\partial r_n^{(i)}} \right|_{\mathbf{r}_n = \mathbf{r}_n^{(i0)}}$, $\left. \frac{\partial u_n(\mathbf{p}_n, \mathbf{p}_{-n})}{\partial r_n^{(i)}} \right|_{\mathbf{r}_n = \mathbf{r}_n^o} = 0$, i.e. $\left. \frac{\partial(\sum_j p_n^{(j)})}{\partial r_n^{(i)}} \right|_{\mathbf{r}_n = \mathbf{r}_n^o} = \frac{1}{u(\mathbf{p}_n^o, \mathbf{p}_{-n})}$;

(ii)   otherwise, $r_n^{(i)o} = 0$,

where

$$\mathbf{r}_n^{(i0)} = [r_n^{(1)o}, r_n^{(2)o}, \dots, r_n^{(i-1)o}, 0, r_n^{(i+1)o}, \dots, r_n^{(K)o}].$$

By transformation of parameters, $\frac{\partial f}{\partial r_n^{(i)}} = \frac{\partial f}{\partial p_n^{(i)}} \Big/ \frac{\partial r_n^{(i)}}{\partial p_n^{(i)}} = \frac{\partial f}{\partial p_n^{(i)}} \frac{1}{R'(\eta_n^{(i)}) \gamma_n^{(i)}}$, where $R'()$ is the first-order derivative of $R()$ and $\gamma_n^{(i)} = \frac{\eta_n^{(i)}}{p_n^{(i)}} = \frac{g_{nn}^{(i)}}{\sum_{j=1, j \neq n}^N p_j^{(i)} g_{jn}^{(i)} + \sigma^2}$. Hence, we have the following equivalent condition for each user. For any subchannel i,

(i)   if $\frac{\sum_{j \neq i} r_n^{(j)}}{p_c + \sum_{j \neq i} p_n^{(j)}} \leq R'(0) \gamma_n^{(i)}$, $\left. \frac{\partial u_n(\mathbf{p}_n, \mathbf{p}_{-n})}{\partial p_n^{(i)}} \right|_{\mathbf{p}_n = \mathbf{p}_n^o} = 0$, i.e.

$$R'(\gamma_n^{(i)} p_n^{(i)o}) \gamma_n^{(i)} = u(\mathbf{p}_n^o, \mathbf{p}_{-n}); \tag{A.1}$$

(ii)   otherwise, $p_n^{(i)o} = 0$.

It is easy to see that the network achieves an equilibrium if and only if the power settings of all users satisfy the above conditions. Theorem 21.3 is readily obtained.

□

A.15   Proof of Lemma 21.4

Proof $p_n^o = f_n(\mathbf{p}_{-n}) = \arg\max_{p_n} u_n(p_n, \mathbf{p}_{-n})$. Since $u_n(0, \mathbf{p}_{-n}) = 0$ and $u_n(p_n, \mathbf{p}_{-n}) > 0$ for any $p_n > 0$, $f_n(\mathbf{p}_{-n}) > 0$ and we have the positivity. Define $I_n = \sum_{j=1, j \neq n}^N p_j g_{jn}$

and $\gamma_n = \frac{g_{nn}}{1+\sigma^2}$. According to (A.1), p_n^o satisfies

$$R'(\gamma_n p_n^o)\gamma_n = u(p_n^o, \mathbf{p}_{-n}) = \frac{R(\gamma_n p_n^o)}{p_c + p_n^o}. \tag{A.1}$$

Substituting $R(\eta) = w\log(1+\eta)$ in to (A.1), we have the following equivalent condition,

$$Q(p_n^o, I) =$$

$$g_{nn}(p_c + p_n^o) - (p_n^o g_{nn} + I + \sigma^2)\log\left(1 + \frac{p_n^o g_{nn}}{I + \sigma^2}\right) \tag{A.2}$$

$$= 0.$$

Hence, $\frac{\partial p_n^o}{\partial I} = -\frac{\partial Q}{\partial I} \Big/ \frac{\partial Q}{\partial p_n^o} = \frac{p_n^o \gamma_n - \log(1+p_n^o \gamma_n)}{g_{nn}\log(1+p_n^o \gamma_n)}$. Since $x > \log(1+x)$ for all $x > 0$, we have $\frac{\partial p_n^o}{\partial I} > 0$. The monotonicity follows immediately. Furthermore,

$$\frac{\partial^2 p_n^o}{\partial I^2} = \frac{\partial \frac{\partial p_n^o}{\partial I}}{\partial I}$$

$$= -\frac{p_n^o(-p_n^o \gamma_n + (1 + p_n^o \gamma_n)\log(1 + p_n^o \gamma_n))}{(I + \sigma^2)(I + \sigma^2 + p_n^o g_{nn})\log(1 + p_n^o \gamma_n)^2}. \tag{A.3}$$

We can easily show that $(1 + x)\log(1 + x) > x$ for all $x > 0$ since $(1 + 0)\log(1 + 0) = 0$ and $(1 + x)\log(1 + x) - x$ has positive first-order derivative when $x > 0$. Thus, $\frac{\partial^2 p_n^o}{\partial I^2} < 0$. Since I is a linear combination of $\mathbf{p}_{-n}$, $f_n(\mathbf{p}_{-n})$ is strictly concave in $\mathbf{p}_{-n}$. Let $F(\alpha) = \alpha f_n(\mathbf{p}_{-n}) - f_n(\alpha \mathbf{p}_{-n})$ and we need to show $F(\alpha) > 0$ for all $\alpha > 1$ to prove the scalability. Note that $F(1) = 0$ and $\frac{\partial^2 F(\alpha)}{\partial \alpha^2} > 0$, it is sufficient to show that $\frac{\partial F(\alpha)}{\partial \alpha}\Big|_{\alpha=1} = f_n(\mathbf{p}_{-n}) - \mathbf{p}_{-n}f_n'(\mathbf{p}_{-n}) > 0$, which is obvious because of the positivity and concavity of $f_n(\mathbf{p}_{-n})$. $\square$

References

[1] 3GPP TS 23.107, Quality of Service (QoS) concept and architecture, V6.2.0, 2004.

[2] European Commission. 2011. *EARTH: Driving the Energy Efficiency of Wireless Infrastructure to its Limits*, https://www.ict-earth.eu/ (last visited: 5 October 2013).

[3] 2011. *GreenTouch*. http://www.greentouch.org/(last visited: 5 October 2013).

[4] A. J. Goldsmith and S. G. Chua. 1997. Variable-rate variable-power MQAM for fading channels. *IEEE Trans. Commun.*, 45(October), 1218–1230.

[5] C. Xiong, G. Li, S. Zhang, Y. Chen, and S. Xu. 2011. Energy- and spectral-efficiency tradeoff in downlink OFDMA networks. *IEEE Trans. Wireless Commun.*, 10(11), 3874–3886.

[6] C. Xiong, G. Li, Y. Liu, Y. Chen, and S. Xu. 2013. Energy-efficient design for downlink OFDMA with delay-sensitive traffic. *IEEE Trans. Wireless Commun.*, 12(6), 3085–3095.

[7] G. Cao and M. Singhal. 2000. An adaptive distributed channel allocation strategy for mobile cellular networks. *J. Parallel and Dist. Comput.*, 60, 451–473.

[8] G. Song and Y. (G.) Li. 2005a. Cross-layer optimization for OFDM wireless networks - part I: theoretical framework. *IEEE Trans. Wireless Commun.*, 4(2), 614–624.

[9] G. Song and Y. (G.) Li. 2005b. Cross-layer optimization for OFDM wireless networks - part II: algorithm development. *IEEE Trans. Wireless Commun.*, 4(2), 625–634.

[10] G. Sun, J. Chen, W. Guo, and K. J. R. Liu. 2005. Signal processing techniques in network-aided positioning: a survey of state-of-the-art positioning designs. *IEEE Sig. Processing Mag.*, 22(July), 12–24.

[11] K. Bai and J. Zhang. 2006. Opportunistic multichannel Aloha: distributed multi-access control scheme for OFDMA wireless networks. *IEEE Trans. Veh. Tech.*, 55(March), 848–855.

[12] M. Hellebrandt and R. Mathar. 1999. Location tracking of mobiles in cellular radio networks. *IEEE Trans. Veh. Tech.*, 48(September), 1558–1862.

[13] M. Hellebrandt, R. Mathar, and M. Scheibenbogen. 1997. Estimating position and velocity of mobiles in a cellular radio network. *IEEE Trans. Veh. Tech.*, 46(February), 65–71.

[14] R. Prakash, N. Shivaratri, and M. Singhal. 1999. Distributed dynamic fault-tolerant channel allocation for mobile computing. *IEEE Trans. Veh. Tech.*, 48(November), 1874–1888.

[15] T. Nandagopal, T. Kim, X. Gao, and V. Bhargavan. 2000. Achieving MAC layer fairness in wireless packet networks. *Proc. of the ACM MobiCom'00* (August), pages 87-98.

[16] W. Dinkelbach. 1967. On nonlinear fractional programming. *Management Science*, 13(7), 492–498.

[17] ABI Research. 2008. Mobile networks go green-minimizing power consumption and leveraging renewable energy. Tech Report. ABI Research, New York.

[18] 3GPP, TR 25.913. *Requirements for Evolved UTRA and Evolved UTRAN*. http://www.3gpp.org/DynaReport/25913.htm.

[19] 3GPP TS 25.321. 2007. 3rd Generation Partnership Project; Technical Specification Group Radio Access Network; Medium Access Control (MAC) Protocol Specification. July. http://www.3gpp.org/DynaReport/25321.htm.

[20] 3GPP TSG-RAN R1-050764. 2005. Inter-cell interference handling for EUTRA. Ericsson. (August).

[21] 3GPP TSG RAN R1-050841. 2005. Further analysis of soft frequency reuse scheme. Huawei. (September).

[22] A. E. Gamal, C. Nair, B. Prabhakar, E. Uysal-Biyikoglu, and S. Zahedi. 2002. Energy-efficient scheduling of packet transmissions over wireless networks. *Proc. IEEE INFO-COM 2002*, 3(June), pages 1773–1782.

[23] A. Gjendemsjø, D. Gesbert, G. E. Øien, and S. G. Kiani. 2006. Optimal power allocation and scheduling for two-cell capacity maximization. *Proc. RAWNET (WiOpt)* (April).

[24] A. Goldsmith. 2005. *Wireless Communications*. Cambridge University Press.

[25] A. J. Goldsmith and P. P. Varaiya. 1997. Capacity of fading channels with channel side information. *IEEE Trans. Inf. Theory*, 43(6), 1986–1992.

[26] A. J. Paulraj, D. A. Gore, R. U. Nabar, and H. Bolcskel. 2002. An overview of MIMO communications - a key to gigabit wireless. *Proc. of the IEEE*, 92(2), 198–218.

[27] A. Kumar, D. Manjunath, and J. Kuri. 2008. *Wireless Networking*. Morgan Kaufmann.

[28] A. Qayyum, L. Viennot, and A. Laouiti. 2002. Multipoint relaying: an efficient technique for flooding in mobile wireless networks. *Proc. of 35th Annual Hawaii Int. Conf. on Sys. Sci.*, pages 3866– 3875.

[29] A. Raghunathan, N. Jha, and S. Dey. 1998. *High-Level Power Analysis and Optimization*. Norwell, MA: Kluwer Academic Publishers.

[30] A. Soysal and S. Ulukus. 2009. Optimality of beam-forming in fading MIMO multiple access channels. *IEEE Trans. Commun.*, 57(4), 1171–1183.

[31] A. T. Koc, S. Talwar, A. Papathanassiou, R. Yang, N. Himayat, and H. Yin. 2008. *IEEE C802.16m-08/666r2: Uplink Power Control Recommendations for IEEE 802.16m*. (July).

[32] I. F. Akyildiz, J. McNair, L. Carrasco, R. Puigjaner, and Y. Yesha. 1999. Medium access protocols for multimedia traffic in wireless networks. *IEEE Network*, 13(4), 39–48.

[33] E. Altman, T. Basar, T. Jimenez, and N. Shimkin. 2000. Competitive routing in networks with polynomial cost. *Proc. IEEE INFOCOM*. (March), 1586–1593.

[34] A. Ashwin. *Green Communications: Annotated Review and Research Vision*. Virginia Tech.

[35] M. Andrews, K. Kumaran, K. Ramanan, A. Stolyar, and P. Whiting. 2001. Providing quality of service over a shared wireless link. *IEEE Commun. Magazine* (February), 150–154.

[36] M. Andrews. 2004. Instability of the proportional fair scheduling algorithm for HDR. *IEEE Trans. Wireless Commun.*, 3(5), 1422–1426.

[37] M. Andrews, S. Borst, F. Dominique, P. Jelenkovic, K. Kumaran, K. G. Ramakrishnan, and P. Whiting. 2000. Dynamic bandwidth allocation algorithms for high-speed data wireless networks. Tech. report. Bell Labs Technical Memorandum.

[38] Atheros Communications. 2003. *Power Consumption and Energy Efficiency: Comparisons of WLAN Products*. White Paper.

[39] B. Debaillie, B. Bougard, 1, G. Lenoir, G. Vandersteen, and F. Catthoor. 2006. Energy-scalable OFDM transmitter design and control. *Proc. the 43rd Annual. Conf. Design Automation*, San Francisco (July), pages 536–541.

[40] B. G. Lee, D. Park, and H. Seo. 2009. *Wireless Communications Resource Management*. John Wiley & Sons (Asia).

[41] B. H. Calhoun, D. C. Daly, N. Verma, D. Finchelstein, D. D.Wentzloff, A.Wang, S. H. Cho, and A. P. Chandrakasan. 2005. Design considerations for ultra-low energy wireless microsensor nodes. *IEEE Trans. Comput.*, 54(6), 727–740.

[42] B. Prabhakar, E. U. Biyikoglu, and A. E. Gamal. 2001. Energy-efficient transmission over a wireless link via lazy packet scheduling. *Proc. IEEE INFOCOM 2001*, 1(April), 386–394.

[43] J. Bennett and H. Zhang. 1996. Hierarchical packet fair queueing algorithms. *Proc. SIGCOMM.* (August), 143–156.

[44] D. Bertsekas and R. Gallager. 1987. *Data Networks*. Englewood Cliffs, NJ: Prentice-Hall.

[45] S. Borst. 2003. User-level performance of channel-aware scheduling algorithms in wireless data networks. *Proc. IEEE INFOCOM 2003* (March), pages 321–331.

[46] C. Drane, M. Macnaughtan, and C. Scott. 1998. Positioning GSM telephones. *IEEE Communications Magazine*, 36(April), 45–54.

[47] C. Isheden and G. P. Fettweis. 2011. Energy-efficient link adaptation with transmitter CSI. *Proc. IEEE WCNC 2011* (March), 1381–1386.

[48] C. Schurgers. 2002. Energy-aware wireless communications. Ph.D. thesis, University of California Los Angeles.

[49] Y. Cao and V. O. K. Li. 2001. Scheduling algorithms in broadband wireless networks. *Proceedings of the IEEE*, 89(1), 76–87.

[50] E. Castillo. 1988. *Extreme Value Theory in Engineering*. New York: Academic Press.

[51] C.-J. Chen and L.-C. Wang. 2004. A unified capacity analysis for wireless systems with joint antenna and multiuser diversity in Nakagami fading channels. *Proc. IEEE Int. Conf. on Commun.*(June).

[52] J. Chuang and N. Sollenberger. 2000. Beyond 3G: wideband wireless data access based on OFDM and dynamic packet assignment. *IEEE Commun. Magazine* (July), 78–87.

[53] Cisco. Cisco Aironet 802.11A/B/G Wireless CardBus Adapter data sheet.

[54] D. Bertsekas and R. Gallager. 1992. *Data Networks*. 2nd edn. Englewood Cliffs, NJ: Prentice Hall.

[55] D. Fudenberg and J. Tirole. 1991. *Game Theory*. Cambridge, MA: MIT Press.

[56] D. Gesbert and S. Alouini. 2004 (June). How much feedback is multi-user diversity really worth. *Proc. IEEE ICC 2004*, 1, 234–238.

[57] D. Gesbert, S. G. Kiani, A. Gjendemsjø, and G. E. Øien . 2007. Adaptation, coordination, and distributed resource allocation in interference-limited wireless networks. *Proc. of the IEEE*, 95(12), 2393–2409.

[58] D. Goodman and N. Mandayam. 2000. Power control for wireless data. *Personal Communications, IEEE*, 7(2), 48–54.

[59] D. Tse and P. Viswanath. 2005. *Fundamentals of Wireless Communication*. Cambridge University Press.

[60] J. G. Dai. 1995. On positive Harris recurrence of multiclass queueing networks: a unified approach via fluid limit models. *Annals Applied Probab.*, 5, 49–77

[61] H. A. David. 1970. *Order Statistics*. New York: John Wiley & Sons.

[62] Decina, M. and Toniatti, T. 1992. Bandwidth allocation and selective discarding for a variable bit rate video and bursty data calls in ATM networks. *Int'l J. Digital Analog Commun. Systems*, 5(2), 85–96.

[63] Draft IEEE std 802.16e/D9. 2005. IEEE standard for local and metropolitan area networks, part 16: air interface for fixed and mobile broadband wireless access systems. June.

[64] E. L. Hahne. 1991. Round-robin scheduling for max-min fairness in data networks. *IEEE J. Sel. Areas Commun.*, 9(7), 1024–1039.

[65] E. Uysal-Biyikoglu, B. Prabhakar, and A. E. Gamal. 2002. Energy-efficient packet transmission over a wireless link. *IEEE/ACM Trans. Networking*, 10(7), 487–499.

[66] E. Wolfstetter. 1999. *Topics in Microeconomics: Industrial Organization, Auctions, and Incentives*. Cambridge University Press.

[67] E. P. Kim and N. R. Shanbhag. 2011. An energy-efficient multiple-input multiple-output (MIMO) detector architecture. *2011 IEEE Workshop on Signal Processing Systems (SiPS)* (October).

[68] Ericsson White Paper 2007. *Sustainable Energy Use in Mobile Communications*, EAB-07:021801 Ericsson AB 2007 (October)

[69] A. Eryilmaz, R. Srikant, and J. Perkins. 2005. Stable scheduling policies for fading wireless channels. *IEEE/ACM Trans. Networking*, 13(2), 411–424.

[70] F. Burghardt, S. Mellers, and J. Rabaey. 2002. *The PicoRadio Test Bed*. (December).

[71] F. Meshkati, H. V. Poor, S. C. Schwartz, and N. B. Mandayam. 2006. An energy efficient approach to power control and receiver design in wireless networks. *IEEE Trans. Commun.*, 5(1), 3306–3315.

[72] F. P. Kelly, A. Maulloo, and D. Tan. 1998. Rate control for communication networks: shadow prices, proportional fairness and stability. *J. Operat. Res. Soc.*, 49, 237–252.

[73] A. Federgruen and H. Groenevelt. 1986. The greedy procedure for resource allocation problems: necessary and sufficient conditions for optimality. *Operations Research*, 34(6), 909–918.

[74] G. P. Fettweis and E. Zimmermann. 2008. ICT energy consumption-trends and challenges. *Proc. 11th Int. Symp. on Wireless Personal Multimedia Communications (WPMC '08)*, pp. 1–6.

[75] F. M. Gardner. 1979. *Phase Lock Techniques*. New York: John Wiley & Sons.

[76] G. Caire, G. Taricco, and E. Biglieri. 1999. Optimum power control over fading channels. *IEEE Trans. Inf. Theory.*, 45(5), 1468–1489.

[77] G. Ganesan, G. Song, and Y. (G.) Li. 2005. Asymptotic throughput analysis of distributed multichannel random access schemes. *Proc. IEEE ICC 2005*, pages 3637–3641.

[78] G. J. Foschini and Z. Miljanic. 1993. A simple distributed autonomous power control algorithm and its convergence. *IEEE Trans. Veh. Tech.*, 42, 641–646.

[79] G. L. Stüber. 2001. *Principles of Mobile Communication*. Norwell, MA: Kluwer Academic Publishers.

[80] G. P. Fettweis and E. Zimmermann. 2008. ICT energy consumptiontrends and challenges. *Proc. 11th Int. Symp. Wireless Personal Multimedia Commun.* (September).

[81] G. W. Miao and Z. Niu. 2006. Practical feedback design based OFDM link adaptive communications over frequency selective channels. *Proc. IEEE ICC 2006*. (June), 4624–4629.

[82] J. Galambos. 1978. *The Asymptotic Theory of Extreme Order Statistics*. New York: John Wiley & Sons.

[83] A. J. Goldsmith and S. G. Chua. 1997. Variable-rate variable-power MQAM for fading channel. *IEEE Trans. Commun.*, 45(10), 1218–1230.

[84] A. J. Goldsmith and M. Effros. 2001. The capacity region of broadcast channels with intersymbol interference and colored Gaussian noise. *IEEE Trans. Inf. Theory*, 47(1), 219–240.

[85] D. J. Goodman and N. B. Mandayam. 2000. Power control for wireless data. *IEEE Personal Commun.*, 7(April), 48–54.

[86] H. Dai, A. F. Molisch, and H. V. Poor. 2004. Downlink capacity of interference limited MIMO systems with joint detection. *IEEE Trans. Wireless Commun.*, 3(2), 442–453.

[87] H. Kim, C. Chae, G. Veciana, and R. Heath. 2009. A cross-layer approach to energy efficiency for adaptive MIMO systems exploiting spare capacity. *IEEE Trans. Wireless Commun.*, 8(8), 4264–4275.

[88] H. Schoeneich and P. A. Hoeher. 2003. Single antenna interference cancellation: iterative semi-blind algorithm and performance bound for joint maximum likelihood interference cancellation. *Proc. IEEE Global Commun. Conf. 2003*, pages 1716–1720.

[89] N. B. Haaser and J. A. Sullivan. 1971. *Real Analysis*. New York: Van Nostrand Reinhold.

[90] L. M. C. Hoo, J. Tellado, and J. M. Cioffi. 2000. FDMA-based multiuser transmit optimization for broadcast channels. *Proc. IEEE Wireless Commun. Networking Conf.*, 2(September), 597–602.

[91] I. Chih-Lin. 2012. *Green Evolution of Mobile Communications (CMCC Perspective)*.

[92] I. E. Telatar. 1999. Capacity of multi-antenna Gaussian channels. *Europ. Trans.Telec.*, 10(November), 585–595.

[93] IEEE. 2004. IEEE 802.16e-2004, part 16: air interface for fixed and mobile broadband wireless access systems - amendment for physical and medium access control layers for combined fixed and mobile operation in licensed bands. November.

[94] IEEE, P802.11D5, Draft Standard IEEE 802.11. 1996. Wireless LAN medium access control (MAC) and physical layer (PHY) Spec. May.

[95] ITU-R Recommendation M.1225. 1997. *Guidelines for Evaluation of Radio Transmission Technologies for IMT-2000*.

[96] J. Chen, K. M. Sivalingam, P. Agrawal, and S. Kishore. 1998. A comparison of MAC protocols for wireless local networks based on battery power consumption. *Proc. IEEE INFOCOM 1998*, 1(March), 150–157.

[97] J. G. Andrews. 2005. Interference cancellation for cellular systems: a contemporary overview. *IEEE Wireless Commun.*, 12(2), 19–29.

[98] J. Jang and K. B. Lee. 2003. Transmit power adaptation for multiuser OFDM systems. *IEEE J. Sel. Areas Commun.*, 21(February), 171–178.

[99] J. Lee, J. Han, and J. Zhong. 2009. MIMO technologies in 3GPP LTE and TE Advanced. *EURASIP Journal on Wireless Communications and Networking*, 2009.

[100] J. Li, B. K. Letaief, and Z. Cao. 2003. Co-channel interference cancellation for space-time coded OFDM systems. *IEEE Trans. Wireless Commun.*, 2(1), 41–49.

[101] J. Lipman, P. Boustead, and J. Judge. 2003. Neighbor aware adaptive power flooding in mobile ad hoc networks. *Int. J. Foundations of Comp. Sci.*, 14(April), 237–252.

[102] J. M. Cioffi. 1999. *A Multicarrier Primer*. Stanford University/Amati T1E1 (November).

[103] J. Mo and J. Walrand. 2000. Fair end-to-end window-based congestion control. *IEEE/ACM Trans. Networking*, 8(5), 556–567.

[104] J. Rabaey, J. Ammer, J. L. da Silva Jr., and D. Patel. 2000. PicoRadio: *Ad-hoc* wireless networking of ubiquitous low-energy sensor/monitor nodes. *IEEE VLSI*, pages 9–12.

[105] J. W. Friedman. 1977. *Oligopoly and the Theory of Games*. Amsterdam: North-Holland.

[106] J. Zander. 1992. Performance of optimum transmitter power control in cellular radio systems. *IEEE Trans. Veh. Tech.*, 41, 57–62.

[107] V. Jacobson. 1998. Congestion avoidance and control. *Computer Communications Review*, 18(4), 314–329.

[108] Z. Jiang, Y. Ge, and Y. Li. 2005. Max-utility wireless resource management for best effort traffic. *IEEE Trans. Wireless Commun.*, 4(1), 100–111.

[109] K. Begain, G. I. Rozsa, A. Pfening, and M. Telek. 2002. Performance analysis of GSM networks with intelligent underlay-overlay. *Proc. 7th Int. Symp. on Comp. and Commun. (ISCC 2002)*, pages 135–141.

[110] K. C. Kiwiel and K. Murty. 2005. Convergence of the steepest descent method for minimizing quasiconvex functions. *J. of Optimization Theory and Applications*, 89(September), 221–226.

[111] K. Feher. 1987. *Advanced Digital Communications*. Englewood Cliffs, NJ: Prentice-Hall.

[112] K. K. Leung, B. McNair, L. J. Cimini, and J. H. Jr. Winters. 2002. Outdoor IEEE 802.11 cellular networks: MAC protocol design and performance. *Proc. IEEE ICC 2002*, 1, 595–599.

[113] K. Kar, S. Sarkar, and L. Tassiulas. 2004. Achieving proportional fairness using local information in Aloha networks. *IEEE Trans. Autom. Control*, 49(October), 1858–1863.

[114] K. Lahiri, A. Raghunathan, S. Dey, and D. Panigrahi. 2002. Battery-driven system design: a new frontier in low power design. *Proc. Intl. Conf. on VLSI Design* (January), 261–267.

[115] K. Xu, M. Gerla, and S. Bae. 2002. How effective is the IEEE 802.11 RTS/CTS handshake in ad hoc networks? *Proc. IEEE Globecom 2002* (November).

[116] F. Kelly 1997. Charging and rate control for elastic traffic. *European Trans. On Telecommunications*, 8, 33–37.

[117] F. Kelly, A. Maulloo, and D. Tan. 1998. Rate control in communication networks: shadow prices, proportional fairness and stability. *Journal of the Operational Research Society*, 49, 237–252.

[118] D. Kivanc, G. Li, and H. Liu. 2003. Computationally efficient bandwidth allocation and power control for OFDMA. *IEEE Trans. Wireless Commun.*, 2(6), 1150–1158.

[119] L. Benini, A. Bogliolo, and G. De Micheli. 2000. A survey of design techniques for system-level dynamic power management. *IEEE Trans. VLSI Syst.*, 8(3), 299–316.

[120] L. M. Feeney and M. Nilsson. 2001 (April). Investigating the energy consumption of a wireless network interface in an *ad hoc* networking environment. *Proc. IEEE INFOCOM 2001*, 3, 1548–1557.

[121] L. P. Qian, Y. J. Zhang, and J. Huang. 2009. Mapel: achieving global optimality for a non-convex wireless power control problem. *IEEE Trans. Wireless Commun.*, 8(3), 1553–1563.

[122] R. Laroia, S. Uppala, and J. Li. 2004. Designing a mobile broadband wireless access network. *IEEE Sig. Processing Mag.* (September), 20–28.

[123] L. Li, and A. J. Goldsmith. 2001. Optimal resource allocation for fading broadcast channels – part I: ergodic capacity. *IEEE Trans. Inf. Theory.*, 47(3), 1083–1102.

[124] Y. (G) Li, L. J. Cimini, Jr., and N. R. Sollenberger. 1998. Robust channel estimation for OFDM systems with rapid dispersive fading channels. *IEEE Trans.Commun.*, 46(July), 902–915.

[125] Y. (G) Li. 2000. Pilot-symbol-aided channel estimation for OFDM in wireless systems. *IEEE Trans. Veh. Tech.*, 49(4), 1207 –1215.

[126] Y. (G) Li., J. C. Chuang, and N. R. Sollenberger. 1999. Transmitter diversity for OFDM system and its impact on high-rate data wireless networks. *IEEE J. Select. Areas Commun.*, 17(7), 1233–1243.

[127] P. Liu, M. Honig, and S. Jordan. 2000. Forward link CDMA resource allocation based on pricing. *Proc. IEEE Wireless Commun. Networking Conf.*, 2(September), 619–623.

[128] P. Liu, R. Berry, and M. Honig. 2003. Delay-sensitive packet scheduling in wireless networks. *Proc. IEEE Wireless Commun. Networking Conf.* (March).

[129] X. Liu, E. Chong, and N. Shroff. 2001. Opportunistic transmission scheduling with resource-sharing constraints in wireless networks. *IEEE J. Select. Areas Commun.*, 19(10), 2053–2064,.

[130] X. Liu, N. B. Shroff, and E. K. P. Chong. 2004. Opportunistic scheduling: an illustration of cross-layer design. *Telecommunications Review*, 16(6), 947–959.

[131] M. A. Haleem and R. Chandramouli. 2005. Adaptive downlink scheduling and rate selection: a cross-layer design. *IEEE J. Sel. Areas Commun.*, 23(January), 1287–1297.

[132] M. C. Necker. 2007. Coordinated fractional frequency reuse. *Proc. 10th ACM Symp. on Mod., Anal., and Sim. of wireless and Mob. Syst.* (October), 296–305.

[133] M. Chiang, P. Hande, T. Lan, and C. W. Tan. 2008. Power control in wireless cellular networks. *Foundations and Trends in Networking*, 2(4), 381–533.

[134] M. Haenggi and D. Puccinelli. 2005. Routing in *ad hoc* networks: a case for long hops. *IEEE Commun. Magazine*, 43(10), 112–119.

[135] M. Pedram. 2001. Power optimization and management in embedded systems. *Proc. ASP-DAC 2001*. (February), 239–244.

[136] M. Pischella and J.-C. Belfiore. 2008. Power control in distributed cooperative OFDMA cellular networks. *IEEE Trans. Wireless Commun.*, 7(5), 1900–1906.

[137] J. K. MacKie-Mason and H. R. Varian. 1995. Pricing congestible network resources. *IEEE J. Select. Areas Commun.*, 13(7), 1141–1149.

[138] S. I. Maniatis, E. G. Nikolouzou, and I. S. Venieris. 2003. QoS issues in the converged 3G wireless and wired networks. *IEEE Commun. Magazine* (August), 44–53.

[139] N. McKeown, A. Mekkittikul, V. Anantharam, and J. Walrand. 1999. Achieving 100% throughput in an input-queued switch. *IEEE Trans.Commun.*, 47(8), 1200–1267.

[140] S. P. Meyn and R. L. Tweedie. 1993. *Markov Chains and Stochastic Stability*. Springer-Verlag.

[141] A. Mordecai, A. 1976. *Nonlinear Programming: Analysis and Methods*. Englewood Cliffs, NJ: Prentice-Hall.

[142] N. Feng, S. C. Mau, and N. B. Mandayam. 2004. Pricing and power control for joint network-centric and user-centric radio resource management. *IEEE Trans. Commun.*, 52(9), 1547–1557.

[143] N. Moezzi-Madani, T. Thorolfsson, J. Crop, P. Chiang, and W. Davis. 2011. An energy-efficient 64-QAM MIMO detector for emerging wireless standards. *Design, Automation & Test in Europe Conference and Exhibition (DATE), 2011*. (March).

[144] S. Nanda, K. Balachandran, and S. Kumar. 2000. Adaptation techniques in wireless packet data services. *IEEE Commun. Magazine* (January), 54–64.

[145] N. Narendran. 2009. *Overview of Recent Technology Trends in Energy-Efficient Lighting*. http://www.elcomaindia.com/tech trends energy lighting.pdf (last visited: 20131022).

[146] M. J. Neely, E. Modiano, and C. E. Rohrs. 2003. Power allocation and routing in multi-beam satellites with time varying channels. *IEEE/ACM Trans. Networking*, 11(1), 138–152.

[147] O. Edfors, M. Sandell, J.-J. van de Beek, and S.K. Wilson, 1998. OFDM channel estimation by singular value decomposition. *IEEE Trans. on Communications*, 46(July), 931–939.

[148] P. D. Straffin. 1993. *Game Theory and Strategy*. Mathematical Association of America.

[149] P. Gupta and A. L. Stolyar. 2006. Optimal throughput allocation in general random-access networks. *Proc. Conf. Information Science and Systems (CISS)*. 1254–1259.

[150] P. Nuggehalli, V. Srinivasan, and R. R. Rao. 2002 (June). Delay constrained energy efficient transmission strategies for wireless devices. *Proc. IEEE INFOCOM 2002*. 3, 1765–1772.

[151] P. P. Pham, S. Perreau, and A. Jayasuriya. 2005. New cross-layer design approach to *ad hoc* networks under Rayleigh fading. *IEEE J. Sel. Areas Commun.*, 23(January), 28–39.

[152] P. Viswanath, D. N. C. Tse, and R. Laroia. 2002. Opportunistic beamforming using dumb antennas. *IEEE Trans. Inf. Theory.*, 48(6), 1277–1294.

[153] Parekh, A. and Gallager, R. 1993. A generalized processor sharing approach to flow control in integrated services networks: The single node case. *IEEE/ACM Trans. Networking*, 1(3), 344–357.

[154] J. Pickands III. 1968. Moment convergence of sample extremes. *Annals of mathematical statistics*, 39(3), 881–889.

[155] Q. Li, G. Li, W. Lee, M. Lee, D. Mazzarese, B. Clerckx, and Z. Li. 2010. MIMO techniques in WiMAX and LTE: a feature overview. *IEEE Commun. Magazine*, 48(5), 86–92.

[156] Q. Li, X. Lin, J. Zhang, and W. Roh. 2009. Advancement of MIMO technology in WiMAX: from IEEE 802.16d/e/j to 802.16m. *IEEE Commun. Magazine*, 47(6), 100–107.

[157] Q. H. Spencer, A. L. Swindlehurst, and M. Haardt. 2004. Zero-forcing methods for downlink spatial multiplexing in multiuser MIMO channels. 52(2), 461–471.

[158] X. Qiu and K. Chawla. 1999. On the performance of adaptive modulation in cellular systems. *IEEE Trans. Commun.*, 47(6), 884–895.

[159] R. D. Yates. 1995. A framework for uplink power control in cellular radio systems. *IEEE J. Select. Areas Commun.* (September), 1341–1347.

[160] R. G. Bartle. 1964. *The Elements of Real Analysis*. New York: John Wiley & Sons.

[161] R. G. Gallager. 1968. *Information Theory and Relaible Communication*. New York: John Wiley & Sons.

[162] R. G. Gallager. 1988. Power limited channels: coding, multiaccess, and spread spectrum. *Proc. Conf. Inform. Sci. and Syst.*, 1(March).

[163] R. J. McEliece and W. E. Stark. 1984. Channels with block interference. *IEEE Trans. Inf. Theory.*, 30(1), 44–53.

[164] R. Mangharam, R. Rajkumar, S. Pollin, F. Catthoor, B. Bougard, L. Van der Perre, and I. Moeman. 2005. Optimal fixed and scalable energy management for wireless networks. *Proc. IEEE INFOCOM 2005*, 1(March), 114–125.

[165] R. Mazumdar, L. G. Mason, and C. Douligieris. 1991. Fairness in network optimal flow control: optimality of product forms. *IEEE Trans. Commun.*, 39(May), 775–782.

[166] R. Tobin. 2003. US army's BLUE radio. *Proc. SPIE, Unattended Ground Sensor Technologies and Applications V.* (April).

[167] Recommendation ITU-R M.1225. 1997. *Guidelines for Evaluation for of Radio Transmission Technologies for IMT-2000*.

[168] W. Rhee and J. M. Cioffi 2000. Increase in capacity of multiuser OFDM system using dynamic subcarrier allocation. *Proc. IEEE Veh.Tech. Conf.*, 2,1085–1089.

[169] R. T. Rockafellar. 1970. *Convex Analysis*. New Jersey: Princeton University Press.

[170] W. Rudin. 1976. *Principles of Mathematical Analysis*. McGraw-Hill.

[171] S. Boyd and L. Vandenberghe. 2004. *Convex Optimization*. Cambridge University Press.

[172] S. Cui, A. Goldsmith, and A. Bahai. 2004a. Energy-efficiency of MIMO and cooperative MIMO techniques in sensor networks. *IEEE J. Sel. Areas Commun.*, 22(August), 1089–1098.

[173] S. Cui, A. Goldsmith, and A. Bahai. 2004b. Joint modulation and multiple access optimization under energy constraints. *Proc. IEEE Globecom 2004*, 1(November), 151–155.

[174] S. K. Jayaweera. 2004. An energy-efficient virtual MIMO architecture based on V-BLAST processing for distributed wireless sensor networks. *Proc. IEEE SECON 2004* (October), 299–308.

[175] S. M. Ross. 1996. *Stochastic Process*. New York: John Wiley & Sons.

[176] S. Tombaz, A. Vastberg, and J. Zander. 2011. Energy- and cost-efficient ultra-high-capacity wireless access. *IEEE Commun. Magazine* (October), 18–24.

[177] S. Verdu. 1998. *Multiuser Detection*. Cambridge University Press.

[178] S. Verdu. 2002. Spectral efficiency in the wideband regime. *IEEE Trans. Inf. Theory.*, 48(6), 1319–1343.

[179] C. U. Saraydar, N. B. Mandayam, and D. J. Goodman. 2001. Pricing and power control in a multicell wireless data network. *IEEE J. Select. Areas Commun.*, 19(10), 1883–1892.

[180] M. Schwartz. 1990. *Information Transmission, Modulation, and Noise*. McGraw-Hill.

[181] V. Shah, N. B. Mandayam, and D. J. Goodman. 1998. Power control for wireless data based on utility and pricing. *Proc. IEEE PIMRC* (September),1427-1432.

[182] S. Shakkottai and A. L. Stolyar. 2002. Scheduling for multiple flows sharing a time-varying channel: the exponential rule. *Analytic Methods in Applied Probability*, 207, 185–202.

[183] C. E. Shannon. 1949. Communication in the Presence of Noise. *Proc. IRE*, 37(January), 10–21.

[184] S. Shenker. 1995. Fundamental design issues for the future Internet. *IEEE J. Select. Areas Commun.*, 13(7), 1176–1188.

[185] G. Song and Y. (G.) Li. 2003 (April). Adaptive subcarrier and power allocation in OFDM based on maximizing utility. *Proc. IEEE Veh.Tech. Conf.*, 2, 905–909 .

[186] G. Song and Y. (G.) Li. 2005. Cross-layer optimization for OFDM wireless network – part II: algorithm development. *IEEE Trans. Wireless Commun.*, 4(2), 625– 634.

[187] G. Song, Y. (G.) Li, L. J. Cimini, and H. Zheng, H. 2004. Joint channel aware and queue-aware data scheduling in multiple shared wireless channels. *Proc. IEEE Wireless Commun. Networking Conf.* (March).

[188] L. Song and N. B. Mandayam. 2001. Hierarchical SIR and rate control on the forward link for CDMA data users under delay and error constraints. *IEEE J. Select. Areas Commun.*, 19(10), 1871–1882.

[189] G. L. Stüber. 2000. *Principles of Mobile Communication*. 2nd edn. Norwell, MA: Kluwer Academic Publishers.

[190] T. Alpcan, T. Basar, R. Srikant, and E. Altman. 2002. CDMA uplink power control as a noncooperative game. *Wireless Networks*, 8, 659–670.

[191] T. Keller and L. Hanzo. 2000. Adaptive multicarrier modulation: a convenient framework for timefrequency processing in wireless communications. *Proc. of the IEEE*, 88(May), 611–640.

[192] L. Tassiulas and A. Ephremides. 1992. Stability properties of constrained queueing systems and scheduling for maximum throughput in multihop radio networks. *IEEE Trans. Automatic Control*, 37(12), 1936–1949.

[193] L. Tassiulas and A. Ephremides. 1993. Dynamic server allocation to parallel queues with randomly varying connectivity. *IEEE Trans. Inf. Theory.*, 39(2), 466–478.

[194] T. S. Rappaport. 1996. *Wireless Communications*. Englewood Cliffs, NJ: Prentice-Hall.

[195] D. Tse and S. Hanly. 1998. Multi-access fading channels – part I: polymatroid structure, optimal resource allocation and throughput capacities. *IEEE Trans. Inf. Theory*, 44(7), 2796–2815.

[196] D. N. Tse. 1997. Optimal power allocation over parallel Gaussian broadcast channel. *Proc. IEEE Int. Symp. on Inform. Theory* (June), 27.

[197] U. C. Kozat, I. Koutsopoulos, and L. Tassiulas. 2006. Cross-layer design for power efficiency and QoS provisioning in multi-hop wireless networks. *IEEE Trans. Wireless Commun.*, 5(1), 3306–3315.

[198] United States Department of Commerce. 2003. *United States Radio Spectrum Frequency Allocations Chart as of 2003*. http://www.ntia.doc.gov/osmhome/allochrt.pdf (last visited: 5 October 2013).

[199] V. Bhaghavan, A. Demers, S. Shenker, and L. Zhang. 1994. MACAW: a media access protocol for wireless LAN's. *Proc. Sigcomm 1994*. (September), 212–225.

[200] P. Viswanath, D. N. C. Tse, and R. L. Laroia. 2002. Opportunistic beamforming using dumb antennas. *IEEE Trans. Inf. Theory.*, 48(6), 1277–1294.

[201] W. C. Jakes. 1974. *Microwave Mobile Communications*. New York: John Wiley & Sons.

[202] W. Ye, J. Heidemann, and D. Estrin. 2002. An energy-efficient MAC protocol for wireless sensor networks. *Proc. IEEE INFOCOM 2002* (June), 1567–1576.

[203] W. Yu and T. Lan. 2007. Transmitter optimization for the multi-antenna downlink with per-antenna power constraints. 55(6), 2646–2660.

[204] X. Wang. 2005. An FDD wideband CDMAMAC protocol with minimum power allocation and GPS-scheduling for wireless wide area multimedia networks. *IEEE Trans. Mobile Computing*, 4(1), 16–28.

[205] Wikipedia. 2013. *Timeline of lighting technology*. http://en.wikipedia.org/wiki/Timeline of lighting technology (last visited: 22 October 2013).

[206] A. Wittneben. 1993. A new bandwidth efficient transmit antenna modulation diversity scheme for linear digital modulation. *Proc. IEEE Int. Conf. on Commun.* (June), 1630–1634.

[207] W. Eberle, B. Bougard, S. Pollin, and F. Catthoor. 2005. From myth to methodology: cross-layer design for energy-efficient wireless communication. *Proc. the 42nd Annual. Conf. Design Automation*, 303–308.

[208] C. Y. Wong, R. S. Cheng, K. B. Letaief, and R. D. Murch. 1999. Multiuser OFDM with adaptive subcarrier, bit, and power allocation. *IEEE J. Select. Areas Commun.*, 17(10), 1747–1758.

[209] X. Qin and R. Berry. 2003. Exploiting multiuser diversity for medium access control in wireless networks. *Proc. IEEE INFOCOM 2003.* (April), 1084–1094.

[210] X. Qin and R. Berry. 2004a. Opportunistic splitting algorithms for wireless networks. *Proc. IEEE INFOCOM 2004.* (April), 1662–1672.

[211] X. Qin and R. Berry. 2004b. Opportunistic splitting algorithms for wireless networks with heterogeneous users. *38th Annual Conference on Information Sciences and Systems (CISS)* (March).

[212] M. Xiao, N. B. Shroff, and E. K. P. Chong. 2003. A utility-based power control scheme in wireless cellular systems. *IEEE/ACM Trans. Networking*, 11(10), 210–221.

[213] Y. (G.) Li and G. L. Stüber. 2006. *OFDM for Wireless Communications*. Springer.

[214] Y. Hu and A. Ribeiro. 2011a. Adaptive distributed algorithms for optimal random access channels. *IEEE Trans. Wireless Commun.*, 10(8), 2703–2715.

[215] Y. Hu and A. Ribeiro. 2011b. Optimal random access for wireless networks in the presence of fading. *Proc. Allerton Conf. on Commun. Control Computing.* (September).

[216] Y. Xiao. 2005. Energy saving mechanism in the IEEE 802.16e wireless MAN. *IEEE Commun. Let.*, 9(7), 595–597.

[217] L. Yang and M.-S. Alouini. 2004. Performance analysis of multiuser selection diversity. *Proc. IEEE Int. Conf. on Commun.* (June).

[218] R. M. Young. 1991. Euler's constant. *Math. Gaz.*, 75, 189–190.

[219] W. Yu and J. M. Cioffi. 2002. FDMA capacity of Gaussian multiple-access channels with ISI. *IEEE Trans. Commun.*, 50(1), 102–111.

[220] C. Zhou, M. L. Honig, and S. Jordan. 2001. Two-cell power allocation for wireless data based on pricing. *39th Annual Allerton Conference* (October).

Index